CARBON CAPTURE AND ITS STORAGE

T0264825

Carbon Capture and its Storage
An Integrated Assessment

Edited by

Simon Shackley
The University of Manchester, UK

and

Clair Gough
The University of Manchester, UK

Routledge
Taylor & Francis Group

LONDON AND NEW YORK

First published 2006 by Ashgate Publishing

2 Park Square, Milton Park, Abingdon, Oxon OX14 4RN
711 Third Avenue, New York, NY 10017, USA

Routledge is an imprint of the Taylor & Francis Group, an informa business

First issued in paperback 2016

British Library Cataloguing in Publication Data
Carbon capture and its storage : an integrated assessment.
 1.Carbon dioxide mitigation 2.Atmospheric carbon dioxide -
 Storage 3.Atmospheric carbon dioxide - Environmental
 aspects 4.Environmental geochemistry
 I.Shackley, Simon II.Gough, Clair
 628.5'3

Library of Congress Cataloging-in-Publication Data
Carbon capture and its storage : an integrated assessment / edited by Simon Shackley and Clair Gough.
    p. cm. --
Includes bibliographical references and index.
ISBN-13: 978-0-7546-4499-6
ISBN-10: 0-7546-4499-5
  1. Carbon dioxide--Environmental aspects. 2. Carbon dioxide sinks. 3. Air quality management. 4. Greenhouse effect, Atmospheric. 5. Greenhouse gases. I. Shackley, Simon. II. Gough, Clair.

TD885.5.C3C357 2006
363.738'746--dc22

                                                                    2006021129
ISBN 978-1-138-25490-9 (pbk)
ISBN 978-0-7546-4499-6 (hbk)

Contents

List of Figures

List of Tables

List of Contributors

Michelle Bentham joined the British Geological Survey in 2000. For the past five years Michelle has been working on the geological storage of carbon dioxide focusing on site selection and storage capacity estimates. She is BGS Project Manager of the EU CASTOR, Green Energy from Coal and GeoCapacity projects and is a Researcher on several EU and UK funded projects such as CO2GeoNet project (European network of excellence on geological storage of CO_2) and the UK Carbon Capture and Storage Consortium (UKCCSC).

Contact details: British Geological Survey, Kingsley Dunham Centre, Keyworth, Nottingham, NG12 5GG, UK; Email: mbrook@bgs.ac.uk; Tel: ++ 44 (0)115 9363048.

Igor Bulatov is a Researcher at the Centre for Process Integration, CEAS, The University of Manchester. He has an MSc and PhD in Chemical Engineering from Mendeleev University of Chemical Engineering, Moscow, Russian Federation. His previous work through a UK Royal Society/NATO Post-Doctoral Fellowship included research on low temperature heat exchanger network design and retrofit. His research interests cover energy saving, CO_2 mitigation and sequestration, heat recovery, cost optimisation and environmental aspects of process design.

Contact details: Centre for Process Integration, School of Chemical Engineering and Analytical Science, The University of Manchester, PO Box 88, Manchester, M60 1QD, UK; E-mail: igor.bulatov@manchester.ac.uk; Tel +44 (0)161 306 4398; Fax +44 (0) 236 7439.

Tim Cockerill has a First in Engineering from Cambridge University and an MSc by research that investigated the Fluid Mechanics and Thermodynamics of the Ranque-Hilsch Vortex Tube. He also has a PhD in the 'Cost Modelling of Bottom Mounted Offshore Wind Energy Converter Systems' from the University of Sunderland, where he was employed as Research Fellow from 1996–2004. He is currently a Lecturer in Wind Energy and Carbon Management at the University of Reading's School of Construction, Management and Engineering.

Contact details: School of Construction Management and Engineering, University of Reading, Whiteknights, PO Box 217, Reading, RG6 6AH, UK; Email: t.t.cockerill@reading.ac.uk; Tel: +44 (0)118-378-8567.

Clair Gough is a Senior Research Fellow with the Tyndall Centre where she has been working on Carbon Capture and Storage since 2001. Her research interests are in public and stakeholder engagement in Integrated Assessment, in particular in the context of climate change mitigation. She is currently engaged in projects under the UK Carbon Capture and Storage Consortium (UKCCSC).

Contact details: Tyndall Centre for Climate Change Research, Pariser Building, University of Manchester, Manchester, M60 1QD, UK; Email: clair.gough@ manchester.ac.uk; Tel: +44 (0) 161 306 3447; Fax: +44 (0) 161 306 3723.

Sam Holloway is a Senior Geologist at the British Geological Survey. His background is in the petroleum geology of the UK and its continental shelf. However, he has worked mainly on the geological storage of carbon dioxide since 1991, when he was a scientist on one of the first studies of the potential for underground CO_2 sequestration: 'A pre-feasibility study for the underground disposal of carbon dioxide', undertaken by BGS for British Coal. Since then he has worked on numerous UK and European carbon dioxide storage projects, including the monitoring of the underground storage of CO_2 at the Sleipner field, North Sea. He is a lead author of the IPCC Special Report on Carbon Dioxide Capture and Storage (2005) and the 2006 IPCC Guidelines for National Greenhouse Gas Inventories and is a member of the UK Carbon Capture and Storage Consortium (UKCCSC).

Contact details: British Geological Survey, Kingsley Dunham Centre, Keyworth, Nottingham, NG12 5GG, UK; Email: shol@bgs.ac.uk; Tel: +44 (0)115 9363190; Fax: +44 (0) 115 0363437.

Karen Kirk graduated from Derby University with a BSc(Hons) in Geology. She has been employed at the British Geological Survey since 1994. For the past five years she has primarily worked on industry and government funded projects focused on the geological storage of carbon dioxide and natural gas. Her current research interests include the storage potential of the Bunter Sandstone Formation in the southern North Sea (GESTCO), cleaner energy from coal technologies (CARNOT), UK sources and sinks and the storage of carbon dioxide from UK power stations.

Contact details: British Geological Survey, Kingsley Dunham Centre, Keyworth, Nottingham, NG12 5GG, UK; Email: klsh@bgs.ac.uk; Tel: +44 (0)115 9363013.

Jiri Klemeš is a Senior Project Officer and Honorary Reader at the Centre for Process Integration, The University of Manchester. He has a PhD in Chemical Engineering (Technical University of Brno, Czechoslovakia), Hon DSc (Kharkiv, Ukraine) and has many years of research and industrial experience in mathematical modelling and process integration.

Contact details: Centre for Process Integration, CEAS, The University of Manchester, PO Box 88, Manchester, M60 1QD, UK; E-mail: j.klemes@manchester. ac.uk; Tel: +44 (0)161 306 4389; Fax: +44 (0)871 247 7774.

Carly McLachlan is based at the Tyndall Centre at the University of Manchester. Following her contribution to the Centre's research on public and stakeholder perceptions of CCS, Carly has recently started her PhD investigating stakeholder perceptions of new and emerging renewable energy technologies.

Contact details: Tyndall Centre for Climate Change Research, Pariser Building, University of Manchester, PO Box 88, Manchester, M60 1QD, UK; Email: c.mclachlan@manchester.ac.uk; Tel: +44(0)1613063764; Fax: +44(0)1613063255.

Ray Purdy is a Senior Research Fellow in environmental law at the Faculty of Laws, University College London. He is also Deputy Director of the faculty's Centre for Law and the Environment. He has a strong interest in climate change and was the managing editor of the journal *Climate Policy* between 2000 and 2003.

Contact details: Centre for Law and the Environment, Faculty of Laws, University College London, Bentham House, Endsleigh Gardens, London, WC1H OEG, UK; Email: uctlrap@ucl.ac.uk; Tel: +44 (0)207 679 4554; Fax: +44 (0)207 679 1440.

Simon Shackley is a founding member of the Tyndall Centre for Climate Change Research at the University of Manchester. Between 2000 and 2005 he served as Co-Manager of the 'Decarbonising the UK' research theme, which culminated in the publication of a new set of energy scenarios for the UK. His research interests include the public and stakeholder perceptions of climate change impacts and of carbon mitigation options, in particular biomass and CO_2 capture and storage.

Contact details: Manchester Business School, University of Manchester, Manchester, M60 1QD, UK; Email: simon.shackley@manchester.ac.uk; Tel: +44 (0)161 306 8781; Fax: +44 (0)161 306 3723.

Preface

Malcolm Wicks MP

Minister for Energy, Department of Trade and Industry, UK Government

The level of CO_2 in the atmosphere has risen by more than a third since the industrial revolution and is currently increasing faster than ever before. Indeed the Intergovernmental Panel on Climate Change (IPCC) has estimated that a continuation of current trends would see CO_2 concentrations exceeding three times pre-industrial levels by 2100. Such an increase would result in the global mean temperature rising by up to 5.8°C, which would cause major climatic changes. The Prime Minister has described climate change as 'probably, long-term, the single most important issue we face as a global community'.

The main cause of this rise in atmospheric CO_2 is of course the increasing use of fossil fuels, driven by our growing need for the energy services they provide. Energy is fundamental to the quality of life enjoyed in modern developed economies: it provides warmth and light for homes and offices, mobility and communications for business and leisure and the power needed to drive industrial processes. Energy is also essential to the continued progress of developing countries in expanding their economies and improving social conditions. Current projections anticipate global energy demand growing by about 60 per cent over the next 25 years with a similar increase in CO_2 emissions.

The UK's Energy White Paper of February 2003, *Our Energy Future – Creating a Low Carbon Economy,* recognised the need for urgent global action to combat climate change while still maintaining the secure, reliable and affordable energy supplies that are essential to economic stability and development. It showed global leadership by putting the UK on a path to a 60 per cent reduction in CO_2 emissions by 2050. More recently the UK reinforced this leadership by making climate change, energy and sustainable development one of the two key themes for the 2005 G8 summit. Amongst other conclusions, this led to a reaffirmation of the commitment to the UNFCCC and its ultimate objective to stabilise greenhouse gas concentrations so as to prevent dangerous anthropogenic interference with the climate system.

There is no one solution to the challenge of continuing to provide the energy services needed by modern economies while making radical reductions in CO_2 emissions. Clearly there is an imperative to use energy as effectively and efficiently as possible, and the development and deployment of technologies to harness non-CO_2 renewable energy sources is also a priority – but more is needed. With over 80 per cent of current world energy supplies coming from fossil fuels there would be substantial advantages to be gained from abatement methods that are central to the established energy supply system. The importance of this is illustrated by International Energy Agency estimates that $16 trillion will need to be invested in the world's energy system over the next 25 years, much of this on fossil technologies. This will put in place capital plant that will operate for 30–40 years, 'locking in' our dependence on fossil fuels.

In June 2005 the DTI launched its strategy for Carbon Abatement Technologies (CATs) for Fossil Fuel Use. CATs is a group of innovative technologies that provide an option for using fossil fuels during the transition to a low carbon economy by delivering emission reductions of between 5 per cent and 85 per cent. Emission reductions of 5–30 per cent can be gained by improving efficiency of combustion plant such as power stations and by co-firing with CO_2 neutral biomass. The larger reduction of up to 85 per cent (and even more with co-firing) can be achieved through carbon capture and storage (CCS), in which the CO_2 produced in large combustion plant is separated and transported to suitable geological formations for injection and permanent storage. The strategy recognises that the UK has a strong industry base for the development of CATs as well as the natural resources for CO_2 storage in depleted oil and gas reservoirs or deep saline aquifers. It is aimed at helping UK business to take a lead in the development and deployment of these technologies both at home and abroad.

CCS involves a chain of technologies and, although most have been used individually for different purposes, there is very limited experience of their full scale integration in a CCS system. There are alternative CCS technology options with different technical and economic merits, different deployment schedules and varying applicability to the UK as well as overseas markets. Also it will require additional frameworks to manage such factors as its consents and licensing, monitoring and verification of the CO_2 stored. Other issues to be addressed include entry into carbon abatement measures such as the EU Emissions Trading Scheme and long term liability for the CO_2 stored. Answers to these issues will require detailed objective analysis and an informed debate on the merits of CCS relative to other abatement options. I therefore welcome the Tyndall Centre's initiative in undertaking a three-year multi-disciplinary study of CCS that has covered not just technical and economic issues, but the crucial areas of citizen and stakeholder perspectives. The resulting book *Carbon Capture and its Storage: An Integrated Assessment* makes a valuable and timely contribution to the growing debate on these issues and I am pleased to commend it to everyone concerned with CO_2 emissions and their reduction.

Acknowledgements

We would like to acknowledge funding support for the research that underlies this book from the Tyndall Centre for Climate Change Research which decided in 2001 that it would support research on the topic of CO_2 capture and storage. The Tyndall Centre is itself funded by the Engineering and Physical Sciences Research Council (EPSRC), Natural Environment Research Council (NERC) and the Economic and Social Research Council (ESRC). Additional funding was provided by E.ON (UK) plc and, for the work in Chapter 5, by the Department of Trade and Industry of the UK government.

We would also like to thank the support and help given by the following individuals, without whom this book would not have been possible: Jon Gibbins, Brian Morris, Adrian Bull, Tim Hill, Kevin Anderson, Richard Pearce, Doug Coleman, Muir Miller, Gregg Butler, John Gale, Martin Angel, Paul Freund, Richard McIlwain, Andy Chadwick, Jonathan Pearce, Caroline Salthouse, Chris Rochelle, David Bowe, Nick Riley, Neil Hurford, Mike Farley, Peter Strutton, Roger Brandwood, Andy Brunt, Tim Melling, Brian Ricketts, Helen Chadwick, Dorian Speakman, Julian Carter, Bill Senior, Guy Wallbanks, Asher Minns, Samantha Jones, Mike Hulme, John Schellnhuber, John Shepherd and Gordon MacKerron. We would also like to thank the members of the public in Manchester and York who participated so actively in our Citizen Panels.

An earlier version of Chapter 5 appeared as Shackley, S., McLachlan, C. and Gough, C. (2005), 'The Public Perception of Carbon Capture and Storage in the UK: Results from Focus Groups and A Survey', *Climate Policy*, **4**(4): 377–398. We are grateful to Earthscan Publishers for permission to re-use some of the material here.

List of Abbreviations

C	carbon
CBM	coal bed methane
CCGT	combined cycle gas turbine
CCS	carbon dioxide capture and storage
CE	capture efficiency
CHP	combined heat and power
CO_2	carbon dioxide
COE	cost of electricity
DEC	decarbonised electricity certificate
DEFRA	Department of Environment, Food and Rural Affairs (UK)
DTI	Department of Trade and Industry (UK)
ECBM	enhanced coal bed methane recovery
EGR	enhanced gas recovery
EMYH	East Midlands, Yorkshire and Humber
ENGO	environmental non-governmental organisation
EOR	enhanced oil recovery
EU	European Union
EU-ETS	European Union Emissions Trading Scheme
FBC	fluidised bed combustion
FGD	flue gas desulphurisation
GDP	gross domestic product
GHG	greenhouse gas
GIS	geographical information system
GMO	genetically modified organism
Gt	giga tonne (10^9)
GW	gigawatt (10^9)
GWh	gigawatt hour
H_2S	hydrogen sulphide
HHV	higher heating value
HMT	Her Majesty's Treasury (UK)
HoC	House of Commons (UK)
IEA	International Energy Agency
IEA-GHG	International Energy Agency – Greenhouse Gas R&D Programme
IGCC	integrated gasification combined cycle
IPCC	Intergovernmental Panel on Climate Change
kW	kilowatt
kWh	kilowatt hour
LHV	lower heating value

MCA	multi criteria assessment
MEA	mono-ethanolamine
Mt CO_2	million tonnes of carbon dioxide
mtoe	million tonnes of oil equivalent
MW	megawatt (10^6)
MWh	megawatt hour
NG-CCGT	natural gas combined cycle gas turbine
NGO	non-governmental organisation
NPV	net present value
NW	North West
OECD	Organisation for Economic Co-operation and Development
OSPAR	Convention for the Protection of the Marine Environment of the North-East Atlantic
PF	pulverised fuel
p/kWh	pence per kilowatt hour
ppmv	parts per million of unit volume
PSA	pressure swing adsorption
R&D	research and development
RD&D	research, development and demonstration
RV	representative value
SAC	Special Area of Conservation
SCR	selective catalytic reduction
SMR	steam methane reformation
SPA	Special Protection Area
SRES	Special Report on Emissions Scenarios
TAR	Third Assessment Report
TW	terawatt (10^{12})
UNCLOS	United Nations Convention on the Law of the Sea
UNFCCC	United Nations Framework Convention on Climate Change

Chapter 1

Introduction

Simon Shackley and Clair Gough

1.1 The Climate Change Problem and the Potential Role of Carbon Dioxide Capture and Storage

It is now widely recognised that large scale reductions in carbon dioxide (CO_2) emissions are required during this century in order to limit the extent of climate change modification. Anthropogenic CO_2 emissions are largely a result of burning fossil fuels (constituting 80 per cent of such emissions). Once in the atmosphere, CO_2 acts as a greenhouse gas, causing: mean global temperatures and sea levels to rise; precipitation patterns to change; the frequency and severity of extreme weather events to be potentially enhanced; and the acidification of the oceans – with widespread implications for global support systems which underpin existing human activities (IPCC 2001 and 2001a; Schellnhuber et al. 2006). Fossil fuel combustion currently accounts for annual emissions into the Earth's atmosphere of about 23 x 10^9 tonnes CO_2 (or 23 Giga tonnes, Gt). This represents a global increase in anthropogenic CO_2 emissions of 70 per cent between 1971 and 2002 (IEA 2004).

Various scenarios have been prepared to explore how anthropogenic CO_2 emissions might change over the next few decades to the end of the twenty-first century. The International Energy Agency's World Energy Outlook (WEO) Reference Scenario (IEA, 2004 and 2004a) predicts that a 1.7 per cent per year increase in CO_2 emissions between 2000 and 2030 (a similar growth rate to that observed over the past 30 years) would result in a further 63 per cent increase in CO_2 emissions from today's level. This assumes that fossil fuel combustion remains the dominant energy source, accounting for more than 90 per cent of the increase in energy use over this period. Even under the World Alternative Policy Scenario, which includes CO_2 reduction measures and a reduced reliance on fossil fuels, the IEA's WEO analysis suggests a 40 per cent increase in global CO_2 emissions from today's level.

The Intergovernmental Panel on Climate Change (IPCC) has conducted extensive work on future CO_2 emissions scenarios, the so-called 'SRES scenarios' (Nakicenovic et al., 2000). It has provided a wide range of future CO_2 emission scenarios to 2100, from emissions only slightly greater than the level in 1990 to six times 1990 levels. The annual and cumulative global CO_2 emissions from several of the SRES scenarios are shown in Figure 1.1, from which it can be seen that scenarios with very different storylines can result in similar emissions levels, whilst scenarios with similar storylines can show extensive divergence in terms of emission levels.

None of the SRES scenarios included CO_2 capture and storage or indeed other climate policy-driven technological developments.

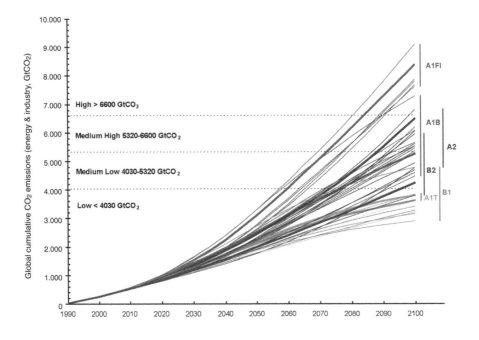

Figure 1.1   Annual and cumulative global carbon dioxide emissions arising from the SRES scenarios (in $GtCO_2$)

Source: Fig. 8.3, page 350, IPCC (2005)

By 2030 most of the IPCC SRES scenarios show an increase in CO_2 emissions of between 1.5 and 3 times the level in 1990. The IPCC emission scenarios translate into an increase in atmospheric CO_2 concentration from 540 ppmv (parts per million of unit volume) to 970 ppmv by the year 2100 compared to the current level of 370ppmv (Watson *et al.*, 2001). Until recently an atmospheric CO_2 concentration of 550 ppmv was regarded by many climate scientists as an appropriate global target for meeting the objective of the United Nations Framework Convention on Climate Change (UNFCCC) as set out in Article 2, namely: 'stabilisation of greenhouse gas concentrations in the atmosphere at a level that would prevent dangerous anthropogenic interference with the climate system'. More recent scientific research suggests that 550 ppmv may well be too high a value and that 400 to 450 ppmv is perhaps a more appropriate target (DEFRA, 2004; Meinshausen, 2006). Even under an optimistic scenario with regards to adoption of low- and zero-carbon energy technologies, therefore, limitation of the build-up of CO_2 in the atmosphere from anthropogenic sources will be a significant challenge over the course of the twenty-first century.

The difficulty of controlling future CO_2 emissions globally is illuminated by the case of China, where approximately one GW of coal powered generation is currently being installed every week, equivalent to the entire capacity of the UK electricity network every year (HoC, 2006). Under one typical scenario, total CO_2 emissions in China are expected to double by 2030 (relative to 2002) to approximately 7 Gt CO_2 per year, compared to an anticipated 4.5 Gt CO_2 per year in the European Union (EU) (which compares with the current EU value of 3.7 Gt CO_2 per year) (Senior, 2005). The increased carbon emissions arising in China result largely from a doubling in the use of coal by 2030 (relative to 2002). By 2030, China is expected in this scenario to have nine times the installed coal power plant capacity as the EU and to be using five times as much coal (Senior, 2005). According to a different study, the *additional* CO_2 emissions from coal-fired power in China and India by 2030 will be 3 Gt CO_2, equivalent to 13 times the UK's current power-sector CO_2 emissions (HoC, 2006).

UK anthropogenic CO_2 emissions are in the order of 0.56 Gt CO_2 per year, less than 2.5 per cent of global emissions. Setting out to demonstrate international leadership on action against climate change, however, the UK Government set a national target of 60 per cent reduction in CO_2 emissions by 2050 in its Energy White Paper (2003). The 60 per cent reduction was derived through a Contraction and Convergence approach[1] (Meyer, 2000) to meet the 550 ppmv atmospheric CO_2 concentration stabilisation target (RCEP, 2000). To meet the 450 ppmv CO_2 target would require a reduction in UK CO_2 emissions of between 80 and 90 per cent (relative to 1990) (Anderson *et al.*, 2005).

The most widely known approaches and technologies for CO_2 emissions reduction are reducing energy demand (e.g. through energy efficiency or behavioural changes), renewable energy technologies and nuclear power. At whatever level of energy use, the ultimate goal must be to establish a sustainable, largely carbon-free, energy supply that is sufficient to satisfy the energy demands of the World's industries, agriculture, transport and domestic usage. Constraining growth in *per capita* energy consumption would greatly assist in meeting this goal, though it is perhaps only a true optimist who would put faith at the present time in our ability to substantially reduce energy demand in post-industrial economies and restrain its growth in industrialising nations. Currently, carbon-free energy technologies are far from being developed and/or deployed sufficiently to meet global demands, nor is there a reasonable prospect that they can match current or future demands in the foreseeable future (Deffeyes, 2005). To re-iterate the point made above, the continued use of fossil fuels and associated emissions of CO_2 probably remain an

---

1 Contraction Convergence is an approach to reducing global carbon emissions through allocating equal *per capita* emission rights by nation, defining an atmospheric CO_2 concentration stabilisation target and a time period over which convergence in per capita emissions and stabilisation is to be achieved. Under this approach, the richer countries have to significantly reduce their emissions over time (by 80–90 per cent), whilst poorer countries can usually grow their per capita emissions to some extent (Anderson and Bows, 2005).

inevitable part of the foreseeable future. Thus it has become urgent that supply side approaches to reducing these emissions from fossil fuel use are developed along side measures to reduce demand and programmes to further develop and deploy non-fossil fuel-based energy technologies.

During the 1990s, a new technology has emerged which offers an additional route to large-scale CO_2 emissions reduction. This is through the capture of CO_2 from large point-sources such as power stations, oil refineries and chemical and metal works and the storage of that CO_2 in suitable geological reservoirs, a technique known as carbon dioxide capture and storage (CCS) (Holloway, 1997; DTI, 2003). Decarbonised energy carriers, such as electricity and hydrogen, can thereby be made from fossil fuels with 80–90 per cent of the CO_2 captured. Such energy carriers could eventually be used for transportation as well as for a myriad of other energy supply applications. If CO_2 is captured from the combustion of biomass, a net reduction of CO_2 from the atmosphere is possible, as biomass crops take up atmospheric CO_2 (known as biological sequestration). This net reduction would allow the continued use of carbon-based liquid fuels in premium applications such as aviation whilst avoiding increased concentration of atmospheric CO_2 (Read and Lermit, 2005; Rhodes and Keith, 2005).

Geological CO_2 storage is now becoming established as a mainstream contender in the portfolio of climate change mitigation measures available. Internationally, it has been the topic of a Special Report from the Intergovernmental Panel on Climate Change (IPCC, 2005) which presents a comprehensive description of the key technologies associated with capture, transport and storage of CO_2 and the implications of the inclusion of CCS within the UNFCCC (United Nations Framework Convention on Climate Change). We have made every attempt not to duplicate the content of the IPCC report here by focusing on the UK context in more detail. In section 1.3, however, we do summarise some of the key findings of the IPCC report at the global scale in order to provide a context for the research we report on in this book.

The British Government has made various statements in support of pursuing CCS further, following on from reports published between 2003 and 2005. In March 2005 the Chancellor, Gordon Brown, announced that he would be looking into providing further incentives for CCS during 2005, followed by the announcement of the Carbon Abatement Technologies Strategy from the DTI which committed £25 million to CCS projects (DTI, 2005). The Chancellor added a further £10 million for CCS technology demonstrations in December 2005 (HoC, 2006). In February 2006 the House of Commons Science and Technology Committee produced an important report on CCS, which strongly supported deployment of CCS in the UK context and the need for government incentives to be put in place (HoC, 2006). Meanwhile, the UK Government published a new Energy Review in 2006 which favoured a new nuclear build programme and anticipated a relatively modest role for additional incentives for encouraging CCS technology.

The research documented in this book is, in essence, an attempt to translate the principles and technologies set out in the IPCC report into practice in the specific context of a national energy system, taking account of the particularities of electricity generating plant and infrastructure, legal and regulatory frameworks, environmental impacts and assessment, stakeholder and public perceptions and priorities and government policy.

1.2   Biological Sequestration, Direct Ocean and On-Shore CO_2 Storage

In this book we have not explicitly examined biological sequestration. The uptake of CO_2 by trees and soil and their potential role in CO_2 reduction has been extensively investigated elsewhere (e.g. IPCC, 2000). It is sufficient to comment here that biological sequestration faces considerable challenges of reversibility, monitoring and verification, wider social impacts and long-term security of the carbon store (ibid. and Brown *et al.*, 2004). CCS has considerable advantages over biological storage on these aspects, provided that rigorous standards are in place and maintained.

Many potential geological storage reservoirs in the UK are located offshore, below the sea bed, and this has sometimes led to geological CCS becoming confused with direct ocean storage. In this latter approach, CO_2 is transferred directly into the deep oceans rather than being stored in geological rock formations. Ultimately (on millennial time-scales) ~80–85 per cent of anthropogenic carbon dioxide emitted to the atmosphere will be taken up by the oceans. Takahashi (2004) estimates that nearly half the anthropogenic carbon emitted since 1800 has entered the oceans already. The deep ocean has an enormous potential capacity of ~38,000Gt and the basis for the concept of direct ocean storage is to short-cut the natural processes whereby carbon dioxide is transferred into the deep ocean. Ocean pH has already declined by 0.1 pH units and Caldeira and Wickett (2003) have estimated that if all fossil fuels were to be burnt ocean pH would eventually drop by ~0.7 units. There is evidence that carbon dioxide emissions are already radically altering the oceans' calcium carbonate system and hence are beginning to have a serious impact on the biota (Feely *et al.*, 2004; Sabine *et al.*, 2004). The oceans turn over on millennial time-scales so if the deep ocean carbon dioxide content is increased more there is the possibility that the deep upwelling water will eventually vent more carbon dioxide back into the atmosphere.

Direct ocean storage is a highly controversial approach because of the large degree of uncertainty in our knowledge of the fate of CO_2 so stored over hundreds and thousands of years. It has not been demonstrated at a pilot or demonstration scale and has not been included for further consideration in this volume. The decision to exclude ocean storage was clear in the UK context because of the ready availability of suitable geological storage sites within the continental shelf making any higher-risk strategy of direct ocean storage unnecessary. The project team also took the decision to concentrate on off-shore storage in geological formations, and not to consider on-shore storage sites. The reason for this decision is that, whilst there are suitable on-shore geological formations in the UK, the risks involved in utilising such storage options are likely to be greater than off-shore storage: in particular, the risks to human health and safety should leakage occur, or intrusion of CO_2 into potable water supplies. Our early work on public perceptions (Gough *et al.*, 2002) also indicated that the public would view on-shore storage much less favourably where there are plentiful sub-sea bed off-shore storage opportunities, as is the case for the UK.

1.3   The Global Context of CCS

The IPCC Special Report (IPCC, 2005) has comprehensively reviewed the research on the potential deployment of CCS at the global scale and below we summarise the key findings, though the reader should refer to the Special Report (in its section 8.3.2.) for more detailed analysis and references:

1. Economic models have explored the role of CCS given different stabilization targets. Not surprisingly, where a more stringent atmospheric CO_2 stabilisation target (e.g. 450 ppmv) is chosen, CCS has a greater role in carbon abatement and is implemented more rapidly.
2. Where coal is more readily available and relatively inexpensive, CCS has a higher take-up than in scenarios in which gas and other less carbon-intensive fuels are more readily available.
3. A further factor that has been explored is the role of international emissions trading schemes. Where efficient trading schemes are in place that are global in scale then the price of a carbon permit should be lower than where emissions trading schemes are more restricted in scope and less efficient. A higher carbon permit price will encourage the more rapid and extensive deployment of CCS technologies.
4. The role of technological learning has also been explored and it has been shown that the deployment of CCS would increase by a factor of 1.5 times if it is assumed that technological learning occurs at a rate similar to that observed with respect to innovation in sulphur removal technologies relative to no change in technology.

The IPCC's Third Assessment Report (TAR) (IPCC, 2001b) explored 76 mitigation scenarios using a number of different energy models. 36 of these scenarios from three different energy models included CCS as a mitigation strategy, assuming a range of stabilisation values from 450 to 750 ppmv. Expressed as the share of CO_2 capture in cumulative carbon emissions reduction between 2000 and 2100, the model results show a very wide range of values from zero to over 90 per cent reliance on CCS (0 to more than 5500 $GtCO_2$). In other words, there is a very high level of uncertainty regarding the deployment of CCS due to the role of assumptions regarding the stabilisation level, costs and availability of fuels, costs of CCS and of other mitigation technologies, and so on. Expressed as an average, the contribution of CCS to CO_2 reduction in the 750 ppmv scenarios is 15 per cent (380 $GtCO_2$) and 54 per cent for the 450 ppmv scenarios (2160 $GtCO_2$). The majority of the results from the six TAR scenarios tend to fall in the 220 to 2200 $GtCO_2$ range for stabilisation targets from 450 ppmv to 750 ppmv. These results, which have been confirmed by subsequent modelling studies, suggest that CCS could have a major role to play in CO_2 mitigation, in particular at the lower stabilisation levels.

A detailed 'CO_2 storage supply curve' has been produced for Europe in which the relationship between the price of CO_2 transportation and storage is related to the cumulative quantity of CO_2 stored. This requires a detailed spatial analysis of the major sources of CO_2 and their relationship to the principal sinks (Wildenborg *et al.*, 2005). The study shows that there are roughly constant costs across a wide range of storage capacity, i.e. once infrastructure for transporting and storing tens of millions of tonnes of CO_2 is in place, then there are few if any additional costs per unit of CO_2 stored if gigatonnes of CO_2 are then to be stored. In other words, the cost of CO_2 transport and storage in Europe (and, incidentally, North America) is likely to have a cap.

The IPCC TAR and Special Report have also examined the issue of the timing of CCS deployment over the course of the twenty-first century. The extent of CCS deployment over the century is highly dependent upon the underlying SRES scenario used in the analysis. If a high growth, high carbon-intensity scenario is employed (for example, scenario A1FI) then the adoption of CCS grows rapidly and continuously over the twenty-first century. On the other hand, if an underlying scenario is used in which growth is slower and there is a greater penetration of less carbon-intensive energy supply (for example, scenario B1), then CCS is not used to any great extent until the second half of the twenty-first century. Alternatively, in some scenarios (e.g. A1B, A2 and B2) CCS is used to a moderate extent into the mid or later part of the present century but then tails off. In these scenarios, CCS is used a bridge between the current fossil-fuel based energy system and an energy system based far more on renewable energy.

Finally, the IPCC Special Report reviews the evidence relating to the broad geographic location of CCS deployment. Dooley *et al.* (2005) identify three categories of country: a) countries where there is ample potential storage resource relative to demand (e.g. USA, Canada and Australia); b) countries where there is limited availability of storage reservoirs relative to demand (e.g. Japan and South Korea); and c) the rest of the world where CCS deployment depends upon both the necessary CO_2 emissions reduction and the availability of reservoirs. It is interesting to note that in terms of large-scale geo-political regions (OECD as of 1990, Reforming Economies (ex-Soviet Union and Eastern Europe), Asia and the Rest of the World (Africa, Latin America and the Middle East), CCS appears to have greater deployment levels in most scenarios in Asia and in the Rest of the World, i.e. in the developing and industrialising countries of today. This is the case for all but the lower-growth, more environmentally-focused scenarios (B1 and B2), in which case deployment of CCS is greater in the industrialised world.

1.4   An Integrated Assessment of CCS in the UK

Having presented briefly the scale of the CO_2 abatement required at a global level and the potential role of CCS within that context, the remainder of this book focuses on the UK scale. Geological carbon storage has the potential to make a significant contribution to the decarbonisation of the UK. Amid concerns over maintaining security, and hence diversity, of energy supply, CCS could allow the continued use of coal, oil and gas whilst avoiding most of the CO_2 emissions currently associated with fossil fuel use. However, as a new technology there remain many uncertainties relating to its viability, effectiveness, affordability and acceptability. We have adopted an Integrated Assessment (IA) approach to addressing these concerns. Integrated Assessment is a process which aims to develop insights beyond those derived from single disciplinary studies (Parson 1995; Risbey *et al.*, 1996; Gough *et al.*, 1998). It is not a substitute for rigorous scientific analysis but an addition to the knowledge gathering process within a complex topic area. The IA process described in this book has enabled us to explore the application of the CCS concept from a range of perspectives to enable a comprehensive evaluation of CCS options in the UK. We have developed and used integrating tools such as scenario generation and multi-criteria assessment, which allows a new synthesis of disciplinary insights and knowledge. The IA therefore allows us to address questions not answerable by a single disciplinary approach. CCS is, of course, only one amongst a range of climate change mitigation options and its assessment must be carried out in a manner that will enable it to be evaluated within this broader context. Although we do not attempt such a direct and comprehensive comparison here, the findings from the project have fed into a holistic assessment of carbon mitigation options for the UK, undertaken by the Tyndall Centre (Anderson *et al.*, 2005).

The results are based on a three year programme of collaborative research organised around the framework illustrated in Figure 1.2. Specific objectives can be summarised as follows:

1. To address major uncertainties associated with specific aspects of carbon capture and storage technologies, in particular:

    - the potential capacity for geological CO_2 storage in the UK;
    - costs of carbon capture and storage through retrofit or new design of power plants, refineries and other large point sources, including pipeline infrastructure,
    - the opinions, perceptions and more formal evaluations of key decision-makers, stakeholders and the public with respect to future potential use of CCS in the UK.

2. To extend the assessment to evaluate the short to medium term conditions for, and implications of, deploying the technologies in two regions of the UK. This has enabled us to identify which sources and sinks of carbon dioxide in those regions have the greatest potential in the near and medium term timescales,

given considerations of cost, capital plant and infrastructure, legal aspects, environmental impacts, geological integrity and public perceptions.

3. To develop integrating frameworks to enable us to address broader implications of implementing CCS. This includes analysis of the concept of storage as a 'bridging' option towards renewable and other zero- and low-carbon energy technologies or whether such groups of technologies imply fundamentally different pathways.

Two case studies were chosen to elaborate the key parameters associated with CCS in the UK in the near and medium term. Case study 1 explores the possibilities for using CO_2 from NW England with storage in the East Irish Sea. Case study 2 analyses options for storing CO_2 from major point sources across the East Midlands and Yorkshire and Humber regions with storage at a number of sites in the southern North Sea.

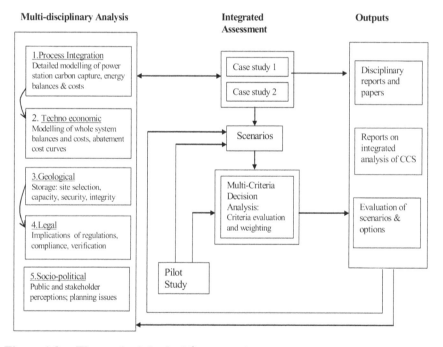

Figure 1.2   The methodological framework

1.5 Structure of the Book

The structure of the book broadly follows the schematic shown in Figure 1.2. In Chapter 2, Holloway *et al.* introduce the concept of geological storage covering issues such as storage site identification, characterisation and selection, storage capacity, long term monitoring and verification of storage, storage safety and security and the integrity of wells in the UK context.

Chapter 3, by Klemeš, Cockerill, Bulatov, Shackley and Gough reviews the costs and effectiveness of capture and recovery of CO_2 for a range of known technologies (such as physical and chemical absorption and adsorption, gas separation/membrane technology, flue gas recycling/oxyfuel boilers) derived from analysis of current literature. The chapter also presents some more novel routes to CO_2 capture and utilisation that are currently less well developed. A techno-economic perspective is then developed for extending the system to encompass infrastructure and transport costs such as pipelines and disposal etc. A techno-economic model has been developed and used for evaluating power plant and pipeline choices in the regional case studies, but due to unavoidable personal circumstances it has not been possible to present the detailed results from the model here.

In Chapter 4, Purdy examines existing legal regimes at international, European Community and national levels, identifying problem areas and areas for proposed change (these include the Kyoto Protocol; International Convention on the Law of the Sea 1982; London Dumping Convention 1972; OSPAR Convention; ESPOO Convention, Habitats Directive, Integrated Pollution Prevention and Control Directive; and UK Climate Change Levy). Issues addressed include where and when these various Conventions apply, the legality of CO_2 storage as an option, requirements for permits or licences and liability, *inter alia*.

Shackley *et al.* assess public perceptions associated with the options for CCS in the UK in Chapter 5. This assessment is based on both Citizen Panels and a larger scale survey. Two Citizens Panels met five times each and called on expert witnesses to present information and opinion relevant to the discussion. Experts included scientists, NGOs and industry spokespersons. A follow-up survey of opinions was designed on the basis of discussions in the panels in order to generalise the findings to a broader public context.

Chapters 6 (Shackley *et al.*) and 7 (Gough *et al.*) present the results from the two case study regions. The case studies incorporate a geological review of the suitability and potential of possible storage reservoirs in the region, a set of scenarios exploring alternative power generation pathways for each region and a Stakeholder Multi-Criteria Assessment (MCA) of these scenarios. The MCA has been used in a heuristic role, similar to the Multi Criteria Mapping approach adopted by Stirling and Mayer (2001) and employed a basic MCA methodology and software tool to illuminate the key issues and trade-offs associated with CCS.

In Chapter 8, Shackley attempts to undertake a comparison of the sustainability of coal CCS with new nuclear power generation. The reason for this comparison is that one of the key issues currently facing UK energy policy makers is how to replace the retiring power generation stock (largely coal and nuclear) in the UK. Whilst nuclear used to offer the only real option of large-scale low-carbon based load generation, fossil CCS is now presenting a viable alternative. Hence, it is vital that we attempt to better understand and compare the two technologies in terms of their economic, environmental and wider social impacts and repercussions.

Finally, in Chapter 9, Shackley and Gough provide a summary and review of the key findings of the book. The Chapter also explores the implications of this research for the future implementation of CCS in the UK context. Future key research needs and critical policy decisions and their timing are presented and discussed. It is intended that these conclusions can contribute to the on-going energy policy debate and process in the UK.

1.6   References

Anderson, K. and Bows, A. (2005), *Growth Scenarios for EU & UK Aviation: Contradictions with Climate Policy*, Tyndall Working Paper 84, Tyndall Centre, Manchester.

Anderson, K., Shackley, S., Mander, S. and Bows, A. (2005), *Decarbonising the UK: Energy for a Climate Conscious Future*, Tyndall Centre Technical Report 33, Tyndall Centre, Manchester.

Brown, K., Adger, N., Boyd, E. and Corbera, E. (2004), *How do CDM Projects Contribute to Sustainable Development?*, Tyndall Centre Technical Report 16, Tyndall Centre, Norwich.

Caldeira, K. and Wickett, M.E. (2003), 'Anthropogenic Carbon and Ocean pH', *Nature*, **425**, p. 365.

Deffeyes, K. S. (2005), *Beyond Oil: The View from Hubbert's Peak*, Hill and Wang, New York.

DEFRA (2004), *Scientific and Technical Aspects of Climate Change, Including Impacts and Adaptation and Associated Costs*, Department for the Environment, Food and Rural Affairs, London, www.defra.gov.uk/environment/climatechange/pdf/cc-science-0904.pdf.

Dooley, J.J., Kim, S.K., Edmonds, J.A., Friedman, S.J., and Wise, M.A. (2005), 'A First Order Global Geologic CO_2 Storage Potential Supply Curve and its Application in a Global Integrated Assessment Model', in E.S. Rubin, D.W. Keith and C.F. Gilboy (eds), *Proceedings of the 7th International Conference on Greenhouse Gas Control Technologies, Volume 1: Peer-Reviewed Papers and Overviews*, Elsevier Science, Oxford, pp. 573–581.

DTI (2003), *Review of the Feasibility of Carbon Dioxide Capture and Storage in the UK*, Department of Trade and Industry, London.

DTI (2005), *A Strategy for Developing Carbon Abatement Technologies for Fossil Fuel Use*, Department of Trade and Industry, London.

Energy White Paper (2003), *Our Energy Future: Creating a Low Carbon Economy*, Cm5761, HMSO, London.

Feely, R.A., Sabine, C.L., Lee, K., Berelson, W., Kleypas, J., Fabry, V.J. and Millero, F.J. (2004), 'Impact of Anthropogenic CO_2 on the $CaCO_3$ System in the Pecans', *Science*, **305**, pp. 362–65.

Gough, C., Castells, N. and Funtowicz, S. (1998), 'Integrated Assessment: An Emerging Methodology for Complex Issues', *Environmental Modelling and Assessment*, **3**(1,2), pp. 19–29.

Gough C. and Shackley, S. (2006), 'Towards a Multi-Criteria Methodology for Assessment of Geological Carbon Storage Options', *Climatic Change*, **74**(1–3), pp. 141–174.

Gough, C., Taylor, I. and Shackley, S. (2002), 'Burying Carbon under the Sea: An Initial Exploration of Public Opinions', *Energy and Environment*, **13**(6), pp. 883–900.

HoC (2006), *Meeting the UK Energy and Climate Needs: The Role of Carbon Capture and Storage*, Volume 1, Report together with formal minutes, House of Commons Science and Technology Committee, HC 578–1, UK Parliament, London.

Holloway, S. (1996) (ed.), *The Underground Disposal of Carbon Dioxide, Final Report of the Joule II Project*, No. CT92-0031, British Geological Survey, Nottingham.

Holloway, S. (1997), 'An Overview of the Underground Disposal of Carbon Dioxide', *Energy Conversion and Management*, **38**, pp. S193–S198.

IEA (2004), *Prospects for CO_2 Capture and Storage*, International Energy Agency, Paris, France.

IEA (2004a), *World Energy Outlook 2004 Edition*, International Energy Agency, Paris, France.

IPCC (2000), *Land Use, Land Use Change and Forestry*, R.T. Watson, I.R. Noble, B. Bolin, N.H. Ravindranath, D.J. Verardo and D.J.Dokken (eds), A Special Report to the IPCC, Cambridge University Press, Cambridge.

IPCC (2001), *Climate Change 2001: The Scientific Basis*, Contribution of Working Group I to the Third Assessment Report of the Intergovernmental Panel on Climate Change, Cambridge University Press, Cambridge.

IPCC (2001a), *Climate Change 2001: Impacts, Adaptation, and Vulnerability*, Contribution of Working Group II to the Third Assessment Report of the Intergovernmental Panel on Climate Change, Cambridge University Press, Cambridge.

IPCC (2001b), *Climate Change 2001: Mitigation*, Contribution of Working Group III to the Third Assessment Report of the Intergovernmental Panel on Climate Change, Cambridge University Press, Cambridge.

IPCC (2005), *IPCC Special Report on Carbon Dioxide Capture and Storage*, prepared by Working Group III of the Intergovernmental Panel on Climate Change, B. Metz, O. Davidson, H.C. de Coninck, M. Loos and L.A. Meyer (eds), Cambridge University Press, Cambridge.

Meinshausen, M. (2006), 'What does a 2°C Target Mean for Greenhouse Gas Concentrations? A Brief Analysis Based Upon Multi-Gas Emission Pathways and Several Climate Uncertainty Estimates', in H.J. Schellnhuber, *et al.* (eds), *Avoiding Dangerous Climate Change*, Cambridge University Press, Cambridge, pp. 265–280.

Meyer, A. (2000), *Contraction and Convergence: The Global Solution to Climate Change*, Schumacher Briefing Number 5, Green Books, Dartington, Devon.

Nakicenovic, N., Alcamo, J., David, G., de Vries, B., Fenhann, J., Gaffin, S., Gregory, K., Grubler, A., Jung, T.Y., Kram, T., La Rovere, E.L., Michaelis, L., Mori, S., Morita, T., Pepper, W., Pitcher, H., Price, L., Riahi, K., Roehrl, A., Rogner, H-H., Sandkovski, A., Schlesinger, M., Shukla, P., Smith, S., Swart, R., van Rooijen, S., Victor, N. and Zhou, D. (2000), *Special Report on Emission Scenarios, A Special Report of Working Group II of the Intergovernmental Panel on Climate Change*, Cambridge University Press, Cambridge.

Parson, E.A. (1995), 'Integrated Assessment and Environmental Policy-making – In Pursuit of Usefulness', *Energy Policy*, **23**(4–5), pp. 463–75.

RCEP (2000), *Energy: The Changing Climate*, 22nd Report, Cm. 4749, The Royal Commission on Environmental Pollution, London.

Read, P. and Lermit, J. (2005), 'Bio-energy with Carbon Storage (BECS): A Sequential Decision Approach to the Threat of Abrupt Climate Change', *Energy*, **30**, pp. 2654–71.

Rhodes, J. and Keith, D. (2005), 'Engineering-economic Analysis of Biomass IGCC with Carbon Capture and Storage', *Biomass and Bioenergy*, **29**, pp. 440–50.

Risbey, J., Kandlikar, M., and Patwardhan, A. (1996), 'Assessing Integrated Assessments', *Climatic Change*, **34**(3–4), pp. 369–95.

Sabine, C.L., Feely, R.A., Gruber, N., Key, R.M., Lee, K., Bullister, J.L., Wanninkhof, R., Wong, C.S., Wallace, D.W.R., Tilbrook, B., Millero, F.J., Peng, T.-H., Kozyr, A., Ono, T. and Rios, A.F., (2004), 'The Oceanic Sink for Anthropogenic CO_2', *Science*, **305**, pp. 367–71.

Schellnhuber, H.J., Cramer, W., Nakicenovic, N., Wigley, T. and Yohe, G. (eds) (2006), *Avoiding Dangerous Climate Change*, Cambridge University Press, Cambridge.

Senior, B. (2005), 'Near-zero Emissions Coal in China: UK/EU/China Cooperation', Presentation to the Power Sector Advisory Group, DTI, 8 November 2005, London.

Shackley, S., Gough C. and Cannell, M. (2002), *Evaluating the Options for Carbon Sequestration*, Tyndall Centre Technical Report 2, Tyndall Centre, Manchester.

Stirling, A. and Mayer, S. (2001), 'A Novel Approach to the Appraisal of Technological Risk: A Multi-Criteria Mapping of a Genetically Modified Crop', *Environment and Planning C: Government and Policy*, **19**(4), pp. 529–55.

Takahashi, T. (2004), 'The Fate of Industrial Carbon Dioxide', *Science*, **305**, pp. 360–2.

Watson, R.T. and the Core Writing Team (2001), *Climate Change 2001: Synthesis Report*, Cambridge University Press, Cambridge.

Wildenborg, T., Gale, J., Hendriks, C., Holloway, S., Brandsma, R., Kreft, E. and Lokhorst, A. (2005), 'Cost Curves for CO_2 Storage: European Sector', in E.S. Rubin, D.W. Keith and C.F. Gilboy (eds), *Proceedings of the 7th International Conference on Greenhouse Gas Control Technologies, Volume 1: Peer-Reviewed Papers and Overviews,* Elsevier Science, Oxford, pp. 603–610.

Chapter 2

Underground Storage of Carbon Dioxide

Sam Holloway, Michelle Bentham, Karen Kirk

1.1   Introduction

The burning of fossil fuels results in the emission of about 23 x 10^9 tonnes CO_2/year to the Earth's atmosphere. Once in the atmosphere CO_2 acts as a greenhouse gas. Most forecasts indicate that global fossil fuel combustion is likely to increase over at least the next two decades, suggesting that CO_2 levels in the Earth's atmosphere are set to increase substantially, which will further enhance the greenhouse effect and may lead to severe climate change. If these forecasts are correct, we need to find a way of reducing CO_2 emissions from fossil fuel combustion. This is a global problem, in which UK emissions reduction can play only a small role, because UK anthropogenic CO_2 emissions are in the order of 0.55–0.56 x 10^9 tonnes CO_2 per year, less than 2.5 per cent of global emissions.

The underground storage of CO_2 is a greenhouse gas mitigation option that reduces CO_2 emissions to the atmosphere by capturing CO_2 at major fossil fuel combustion plants and storing it underground in reservoir rocks. CO_2 would be captured at large stationary point sources such as fossil-fuel-fired power plants, transported, probably by pipeline, to a storage site, where it would be pumped down a well or wells into an underground reservoir rock. Once injected into the reservoir rock it would be held in place by natural geological seals that prevent it moving out of the storage site, just as oil and gas have been retained naturally, in many cases for millions of years (Figure 2.1).

Required Storage Period

If underground storage were to make a contribution to reducing CO_2 levels in the atmosphere, the stored CO_2 would have to be retained until well past the end of the fossil fuel era. After the end of the fossil fuel era, atmospheric CO_2 levels may begin a slow decline as ocean/atmosphere CO_2 levels re-equilibrate (Wilson, 1992; Lenton and Cannell, 2002). Thus the most desirable time frame for storage might be at least thousands of years. Nevertheless, short-term storage of a few hundred years could be valuable in peak-shaving the expected levels of CO_2 in the atmosphere that might otherwise occur towards the end of the fossil fuel era.

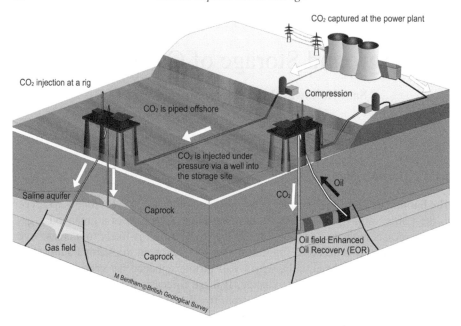

Figure 2.1 Diagram showing carbon dioxide capture and geological storage

Practical Examples of the Underground Storage of CO_2

The underground storage of industrial quantities of CO_2 is not simply a concept. At the Sleipner West gas field in the Norwegian sector of the North Sea, approximately 1 x 10^6 tonnes CO_2 per year are being stored underground (Figure 2.3). This operation (Korbul and Kaddour, 1995) started in 1996 and some 7.5 million tonnes has been stored to date. CO_2 is also being injected underground in enhanced oil recovery (EOR) operations worldwide. The greatest concentration of such projects is in the Permian Basin of west Texas, USA (Hsu *et al.*, 1995), but the best monitored is at Weyburn, in Saskatchewan, Canada (Wilson and Monea, 2004). The longest-running EOR project that has been monitored for leakage started at Rangely, Colorado in 1986 (Klusman, 2003). More recently, commercial-scale CO_2 storage projects have been started at the In Salah gas fields, Algeria (Riddiford, 2003) and reached the construction phase at the Snohvit field in the Barents Sea, offshore Norway (Maldal and Tappel, 2003). Smaller demonstration projects have been undertaken at Nagaoka, Japan (Kikuta *et al.*, 2005), Frio, Texas (Hovorka, 2005) and the K12-B gas field offshore Netherlands (van der Meer, 2005). These projects demonstrate that it is technically possible to place CO_2 in underground storage sites and store it over the medium term (decades). A full list of CO_2 storage projects to date is provided in Table 2.1.

Safe and secure underground storage by man of very large quantities of CO_2 for thousands of years cannot be directly demonstrated because the oldest monitored

Table 2.1 Summary of experience with CO$_2$ storage

Summary of situation at present
<ul><li>Only a few commercial-scale examples storing more than 1 Mt/year CO$_2$ – Sleipner (Norway) and In-Salah (Algeria)</li><li>Many active research demonstrations of 1–100 kt CO$_2$, e.g. CRUST (Netherlands), Nagaoka (Japan), Frio (USA), West Pearl Queen (USA)</li><li>Many imminent research demonstrations, e.g. Teapot Dome (USA), CO$_2$SINK (Germany), Otway (Australia)</li><li>Numerous Enhanced Oil Recovery and acid gas injection projects: more than 70 projects worldwide (59 in Texas), but all onshore; several on a scale of more than 1Mt/y CO$_2$, e.g. Weyburn (Canada), Rangely (USA)</li><li>Large offshore developments planned, e.g. Gorgon (Australia), Snohvit (Norway), Sibilla (Italy), Miller (UK), Carson (California)</li></ul>
Enhanced Oil Recovery (EOR): some examples
<ul><li>Weyburn (Saskatchewan) operating EOR commercially since 2000, producing 122 M bbl extra oil over 15 years, whilst storing 20 Mt CO$_2$ at 1.8 Mt/y. CO$_2$ transported 330km by pipeline</li><li>Teapot Dome (Wyoming) storing 2.6 Mt/y whilst researching monitoring methods</li><li>Rangely has injected 25 Mt CO$_2$ since 1986 and monitoring leakage</li><li>West Pearl Queen monitoring migration of 2100 t CO$_2$</li><li>Plans for CO$_2$ injection into Upper Zakum field of United Arab Emirates by Japanese Oil Development Company</li><li>Plans for storage of 1.5 Mt CO$_2$ at Sibilla, Italy, from the API refinery</li><li>Plans for storage of 1.3 Mt/y CO$_2$ at Miller Field, Scotland with methane reformation and hydrogen-powered electricity generation (BP, Scottish & Southern Energy)</li><li>Plans for CO$_2$ injection at the Draugen and Heldrun offshore oil and gas fields (Norway) with a gas-fired power plant and methanol production at Tjeldbergodden (Shell and Statoil)</li></ul>
Depleted oil fields
<ul><li>Canada has about 30 commercial acid gas injection projects for CO$_2$ and H$_2$S disposal, at rates of 0.003 to 0.06 Mt/y</li><li>Abu Dhabi injects 0.4 Mt/y acid gases</li></ul>
Depleted gas fields and gas reservoirs
<ul><li>The CRUST project (Netherlands) is injecting 20 kt/y CO$_2$ into an offshore sandstone depleted gas reservoir at 3,800 m depth (increasing to 0.5Mt/y CO$_2$)</li><li>In Salah project (Algeria) injects 1 Mt/yr CO$_2$ into a gas reservoir</li></ul>
Saline aquifers
<ul><li>1 Mt/y CO$_2$ injected at Sleipner into an aquifer at 1000m depth</li><li>10 kt CO$_2$ injected into an aquifer at Nagaoka (Japan) in 2003 and monitored</li><li>Plans for 0.7 Mt/yr CO$_2$ storage in an aquifer at Snohvit (Barents Sea)</li><li>Plans for 5 Mt/yr CO$_2$ storage in an aquifer at the Gorgon Field (Australia)</li></ul>
Enhanced coal bed methane (ECBM)
<ul><li>Pilot studies into injection into permeable coals in Mexico (COAL-SEQ)</li><li>Polish pilot studies have shown need to fracture EU coals (RECOPOL)</li><li>Pilot scale injection at Qinshui Basin, eastern China</li><li>Small-scale demonstrations in Japan at 900m depth</li></ul>

Source: Adapted from HoC (2006), pp. 23–24, HMT (2006)

CO_2 injection project (Rangely) has only been injecting CO_2 since 1986. However, there are analogies in nature. The many natural underground CO_2 fields around the world (Studlick *et al.*, 1990, Stevens, 2004) are identical to the more common natural gas fields that contain predominantly methane in every respect apart from their gas composition. Furthermore many of the latter contain varying quantities of CO_2 mixed in with the hydrocarbon gases (Baines and Worden, 2004). Many of these fields of both pure CO_2 and CO_2/hydrocarbon mixtures have existed for thousands to millions of years. This proves that under favourable circumstances CO_2 can be retained underground for geological timescales.

2.2   Underground Storage Concepts

The main concepts that have been put forward for underground storage sites for CO_2 fall into three categories: storage in the saline water-filled pore spaces of underground reservoir rocks, storage in depleted oil and gas fields and storage by adsorption onto the surface of the micropores found in coal seams (otherwise known as coal beds). Storage in caverns and mines cannot make a significant impact on the greenhouse effect because the majority are not leak proof, especially at pressures much greater than atmospheric. Most abandoned mines gradually fill with water, and any gas within them will eventually be forced out. The leak proof mines have alternative uses – for example, storage of documents, natural gas and chemical waste. Solution-mined salt caverns are also unsuitable as they are not stable in the long term because rock salt is a ductile substance that can creep and rupture under the in situ stresses within the Earth's subsurface.

Storage in the Saline Water-Filled Pore Spaces in Porous and Permeable Reservoir Rocks

CO_2 can be stored in geological formations by filling the intergranular pore space within rocks with CO_2. Figure 2.2.1 shows a typical sandstone reservoir rock in a quarry in Eastern England – the darker grey rock is the sandstone, it is overlain by light grey, silty claystone. Figure 2.2.2 shows a close-up of a reservoir sandstone, showing the individual sand grains which are held together by natural mineral cements. These cements do not usually occlude all the space between the grains. This is more clearly shown in Figure 2.2.3, which is a very thin slice of a reservoir sandstone in which the open spaces between the grains (the pore spaces) have been filled with a resin which appears dark grey. The pore spaces in a reservoir rock are connected, allowing fluids to flow into and out of it, such that the rock is permeable as well as porous. In places where such reservoir rocks are found deep underground, their pore spaces are normally filled with water, of varying salinity or, more rarely, with oil or gas (where filled with saline water, such reservoir rocks are sometimes known as saline aquifers). When CO_2 is pumped into a reservoir rock it partially displaces these so-called native pore fluids and fills the pore space (Figure 2.2.4).

1. Outcrop of reservoir sandstone

2. Close up photograph of reservoir sandstone

3. Thin section of reservoir sandstone viewed through a microscope

4. Diagram highlighting porosity and permeability

M Bentham © British Geological Survey

Figure 2.2   Reservoir rocks, porosity and permeability

The process at Sleipner is shown in Figure 2.3. Natural gas from the Sleipner West field contains about 9.5 per cent CO_2. This has to be separated out to get the gas to sales quality. Separation, by amine scrubbing, takes place on the Sleipner T platform. The CO_2 is compressed and sent for storage in the Utsira Sand via a 3 km-long highly deviated well drilled from the Sleipner A platform. Note that the CO_2 storage reservoir is at a shallower level than, and completely separate from, the natural gas production reservoir.

Pressure – Temperature Conditions Underground

The average temperature underground commonly increases downwards, as a result of heat flow from the inside to the outside of the Earth. The rate at which it increases (known as the geothermal gradient) is variable (Tissot and Welte, 1978) but in the sedimentary basins where CO_2 storage is likely to take place it commonly increases by about 20–35°C per kilometre.

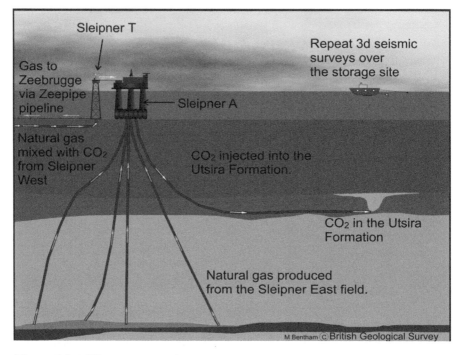

Figure 2.3   The capture and storage process at the Sleipner project

Pressure also increases downwards within the subsurface. Pressure in the pore spaces of sedimentary rocks is commonly close to hydrostatic pressure, that is the pressure generated by a column of water of equal height to the depth of the pore space. This is because the pore space is mostly filled with water and is connected, albeit tortuously, to the ground surface, and the pressure exerted by the overlying rock is supported by the reservoir's rock framework. However, under conditions where the pore space is either not connected to the surface, or not equilibrated to the surface, pressure may be greater than hydrostatic because the fluid within the pore space is supporting part of the weight of the overlying rock.

Underpressure may also exist in the pore fluids of reservoir rocks, either naturally, or as the result of abstraction of fluids such as oil and gas. The physical properties of CO_2 and the ambient temperature and pressure in the storage reservoir define the density at which CO_2 will be stored underground (Holliday *et al.*, 1992). Under typical reservoir conditions, there is a sharp increase in the density and corresponding decrease in the volume of CO_2 at depths between approximately 600 m and 800 m depending on the precise geothermal conditions and pressure (Figure 2.4). This is associated with the phase change from gas to a dense supercritical fluid. Consequently, CO_2 occupies much less space in the subsurface than at the surface; one tonne of CO_2 at a density of 700 kg/m^3 occupies 1.43 m^3, whereas at 0°C and 1 atmosphere one tonne of CO_2 occupies 509 m^3.

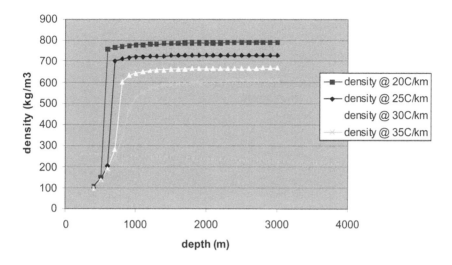

Figure 2.4   Density of CO$_2$ at hydrostatic pressure and typical geothermal gradients

There is no reason why CO$_2$ should not be stored at depths where it would be gaseous rather than in the dense phase, but relatively small masses of CO$_2$ would occupy relatively large volumes of pore space. Also, in onshore areas shallow reservoir rocks commonly have a more important use – groundwater supply.

How is CO$_2$ Retained in Porous and Permeable Reservoir Rocks?

Once CO$_2$ is injected into a reservoir rock, the processes of migration and trapping begin. CO$_2$ is always buoyant compared to water at reservoir conditions, so it tends to migrate upwards through the pore system. However, it may become trapped in four ways:

Structural and stratigraphic trapping   Typical structural traps for buoyant fluids may take the form of domes or otherwise elevated parts of the reservoir rock, which are overlain and confined laterally by so-called cap rocks (Figure 2.1, 2.6). Cap rocks can be divided into two categories; essentially impermeable strata such as thick rock salt layers, known as aquicludes (Figure 2.5) and those with very low permeability such as shales and mudstones, known as aquitards, through which fluids can migrate, albeit extremely slowly.

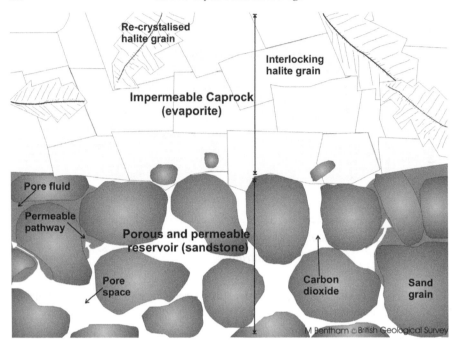

Figure 2.5   Diagram of a halite cap rock overlying a reservoir of sandstone

If the cap rocks contain no faults or other fractures, their effectiveness at retaining gaseous or supercritical CO_2 depends mainly on their capillary entry pressure – the pressure necessary for CO_2 to enter their pore system (Surdam, 1997). This is essentially a function of the size of the pore throats connecting the pores within the cap rock and the interfacial tension between the fluid attempting to enter the rock (CO_2) and the native pore fluid (commonly brine). Capillary entry pressure can be measured in the laboratory and used to predict the sealing capacity of cap rocks. However, it does not provide all the answers because in real situations cap rocks also may contain faults or fractures that could cause them to leak. Internal permeability barriers such as shale interbeds may occur within reservoir rocks and these can trap CO_2 in the same way as permeability barriers at the top of a reservoir formation (Figure 2.6).

Residual saturation trapping   As the CO_2 migrates through the reservoir rock, a fraction of it may be retained in traps formed by internal permeability barriers within the reservoir, and these also make the migration path of the CO_2 through the reservoir more tortuous. Also, CO_2 will be trapped by capillary forces in pores and by adsorption onto grain surfaces along the migration path of the CO_2 within the reservoir (Figure 2.6). This 'residual' CO_2 saturation along the migration path could be in the order of 5–30 per cent (Ennis-King and Paterson, 2001).

Dissolution trapping   During the migration process, CO_2 that comes into contact with the native pore fluid (often called the formation water) will dissolve into it. The solubility of CO_2 in water depends on temperature, pressure and salinity (Czernichowski-Lauriol *et al.*, 1996). For typical subsurface conditions, solubility of CO_2 in brine varies between about 20 and 50 kg/m^3 below 600 m depth. The rate of dissolution will depend on how well the CO_2 mixes with the formation water once it is injected into the reservoir. For many accumulations, dissolution could be slow; in the order of a few thousand years for typical injection scenarios (Ennis-King and Paterson, 2001). However, the CO_2 that does dissolve will only migrate at the (commonly very low) rates at which natural fluid flow occurs within reservoir rocks. If the migration of the CO_2 is very slow and the proposed injection point is a very large distance from the edge of the reservoir, the CO_2 may not reach the edge of the reservoir for millions of years.

Geochemical trapping   Some of the injected CO_2 may also become trapped by chemical reaction with either the formation water or the reservoir rock – the latter will take place only over long time scales (Xu, Apps and Pruess, 2003) – the amount depending on the pore water chemistry, rock mineralogy and the length of the migration path (Gunter *et al.*, 1993). Geochemical reaction of CO_2 with basic aluminosilicate minerals can result in the precipitation of carbonate minerals in the reservoir and thus the long term fixation of carbon as a solid phase.

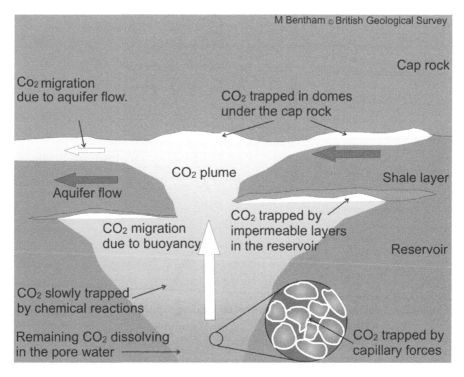

Figure 2.6   The trapping mechanisms that store CO_2 in reservoir rocks

Thus, in the long term, the interaction of five principle mechanisms will determine the fate of the CO_2 in the reservoir. These are: immobilization in traps, immobilization of a residual saturation of CO_2 along the CO_2 migration path, dissolution into the surrounding formation water, geochemical reaction with the formation water or minerals making up the rock framework and, if the seal is not perfect, migration out of the geological storage reservoir.

The amount of CO_2 that can be injected during a particular project or into a particular reservoir is limited by the undesirable effects that could occur. Some of these might be important in the short term, others may occur in much longer timescales, as the result of migration of the injected CO_2. They include: an unacceptable rise in reservoir pressure, conflicts of use of the subsurface (e.g. unintentional interaction with coal mining, or the exploitation of oil and gas), pollution of potable water, e.g. by CO_2 or substances entrained by CO_2 (e.g. hydrocarbons), escape of CO_2 to the outcrop of a reservoir rock and escape of CO_2 via an unidentified migration pathway through the cap rock. Some potential leakage pathways are shown in Figure 2.7.

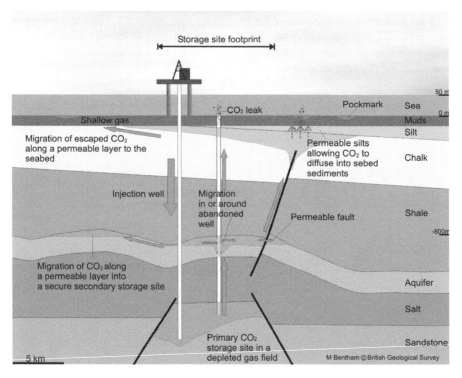

Figure 2.7   Diagram of some potential leakage pathways for CO_2 stored underground

When CO_2 is pumped into a reservoir rock, displacement of the native pore fluid will only occur if the injection pressure is greater than the native pore fluid pressure. If the permeability of the rock is low or there are barriers to fluid flow within the rock, for example faults that compartmentalize the reservoir, CO_2 injection may cause a significant increase in pressure in the pore spaces, especially around the injection well (van der Meer, 1992). This may limit both the amount of CO_2 that can be injected into a reservoir and the rate at which it can be injected. For example, in Alberta, the maximum allowable injection pressure is 90 per cent of the fracture pressure at the top of the reservoir (Law, 1996). This factor could make heavily compartmentalised reservoirs unsuitable for CO_2 injection.

Storage in Depleted or Abandoned Oil and Gas Fields

Oil and gas fields are natural underground traps for buoyant fluids. In many cases there is geological evidence that the oil or gas has been trapped in them for hundreds of thousands or millions of years. In such cases, they will not leak in the geologically short term (a few hundred to a few thousand years) as long as their exploitation by man has not damaged the trap and the cap rock is not adversely affected by the injection of CO_2.

CO_2 is widely used for enhancing oil recovery in depleted oil fields (Taber, 1994) so it should be possible to store CO_2 in such fields and increase oil production at the same time (e.g. Bondor, 1992; Holt, Jensen and Lindeberg, 1995; Stevens, Kuuskraa and Gale, 2001). The production of additional oil would at least partially offset the cost of CO_2 sequestration. The carbon balance of EOR is field-specific and highly dependent on the aims of the project. For current EOR projects in the Permian Basin, Texas, the CO_2 has to be purchased, so common practice is to recover significant amounts of CO_2 from the reservoir at the end of the project by reducing the reservoir pressure, so it can be re-used in further projects. In such cases, some 2000 standard cubic feet of CO_2 or less remain stored underground for every barrel of additional oil recovered (Ruether *et al.*, no date). Therefore approximately 3.95 g carbon is produced for every 1 g carbon stored. However, if there was a value in leaving the CO_2 in the ground, the reservoir pressure might not be reduced at the end of the project and significantly greater amounts of CO_2 might be stored per barrel of oil recovered. Net CO_2 purchased in EOR projects (i.e. stored assuming no leakage and no CO_2 recovery at the end of the project) is about 5600 to 6000 standard cubic feet per barrel additional oil recovered (Hsu 1995; Taber 1993). In this scenario, roughly 1.36 g carbon would be produced for every 1 g carbon stored. It might be possible to store about 8900 standard cubic feet CO_2 per barrel additional oil recovered if there was an incentive to use a higher ratio of CO_2 to water in the EOR operations and not depressurize the field at the end of operations (Ruether *et al.*, *op cit.*). This could result in 0.89 g carbon produced for every 1 g carbon stored, i.e. net carbon storage.

Some of the CO_2 used in EOR projects is anthropogenic, e.g. at Encana's Weyburn field in Saskatchewan anthropogenic CO_2 from a coal gasification plant in North Dakota (Wilson *et al.*, 2001; Wilson and Monea, 2004) is used. The progress of this CO_2 flood is being monitored from a CO_2 sequestration perspective. It will permanently sequester about 18 million tonnes of CO_2 over the lifetime of the project. The Rangely EOR project in Colorado has also been monitored to determine whether CO_2 is leaking from the reservoir to the ground surface (Klusman, 2003). Further opportunities for EOR abound, especially if recent increases in the price of oil are maintained. There is undoubtedly significant potential in many of the world's major onshore oil provinces, for example the Middle East. However, the relatively small amount of CO_2 sequestered in such projects indicates that EOR would have to take place on a massive scale to have a significant impact on global CO_2 emissions to the atmosphere (Stevens, Kuuskraa and Gale, 2001).

From a UK perspective, there is significant potential for EOR in the North Sea (Blunt *et al.*, 1995) but there are cost and technical issues, such as the need for CO_2-compliant infrastructure and the larger distance between wells in offshore fields that need to be addressed (Espie *et al.*, 2001). Offshore EOR has not yet been implemented anywhere in the world, so there is significant financial risk, which possibly could be offset by tax adjustment and carbon credits for the CO_2 stored. Offshore EOR would represent a significant benefit to the UK, both in terms of increased security of energy supply and enhancement of sustainability by making the best use of resources. The majority of the UK's oil fields are located in the Northern and Central North Sea, hundreds of kilometres from the major point sources of CO_2 (Figure 2.8). An additional advantage of the implementation of EOR in these fields would be the development of an infrastructure that could be used to store CO_2 in the saline aquifers of the area. This is unlikely to develop without some kind of stimulus because of the long transport distances involved.

When natural gas is produced from a gas field, the production wells are opened and the pressure is simply allowed to deplete, usually without any fluid being injected to maintain the pressure. Thus, depending on the rate of water inflow into the porosity that comprises the gas reservoir, a large volume of pressure-depleted pore space may be available for CO_2 storage when the gas has been produced. In many cases there is little or no water flow into a gas reservoir. Therefore it may be possible to store underground a volume of CO_2 equal to the underground volume of the gas produced. Furthermore, there is a possibility that CO_2 injection could enhance natural gas production towards the end of field life. CO_2 is denser than methane at reservoir conditions and if it was injected at the base of the field, the pressure increase might help drive methane towards the production wells. However, some mixing of CO_2 and methane would likely take place and any CO_2 that as produced along with the methane would have to be separated from it to get the gas to sales quality. This would reduce the economic benefits so enhanced gas recovery is likely to result in only small fractions of incremental production.

The majority of the UK's gas fields (Figure 2.8) are situated in the Southern North Sea Basin and East Irish Sea Basin (Figure 2.9). They have very substantial potential CO_2 storage capacity and are situated relatively close to some of the major point sources of CO_2.

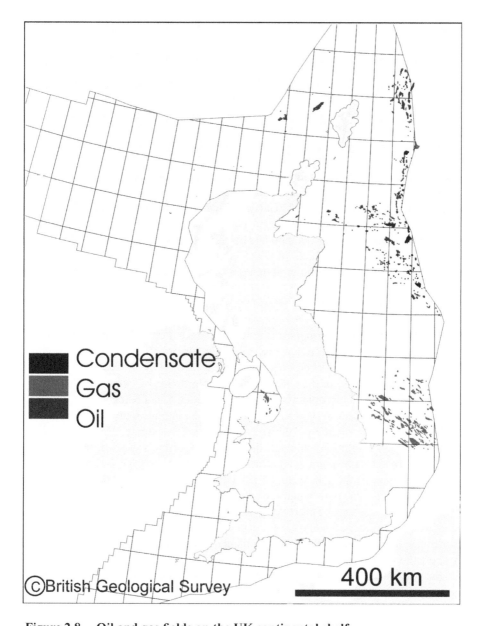

Figure 2.8   Oil and gas fields on the UK continental shelf

What is the Global and UK CO_2 Storage Capacity in Oil and Gas Fields and Saline Water-bearing Reservoir Rocks?

The availability of sufficient storage capacity is one of the critical parameters that could decide whether the underground sequestration of CO_2 can be a major contributor to solving this century's greenhouse problem. The storage capacity of oil and gas fields is relatively well defined being based on the principle that a proportion of the pore space occupied by the recoverable reserves of a field is, or will be, available for the storage of CO_2. As the pore volume of the field is well known, the mass of CO_2 that could be stored in the total pore volume provides an upper bound, which can be discounted to take account of factors that might reduce the storage capacity of oil or gas fields. The global CO_2 storage capacity of oil and gas fields has been estimated to be 923 Gt (Gt = Gigatonnes = 10^9 tonnes) (IPCC, 2005), equivalent to about 40 years of current global anthropogenic CO_2 emissions. The CO_2 storage capacity of the UK's oil and gas fields is estimated to be approximately 6.2 Gt (Gibbins *et al.*, 2006), all of which is offshore. With the exception of Wytch Farm, none of the UK's onshore oil and gas fields is sufficiently large to provide a significant storage opportunity.

No widely accepted methodology for calculating the aquifer storage capacity of reservoir formations has yet been developed, and it is difficult to marshal the minimum data and other resources necessary to make a crude estimate, even in the UK, where data is comparatively easy to obtain. Nonetheless estimates of aquifer storage capacity have been produced. Holloway and Baily (1996) calculated the aquifer storage capacity of the UK sector of the North Sea by two methods, both of which oversimplify the problem. Firstly they assumed that a percentage of the pore volume of all potential reservoir formations would be in traps for buoyant fluids, and a proportion of that volume would be available for CO_2 storage. This is essentially a conservative assumption, although the injectivity of the various formations was not considered. This resulted in an estimate of approximately 8.6 Gt for the aquifer storage potential of the UK sector of the North Sea. Secondly they assumed that a fraction of the total pore volume of all potential reservoir formations would be available for CO_2 storage. This resulted in an estimate of up to 240 Gt CO_2. It is now recognized that the methodology and data on which these estimates were based is inadequate – albeit the best that was available at the time – and the upper estimate is likely too high.

Subsequently, Holloway *et al.* (in press) have estimated the aquifer CO_2 storage capacity of the southern North Sea alone to be up to 14.2 Gt CO_2 and Kirk (2005) has estimated that of the East Irish Sea Basin to be up to 0.6 Gt CO_2; in both cases the storage capacity could possibly be very significantly less, not least because the estimate takes no account of potential leakage. However, given the large number of potential reservoir formations in the sedimentary basins surrounding the UK, it is considered likely that the total CO_2 storage capacity of the UK comfortably exceeds 20 Gt (Gibbins *et al.*, 2006).

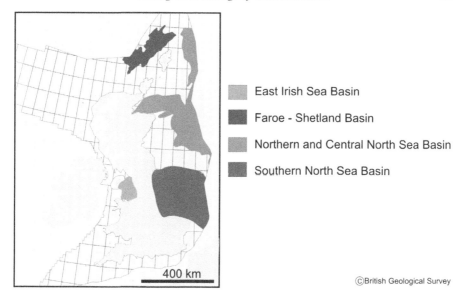

East Irish Sea Basin

Faroe - Shetland Basin

Northern and Central North Sea Basin

Southern North Sea Basin

400 km

©British Geological Survey

Figure 2.9   The main oil and gas-bearing sedimentary basins on the UK continental shelf

Storage in Coal Beds

Coal beds (otherwise known as coal seams) can be reservoirs for gases. Coal contains a natural system of orthogonal fractures known as the cleat, which imparts some permeability to it (Figure 2.10), and although it does not contain significant conventional porosity it contains micropores in which a natural gas known as coalbed methane (CBM) can occur. This usually consists of >90 per cent methane plus small amounts of higher hydrocarbons, CO_2 and N_2. The gas molecules are adsorbed onto the surfaces of the micropores. They are very closely packed and so UK bituminous coals can adsorb up to about 20 m^3 methane/tonne of coal (Creedy, 1991).

The gas molecules in the coal micropores are held in place by electrostatic forces. These are much weaker than chemical bonds and sensitive to changes in temperature and pressure. If the temperature is raised, or the pressure lowered, gas will desorb from the coal (Davidson, Sloss and Clarke, 1995).

Commercial CBM fields exist in the United States, e.g. in the San Juan Basin (Colorado/New Mexico) and Warrior Basin (Alabama) and also in Australia. However only a minority of coalfields are suitable for commercial CBM recovery using present technology, because economic production is only possible from coal beds with exceptional permeability. If there is sufficient permeability within a coal bed, CBM production is achieved by drilling a well into the coal bed, sealing it off from the surrounding strata and pumping water out of the cleat to lower the pressure within the coal bed. This causes the gases to desorb from the sorption sites and move into the cleat system, where the pressure gradient causes them to move towards the production borehole.

Figure 2.10 Cleat in coal

CO_2 has a greater affinity to be adsorbed onto coal than methane. Thus, if CO_2 is pumped into a coal seam, not only will it be stored by becoming adsorbed onto the coal, it may displace any methane at the adsorption sites (Gunter *et al.*, 1996). This theory forms the basis of the concept of enhanced coalbed methane production by the injection of CO_2 into coal seams. In this process CO_2 would be injected into coal seams. Once adsorbed, the CO_2 would be held in place and would not leak to the surface unless the pressure on the coal is reduced or the temperature increased. Thus it would be effectively isolated from the atmosphere. The displaced methane would be recovered at a production well and its sale would offset some of the costs of CO_2 storage.

Enhanced coalbed methane production experiments have been conducted in the San Juan Basin by Burlington Resources (Reeves, 2005). Over 370,000 tonnes of CO_2 have been injected into the Fruitland coal seams since 1996. The results of these experiments were encouraging; a detailed reservoir simulation suggests that CO_2 injection increased the recovery of methane from 77 per cent to 95 per cent of the gas originally in place in the coal. However, the injectivity declined by about 60 per cent during the injection period because the adsorption of CO_2 causes the coal matrix to swell.

Nitrogen can also be used to enhance coalbed methane production (Reeves, 2005). Nitrogen injection reduces the partial pressure of methane and thus encourages methane to desorb from the coal matrix. N_2 injection experiments in the San Juan Basin were highly successful, producing a large increase in methane production in a relatively short time. So it may be possible to enhance coalbed methane production by injecting flue gas (principally a mixture of N_2 and CO_2 with small amounts of nitrogen oxides and sulphur gases) into the coal beds. However, the separation of N_2 and CO_2 from CH_4 is likely to be costly, so once breakthrough of N_2 and CO_2 into the production wells occurs, the process may no longer be economic. Therefore this technique is likely to be used for only a short period at the end of a coalbed methane production project.

Controlled experiments to test enhanced coalbed methane (ECBM) production using CO_2 as a stimulant are under way in Europe (Pagnier *et al.*, 2005), Alberta (Wong and Gunter, 1999) and Japan (Yamaguchi, 2005). However, the methane in coal represents only a small proportion of the energy value of the coal, and the remaining energy would be sterilised if the coal was used as a CO_2 storage reservoir, i.e. the coal could not be mined or gasified underground without releasing the CO_2 to the atmosphere. Moreover it would be important to prevent fugitive CH_4 emissions because CH_4 is a much more powerful greenhouse gas than CO_2.

From a UK perspective it is considered unlikely that CO_2 storage in coal seams will play a significant role in CO_2 storage in the near term, because such evidence as is in the public domain indicates that most UK coals have low permeability (Creedy *et al.*, 2001) and, to date, no coalbed methane production has been established. However, there are very large untouched coal resources in the UK if suitable technology can be developed (Jones *et al.*, 2004).

2.3   Safety and Security of CO_2 Storage

Health Effects of CO_2

CO_2 can be toxic, causing asphyxiation at high concentrations and acidosis at lower concentrations. As CO_2 is colourless and odourless, people (and animals) can be totally unaware of entering areas where the concentrations are dangerously high, and may collapse before being able to make their escape. CO_2 is heavier than air, so it accumulates in depressions and in poorly ventilated enclosed spaces such as cellars and basements. Concentrations of 100,000 ppm (i.e. 10 per cent) are directly toxic (Vendrig *et al.*, 2003). The gas is an asphyxiant, a cerebral vasodilator rapidly causing circulatory failure, coma and then death. The symptoms resulting from acidosis (lowering of the pH of the blood) include headache, nausea, visual disturbance and laboured breathing (dyspnoea). These are normally experienced at concentrations of >15,000 ppm (1.5 per cent CO_2). A few minutes exposure at concentrations of 7–10 per cent is sufficient to cause loss of consciousness and less than a minute's exposure to concentrations of 15–30 per cent can be fatal. West *et al.* (2005) provide an overview natural elevated concentrations and fluxes of CO_2 and the effects of CO_2 exposure on selected organisms.

However, the numerous carbon dioxide springs in Italy illustrate how leakage at rates of 150 tonnes per day can be of little consequence even in regions of high population density. Experience shows that the engineering and routing requirements to optimise the safety of the operation of carbon dioxide pipelines are not going to impose insuperable problems and are likely to be less problematical than the current networks of pipelines transmitting natural gases and hazardous liquids over long distances. Leakage out of the repositories themselves will mimic natural venting of carbon dioxide.

Submarine leakage is of far less concern to mankind. Much of any carbon dioxide released will dissolve in the seawater where the high bicarbonate content will buffer any pH changes. Carbon dioxide is vented from seeps in the seabed, both in shallow water and in deep water with very little apparent influence on marine life. 'Bubbles' of liquid carbon dioxide have been filmed by ROV (Remotely Operated Vehicles) issuing from hydrothermal vents at depths of 900m off the Marianas. Whereas bubbles of gaseous CO_2 appear to have little effect on plankton, *in situ* experiments on the emplacement of liquid carbon dioxide on the seabed at depth have shown that liquid CO_2 is extremely toxic to plankton that ventures into its immediate vicinity. Submarine leakage is likely to have a localised environmental impact, which in the dynamic marine environment will be extremely difficult to detect. Turley *et al.* 2005 produced an extensive review of the literature relating to the potential environmental impacts of CO_2 to the marine environment. The overlying seawater will shield personnel on platforms at the sea surface. It has to be concluded that submarine disposal will be less risky to humankind, since small leaks will not lead to localised accumulations.

Safety and Security of Long-Term Storage

Although the safe and stable storage of CO_2 in the subsurface can be demonstrated over the medium term (decades) the question of whether it can be assured over the long term (centuries to millennia) is probably the most important issue facing the underground storage of CO_2 at present, because this is likely to have a high impact on public acceptability and regulation.

To ensure safe and stable containment of the injected CO_2, a rigorous risk assessment process is required. One approach is to identify all the Features, Events and Processes (FEPs) that could affect the storage site (Savage, 2005) and then assess the risks associated with these. Detailed geological characterization of the selected site and surrounding area should take place prior to CO_2 storage. This should be used to help with the risk assessment process, for example by building geological models of the site, to provide information about the volume of the storage reservoir and any potential migration paths out of it. The geological data and models should also be used to construct numerical reservoir models which can be used to simulate the injection of CO_2 at the site and determine the likelihood, potential magnitude, timing and location of any CO_2 migration out of the storage reservoir or to the ground surface or sea bed. This in turn should provide the basis for a monitoring plan and, if considered necessary, remediation plans. Baseline monitoring surveys should also be acquired prior to injection.

Once injection starts, long term monitoring would be needed to validate storage. Some types of data such as the mass of CO_2 injected, need to be monitored continuously whereas other data, such as the distribution of CO_2 within the reservoir as imaged by seismic surveys, may only need to be acquired intermittently. Seismic reflection surveys, seismic attribute studies, gravity surveys, infra-red CO_2 detection equipment and data and samples acquired from wells are amongst the techniques being used for monitoring at present (Pearce *et al.*, 2005). Monitoring data should be history matched to predictions from the numerical simulation models to check whether the site is performing as predicted. If significant discrepancies are found, more geological data should be acquired and/or the models adjusted as necessary.

Once injection ends, it is considered likely that monitoring would continue for a significant period, until the operator and regulator are satisfied that the site is performing, and will continue to perform, as predicted. Site closure would then follow.

A risk assessment of CO_2 storage and enhanced oil recovery at the Forties oilfield, North Sea (Cawley *et al.*, 2005) concluded that the risk of leakage from the natural geological system was negligible. However, the risk of CO_2 escape via the pre-existing wells or any new wells in the area could not be fully addressed.

Impacts of CO_2 Leakage from Underground or Sub-Seabed CO_2 Storage

There are many areas in the world where CO_2 naturally emanates from the subsurface (Holloway, 1997; Raschi *et al.*, 1997; Czernichowski-Lauriol *et al.*, 2003; Pearce, 2004; Shipton *et al.*, 2005) or is stored in the subsurface (Pearce *et al.*, 1996; Stevens and Fox, 2001; Watson, Zwingmann and Lemon, 2003; Stevens, 2005). Many such natural CO_2 emission provinces are active volcanic areas, particularly areas of hydrothermal activity such as Yellowstone national park in Wyoming, USA. But they also occur in sedimentary basins similar to those in which it is envisaged CO_2 storage might take place, for example the French carbo-gaseous province and the Paradox Basin, USA. In volcanic regions, significant CO_2 emissions take place during volcanic eruptions but these have little bearing on the safety of man-made CO_2 storage sites. Other CO_2 emissions in volcanic and hydrothermal areas can also be sudden and violent because in such areas large amounts of CO_2 are commonly found along with high temperatures and steam at shallow depths.

Natural emissions from carbo-gaseous provinces in sedimentary basins, where high temperatures at shallow depths are not commonly found, are therefore more likely to be useful as analogues for leaks from man-made CO_2 storage facilities than those from volcanic or hydrothermal areas. CO_2 does not emerge in an evenly dispersed manner throughout these provinces, it emerges through small, distributed emission sites such as carbonated springs, geysers and dry emission sites where

CO_2 emerges from the ground as a gas, known as mofettes. Although fluxes from emission points in sedimentary basins are generally low, they can cause danger if CO_2 builds up in confined spaces such as buildings. Consequently, this possibility should be included in risk assessments for CO_2 storage sites and buildings alarmed for increased CO_2 levels.

At man-made CO_2 storage sites, a major well failure in the injection period, when reservoir pressure was relatively high, could theoretically pose the danger of the development of a major cloud of CO_2 at the ground surface (Cox *et al.*, 1996). However, a well blow-out in a natural CO_2 field has occurred, was successfully controlled and did not cause significant damage to man or the natural environment (Lynch *et al.*, 1985).

Leaks from offshore storage sites to the sea bed may prove more difficult to remediate due to problems of access. They could also impact adversely on marine flora and fauna as they are likely to result in a plume of sea water with reduced pH downstream of the leak point. Natural CO_2 emissions from the sea bed are known, e.g. in the Tyrrhenian Sea (Italiano *et al.*, 2001). The environmental impact of these leaks is not presently known.

Insights into safety and security of storage can also be gained from the study of engineering analogues for CO_2 storage and leakage such as natural gas storage facilities in aquifers (Perry, 2005). Methodologies exist for determining storage security in natural gas storage projects, but these are generally significantly smaller than conjectured CO_2 storage schemes, and always confine the gas within a structural trap.

There are three commonly cited natural disasters that have resulted in multiple deaths from asphyxiation by CO_2. These are the 1984 Lake Monoun event, the 1986 Lake Nyos event (both in Cameroon), and the 1979 event at the Dieng volcanic complex in Indonesia. Of these, the Lake Nyos disaster (Kling *et al.*, 1987; Sigurdsson *et al.*, 1987) is probably the most infamous. A brief description of it is included here, not because it necessarily has great direct relevance to putative leakage from man-made underground CO_2 storage sites, but rather to illustrate the low likelihood of such an event occurring as a result of purposeful storage, and that successful remediation has taken place at Lake Nyos that should prevent further disasters.

Sometime during the late evening of August 21, 1986, a huge mass of concentrated CO_2 was emitted from Lake Nyos, a volcanic crater lake in Cameroon. A lethal concentration of the gas reached a height of 120 m above the lake surface, and the total volume of CO_2 in the lethal gas cloud may have been up to 0.63 km^3; a mass of approximately 1.24 Mt CO_2. It flowed out of the spillway at the northwest end of the lake and down the topographic slope, along two valleys. It killed 1746 people in a thinly populated area, and all animal life along its course as far as 14 km from the crater. This disaster was caused by a so-called 'limnic eruption' – a sudden release of CO_2 caused by the overturn of the 220 m deep lake, the lower part of which had become saturated with CO_2 of volcanic origin, caused by a slow leak of CO_2 into the lake waters from below (Kanari, 1989).

The CO_2 leaking into the lake dissolved in the water in the lower part of the lake increasing the water density. This resulted in the lake waters becoming density-stratified and the lower, dense horizon becoming saturated with CO_2. Thus any significant disturbance of the lake waters had the potential to cause part of the CO_2-saturated lower lake waters to rise and emit bubbles of CO_2. Once a train of rising bubbles became established this would lift more of the lower lake waters towards the surface and a self-accelerating increase in the rate of the degassing process would be established until a major and very rapid degassing event occurred. Zhang (1996) presents a thermodynamic and fluid dynamic understanding of this process. The trigger for the overturn of Lake Nyos is not known, but it may have been a long period of cool days that allowed cold surface water to build up and then sink, disturbing the density stratification.

Clearly the likelihood that an accident comparable to the Lake Nyos natural disaster could occur as a result of leakage from a man-made underground CO_2 storage facility must be considered. However, it should be noted that the topography around Lake Nyos appears to provide ideal conditions for the emitted CO_2-rich gas cloud to remain concentrated rather than disperse. The CO_2 held in the lake waters was probably released in a few hours at most, and would have hugged the ground rather than dispersing. High crater walls surround the lake on the east and west sides, and the natural water spillway in the northwest corner of the lake provides a natural outlet for the CO_2 into a valley system, where it would remain confined. The sudden emissions of concentrated CO_2 from crater lakes in Cameroon are the result of slow emissions of carbon dioxide into relatively small, deep lakes.

It would be relatively simple to determine whether any such lakes occur in the vicinity of a proposed CO_2 storage site and, if necessary monitor them. Most lakes outside the tropics overturn seasonally, as a result of temperature changes in the surface waters, and so there may be less potential for stratification outside the tropics. Thus the possibility of an analogous event resulting from the leakage of CO_2 from a storage reservoir could easily be excluded. Furthermore, Lake Nyos is being degassed at the moment, precisely to prevent a recurrence of the tragedy (Kling *et al.*, 2005), and its shores are monitored and alarmed for increased CO_2 levels. A similar strategy could be adopted for any lake into which CO_2 might leak from a man-made CO_2 storage facility.

Carbon Dioxide Vented by Fumaroles

Carbon dioxide is a common constituent of the gases vented by fumaroles. Throughout the East African Rift valley there are many fumaroles that are almost continuously venting nearly pure CO_2. Many of these vents lie at the bottom of depressions, which during windless conditions in the early morning can become filled with carbon dioxide. These hollows are death traps for game that accidentally ventures into them, then the smell of the decaying corpses attracts in scavengers that also succumb. These are described locally as 'elephant graveyards' (Lockwood and Tuttle, 1991). In the Virunga volcanic range on the borders of Congo and Rwanda, vents from fumaroles have recently suffocated several people in the vicinity the refugee camp at Goma. There is also a fear that a build up of CO_2 in the deep waters of Lake Kivu may lead to a repeat of the Lake Nyos disaster.

In Southern Italy there are over 150 vents, or carbon dioxide springs, which release ~150 tonnes CO_2 per day. Mount Etna itself is reported to vent 35,000 tons CO_2/day. In total these Italian vents are estimated to release ~5–10 per cent of the estimated magmatic CO_2 discharged globally to the atmosphere by active volcanoes (Cardellini *et al.,* 2000). Responses of the vegetation around these springs have been extensively studied, using the carbon dioxide springs as natural experimental analogues for global rises in atmospheric CO_2 concentrations (Raschi *et al.*, 1997; Scholefield *et al.*, 2004; Rapparini *et al.*, 2004). A survey of two vents in Tuscany showed an accumulation of dead animals (bats, mice, rabbits, rats, cats and birds, even a 2cm thick crust of flies); only at one was there any sign of living animals – a spider which had woven webs across the vent holes (Bridges *et al.*, 2000). Some species of plant flourish in the vicinity of the springs, and show enhanced productivity. One positive effect of the heightened CO_2 concentrations is the inhibition of the release of chloroplast-derived isoprenoids (Rosentiel, 2003; Scholefield *et al.*, 2004), which otherwise would increase local ozone concentrations. It is worth noting that local human activity is almost totally unaffected by the presence of these CO_2 vents; hence small leaks of carbon dioxide from pipelines need not necessarily cause serious disruption to human activity.

Long Term Liability for CO_2 Storage

Finally, little is known about the long-term CO_2 storage issues. The required storage period is greater than the likely lifetime of any corporation. This raises issues of ownership, monitoring and liability for leaks or man-made breaches of the storage integrity into the distant future. Because of the longevity of storage, it seems inevitable that ownership and liability would, at some stage, be transferred to the State.

2.4   References

Baines, S.J. and Worden, R.H. (2004), 'The Long Term Fate of CO_2 in the Subsurface: Natural Analogues for CO_2 Storage', in S.J. Baines and R.H. Worden (eds), *Geological Storage of Carbon Dioxide*. Geological Society, London, Special Publications, **233**, pp. 59–85.

Blunt, M., Fayers, J.F. and Orr, F.M. (1995), 'Carbon Dioxide in Enhanced Oil Recovery', *Energy Conversion and Management*, **34**, pp. 1197–1204.

Bondor, P.L. (1992), 'Applications of Carbon Dioxide in Enhanced Oil Recovery', *Energy Conversion and Management*, **33**(5–8), pp. 579–586.

Cawley, S., Le Gallo, Y., Carpentier, B., Holloway, S., Bennison, T., Wickens, L., Wikramaratna, R., Saunders, M., Kirby, G.A., Bidstrup, T., Ketzer, J.M., Arkley,

S.L.B. and Browne, M.A.E. (2005), 'The NGCAS Project - Assessing the Potential for EOR and CO$_2$ Sequestration at the Forties Oilfield, Offshore UK', in S.M. Benson (ed.), *Carbon Dioxide Capture for Storage in Deep Geologic Formations – Results from the CO$_2$ Capture Project, Volume Two: Geologic storage of Carbon Dioxide with Monitoring and Verification*, Elsevier Science, Oxford, pp. 713–50.

Cox, H, Heederik, J.P., van der Meer, L.G.H. and van der Straaten, R. (1996), 'Chapter 5: Safety and Stability of Underground Storage', in S. Holloway (ed.), *The Underground Disposal of Carbon Dioxide, Final Report of Joule 2 Project*, No. CT92-0031, British Geological Survey, Nottingham.

Creedy, D.P. (1991), 'An Introduction to Geological Aspects of Methane Occurrence and Control in British Deep Coal Mines', *Quarterly Journal of Engineering Geology*, **24**, pp. 209–20.

Creedy, D.P., Garner, K., Holloway, S. and Ren, T.X. (2001), *A Review of the Worldwide Status of Coalbed Methane Extraction and Utilisation*, DTI Cleaner Coal Technology Programme Report No. COAL R210, DTI/Pub URN 01/1040, DTI, London.

Czernichowski-Lauriol, I., Sanjuan, B., Rochelle, C., Bateman, K., Pearce J.M. and Blackwell, P. (1996), 'Inorganic Geochemistry', in S.Holloway (ed.), *The Underground Disposal of Carbon Dioxide, Final Report of Joule 2 Project*, No. CT92-0031, British Geological Survey, Nottingham, pp. 183–276.

Czernichowski-Lauriol, I., Pauwels, H., Vigouroux, P. and Le Nindre, Y-M. (2003), 'The French Carbo-Gaseous Province: An Illustration of Natural Processes in CO$_2$ Generation, Migration, Accumulation and Leakage', in J. Gale, and Y. Kaya (eds), *Proceedings of the 6th International Conference on Greenhouse Gas Control Technologies*, Volume 1, Pergamon, Oxford, pp. 411–16.

Davidson, R.M., Sloss, L.L. and Clark, L.B. (1995), *Coalbed Methane Extraction*, IEA Coal Research Report No. IEACR/76, IEA Coal Research, London.

Ennis-King, J. and Paterson, L. (2001), 'Reservoir Engineering Issues in the Geological Disposal of Carbon Dioxide', in D. Williams, B. Durie, P. McMullan, C. Paulson and A. Smith (eds), *Proceedings of the 5th International Conference on Greenhouse Gas Control Technologies*, CSIRO Publishing, Collingwood, pp. 290–295.

Espie, A.A. (2001), 'Options for Establishing a North Sea Geological Storage Hub', in D. Williams, B. Durie, P. McMullan, C. Paulson and A. Smith (eds), *Proceedings of the 5th International Conference on Greenhouse Gas Control Technologies*, CSIRO Publishing, Collingwood, pp. 266–271.

Gibbins, J., Haszeldine, S., Holloway, S., Pearce, J.M., Oakey, J., Shackley, S and Turley, C. (2006), 'Scope for Future CO$_2$ Emission Reductions through the Deployment of Carbon Capture and Storage Technologies', in H.J. Schellnhuber, W. Cramer, N. Nakicenovic, T. Wigley and G. Yohe (eds), *Avoiding Dangerous Climate Change*, Cambridge University Press, Cambridge.

Gunter, W.D., Gentzis, T., Rottenfusser, B.A. and Richardson, R.J.H. (1996), 'Deep Coalbed Methane in Alberta, Canada: A Fuel Resource with the Potential of Zero Greenhouse Gas Emissions', *Proceedings of the Third International Conference on Carbon Dioxide Removal*, Cambridge, MA., September 9–11, 1996, pp. 217–22.

Gunter, W.D, Perkins, E.H. and McCann, T.J. (1993), 'Aquifer Disposal of CO_2-Rich Gases: Reaction Design for Added Capacity', *Energy Conversion and Management*, **34**, pp. 941–48.

Holliday, D.W., Williams, G.W., Holloway, S., Savage, D. and Bannon, P. (1991), *A Preliminary Study of the Feasibility of Underground Disposal of Carbon Dioxide*, BGS Technical Report No. WE/91/20, British Geological Survey, Nottingham.

Holloway, S. (1997), 'Safety of the Underground Disposal of Carbon Dioxide,' *Energy Conversion and Management*, **38**, Suppl., pp. S241–S245.

Holloway, S. and Baily, H.E. (1996), 'The CO_2 Storage Capacity of the United Kingdom', in S. Holloway (ed.), *The Underground Disposal of Carbon Dioxide, Final Report of Joule 2 Project*, No. CT92–0031, British Geological Survey, Nottingham, pp. 92–105.

Holloway, S., Vincent, C.J., Bentham, M.S. and Kirk, K.L. (in press), 'Top-Down and Bottom-Up Estimates of CO_2 Storage Capacity in the UK Sector of the Southern North Sea Basin', *Environmental Geoscience*, in press.

Holt, T., Jensen, J-I. and Lindeberg, E. (1995), 'Underground Storage of CO_2 in Aquifers and Oil Reservoirs', *Energy Conversion and Management*, **36**(6–9), pp. 535–38.

Hovorka, S., Doughty, C. and Holtz, M.H. (2005), 'Testing Efficiency of CO_2 Storage in the Subsurface: Frio Brine Pilot Project', in M. Wilson, T. Morris, J. Gale, and K. Thambimuthu (eds), *Proceedings of 7th International Conference on Greenhouse Gas Control Technologies. Volume 2: Contributed Papers and Panel Discussion*, Elsevier Science, Oxford, pp. 1361–66.

Hsu, C-F, Koinis, R.L. and Fox, C.E. (1995), 'Technology, Experience Speed CO_2-flood Design', *Oil and Gas Journal*, **93** (43), pp. 51–59.

IPCC (2005), *IPCC Special Report on Carbon Dioxide Capture and Storage*, prepared by Working Group III of the Intergovernmental Panel on Climate Change, B. Metz, O. Davidson, H.C. de Coninck, M. Loos and L.A. Meyer (eds), Cambridge University Press, Cambridge.

Italiano, F., Favara, R., Etiope, G. and Favali, P. (2001), 'Submarine Emissions of Greenhouse Gases From Hydrothermal And Sedimentary Areas', *Water-rock Interaction 2001*, Swets and Zeitlinger, Lisse, ISBN 90 2651 824 2.

Jones, N.S., Holloway, S. Smith, N.J., Browne, M.A.E., Creedy, D.P., Garner, K. and Durucan, S. (2004), *UK Coal Resource for New Exploitation Technologies*, DTI Report Number COAL R271, DTI/Pub URN 04/1879. Department of Trade and Industry, London.

Kanari, K. (1989), 'An Inference on the Process of Gas Outburst from Lake Nyos, Cameroon', *Journal of Volcanology and Geothermal Research*, **39**, pp. 135–49.

Kikuta, K., Hongo, S., Tanase, D. and Ohsumi, T. (2005), 'Field test of CO_2 Injection in Nagaoka, Japan', in M. Wilson, T. Morris, J. Gale, and K. Thambimuthu (eds), *Proceedings of 7th International Conference on Greenhouse Gas Control Technologies. Volume 2: Contributed Papers and Panel Discussion*, Elsevier Science, Oxford, pp. 1367–1372.

Kirk, K.L. (2005), 'Potential for Storage of Carbon Dioxide in the Rocks Beneath the East Irish Sea', *British Geological Survey Report* No. CR/05/127N, British Geological Survey, Nottingham.

Kling, G.W., Clark, M.A., Compton, H.R. Devine, J.D., Evans, W.C., Humphrey, A.M., Koenigsberg, E.J., Lockwood, J.P., Tuttle, M.L. and Wagner, G.N. (1987), 'The 1986 Lake Nyos Gas Disaster in Cameroon, West Africa', *Science*, **236**, pp. 169–75.

Kling, G.W., Evans, W.C., Tanylike, G., Kusakabe, M., Ohba, T., Yoshida, Y. and Hell, J.V. (2005), 'Degassing Lakes Nyos and Monoun: Defusing Certain Disaster', *Proceedings of the National Academy of Sciences*, **102**(40), pp. 14185–90.

Klusman, R.W. (2003), 'A Geochemical Perspective and Assessment of Leakage Potential for a Mature Carbon Dioxide Enhanced Oil Recovery Project as a Prototype for Carbon Dioxide Sequestration: Rangely Field, Colorado', *Bulletin of the American Association of Petroleum Geologists*, **87**, 1485–507.

Korbul, R. and Kaddour, A. (1995), 'Sleipner Vest CO_2 Disposal – Injection of Removed CO_2 into the Utsira Formation', *Energy Conversion and Management*, **36**(6–9), pp. 509–12.

Law, D. (1996), 'Injectivity Studies', in B. Hitchon (ed.), *Aquifer Disposal of Carbon Dioxide: Hydrodynamic and Mineral Trapping – Proof of Concept*, Geoscience Publishing, Alberta, pp. 59–92.

Lenton, T.M. and Cannell, M.G.R. (2002), 'Mitigating the Rate and Extent of Global Warming', *Climate Change*, **52**(3), pp. 255–262.

Lynch, R.D. McBride, E.J., Perkins, T.K. and Wiley, M.E. (1985), 'Dynamic Kill of an Uncontrolled Well', *Society of Petroleum Engineers*, Paper 11378.

Maldal, T. and Tappel, I.E. (2003), 'CO_2 Underground Storage for Snohvit Gas Field Development', in J. Gale, and Y. Kaya (eds), *Proceedings of 6th International Conference on Greenhouse Gas Control Technologies*, Volume 1, Pergamon, Oxford, pp. 601–606.

Pagnier, H. and 24 others (2005), 'Field Experiment of ECBM-CO_2 in the Upper Silesian Basin of Poland (Recopol)', in M. Wilson, T. Morris, J. Gale and K. Thambimuthu (eds), *Proceedings of 7th International Conference on Greenhouse Gas Control Technologies, Volume 2: Contributed Papers and Panel Discussion*, Elsevier Science, Oxford, pp. 1391–97.

Pearce, J.M. (ed.), (2004), *Natural Analogues for the Geological Storage of CO_2*, BGS Technical Report, British Geological Survey, Nottingham.

Pearce, J.M., Chadwick, R.A., Bentham, M.S., Holloway, S. and Kirby, G.A. (2005), *Technology Status Review – Monitoring Technologies for the Geological Storage of CO_2*, DTI Report No. COAL R285, DTI/Pub URN 05/1033, Department of Trade and Industry, London.

Pearce, J.M., Holloway, S., Wacker, H., Nelis, M.K., Rochelle C. *et al.*, (1996), 'Natural Occurrences as Analogues for Carbon Dioxide Disposal', *Energy Conversion and Management*, **37**(6–8), pp. 1123–28.

Perry, K.F. (2005), 'Natural Gas Storage Industry Experience and Technology: Potential Application to CO_2 Geological Storage', in D.C. Thomas and S.M. Benson (eds), *Carbon Dioxide Capture for Storage in Deep Geologic Formations, Volume 2*, Elsevier Science, Oxford, pp. 815–25.

Raschi, A., Miglietta, F., Tognetti, R., and van Gardingen, P.R. (eds), (1997), *Plant Responses to Elevated CO₂*, Cambridge University Press, Cambridge.

Reeves, S.R. (2005), 'The Coal-Seq Project: Key Results from Field, Laboratory and Modeling Studies', in M. Wilson, T. Morris, J. Gale, and K. Thambimuthu (eds), *Proceedings of 7th International Conference on Greenhouse Gas Control Technologies. Volume 2: Contributed Papers and Panel Discussion*, Elsevier Science, Oxford, pp. 1399–403.

Reuther, J., Dahowski, R., Ramezan, M. and Balash, P. (date unknown), *Gasification-Based Power Generation with CO₂ Production for Enhanced Oil Recovery*, http://www.netl.doe.gov/coal/gasification/pubs/pdf/35.pdf.

Riddiford, F.A., Tourqui, A., Bishop, C.D., Taylor, B. and Smith, M. (2003), 'A Cleaner Development: The In Salah Gas Project, Algeria', in J. Gale and Y. Kaya (eds), *Proceedings of 6th International Conference on Greenhouse Gas Control Technologies, Volume 1*, Pergamon, Oxford, pp. 595–600.

Savage, D. (2005), 'Development of a FEP Database for Geological Storage of Carbon Dioxide', in E.S. Rubin, D.W. Keith and C.F. Gilboy (eds), *Proceedings of 7th International Conference on Greenhouse Gas Control Technologies, Volume 1: Peer-Reviewed Papers and Overviews*, Elsevier Science, Oxford, pp. 701–709.

Shipton, Z. K., Evans, J.P., Dockrill, B., Heath, J.M., Williams, A., Kirchner, D. and Kolesar, P.T. (2005), 'Natural Leaking CO₂-Charged Systems as Analogs for Failed Geologic Storage Reservoirs', in S.M. Benson (ed.), *Carbon Dioxide Capture For Storage in Deep Geologic Formations – Results from the CO₂ Capture Project, Volume 2: Geologic Storage of Carbon Dioxide with Monitoring and Verification*, Elsevier Science, Oxford, pp. 699–712.

Sigurdsson, H., Devine, J.D., Tchoua, F.M., Pressor, T. S., Pringle, M.K.W., and Evans, W. C. (1987), 'Origin of the Lethal Gas Burst from Lake Monoun, Cameroon', *Journal of Volcanology and Geothermal Research*, **31**, pp. 1–16.

Stevens, S.H. (2005), 'Natural CO2 Fields as Analogs for Geologic CO₂ Storage', in D.C. Thomas and S.M. Benson (eds), *Carbon Dioxide Capture For Storage in Deep Geologic Formations – Results from the CO₂ Capture Project, Volume 2: Geologic Storage of Carbon Dioxide with Monitoring and Verification*, Elsevier Science, Oxford, pp. 687–697.

Stevens, S. and Fox, C.E. (2001), 'McElmo Dome and St Johns Natural CO₂ Deposits: Analogs for Geologic Sequestration,' in D. Williams, B. Durie, P. McMullan, C. Paulson and A. Smith (eds), *Proceedings of the 5th International Conference on Greenhouse Gas Control Technologies*, CSIRO Publishing, Collingwood, pp. 317–321.

Stevens, S.H., Kuuskraa, V.A. and Gale, J. (2001), 'Sequestration of CO₂ in Depleted Oil and Gas Fields: Global Capacity and Barriers', in D. Williams, B. Durie, P. McMullan, C. Paulson and A. Smith (eds), *Proceedings of the 5th International Conference on Greenhouse Gas Control Technologies*, CSIRO Publishing, Collingwood, pp. 278–283.

Studlick, J. R. J., Shew, R. D., Basye, G. L. and Ray, J. R. (1990), 'A Giant Carbon Dioxide Accumulation in the Norphlet Formation, Pisgah Anticline, Mississippi', in J.H Barwis, J.G. McPherson and J.R.J. Studlick (eds), *Sandstone Petroleum Reservoirs*, Springer-Verlag, New York, pp. 181–203.

Surdam, R.C. (ed.) (1997), *AAPG Memoir 67, Seals, Traps and the Petroleum System*, American Association of Petroleum Geologists, Tulsa.

Taber, J.J. (1994), 'A Study of Technical Feasibility for the Utilization of CO_2 for Enhanced Oil Recovery', in P.W.F. Riemer (ed.), *The Utilization of Carbon Dioxide from Fossil Fuel Fired Power Stations, Appendix B*, IEA Greenhouse Gas R and D Programme, Cheltenham.

Tissot, B.P. and Welte, D.H. (1978), *Petroleum Formation and Occurrence: A New Approach to Oil and Gas Exploration*, Springer-Verlag, Berlin.

Turley, C., Nightingale, P., Riley, N., Widdicombe, S., Joint, I., Gallienne, C., Lowe, D., Goldson, L. Beaumont, N., Mariotte, P. Groom, S., Smerdon, G. Rees, A. Blackford, J., Owens, N., West, J., Land, P. and Woodason, E., (2004), *Literature Review: Environmental Impacts of a Gradual or Catastrophic Release of CO_2 into the Marine Environment Following Carbon Dioxide Capture and Storage*, DEFRA, London.

van der Meer, L.G.H. (1992), 'Investigations Regarding the Storage of Carbon Dioxide in Aquifers in the Netherlands', *Energy Conversion and Management*, **33**(5–8), pp. 611–18.

van der Meer, L.G.H., Hartman, J., Geel, C. and Kreft, E. (2005), 'Re-injecting CO_2 into an Offshore Gas Reservoir at a Depth of nearly 4000m Subsea', in E.S Rubin, D.W. Keith and C.F. Gilboy (eds), *Proceedings of 7th International Conference on Greenhouse Gas Control Technologies, Volume 1: Peer-Reviewed Papers and Overviews*, Elsevier Science, Oxford, pp. 521–29.

Watson, M.N., Zwingmann, N. and Lemon, N.M. (2003), 'The Ladbroke Grove – Catnook Carbon Dioxide Natural Laboratory: A Recent CO_2 Accumulation in a Lithic Sandstone Reservoir', in J. Gale and Y. Kaya (eds), *Proceedings of 6th International Conference on Greenhouse Gas Control Technologies, Volume 1*, Pergamon, Oxford, pp. 417–22.

Wilson, M., Moberg, R., Stewart, B. and Thambimuthu, K. (2001), 'CO_2 Sequestration on Oil Reservoirs – A Monitoring and Research Opportunity', in D. Williams, B. Durie, P. McMullan, C. Paulson and A. Smith (eds), *Proceedings of the 5th International Conference on Greenhouse Gas Control Technologies*, CSIRO Publishing, Collingwood, pp. 243–49.

Wilson, M. and Monea, M. (2004), *IEA GHG Weyburn CO_2 Monitoring and Storage Project Summary Report 2000–2004*, Petroleum Technology Research Centre, Regina.

Wilson, T.R.S. (1992), 'The Deep Ocean Disposal of Carbon Dioxide', *Energy Conversion and Management*, **33**(5–8), pp. 627–33.

Wong, S and Gunter W.D. (1999), 'Testing CO_2 Enhanced Coalbed Methane Recovery', *Greenhouse Issues*, IEA Greenhouse Gas R&D Programme, **45**, pp. 1–3.

Xu, T., Apps, J.A. and Pruess, K. (2003), 'Reactive Geochemical Transport Simulation to Study Mineral Trapping for CO_2 Disposal in Deep Arenaceous Formations', *Journal of Geophysical Research*, **108**(B2), pp. 2071–84.

Yamada, Y. and Ikeda, T. (1999), 'Acute Toxicity of Lowered pH to some Oceanic Zooplankton', *Plankton Biology and Ecology*, **46**(1), pp. 62–67.

Yamaguchi, S., Ohga, K., Fujioka, M. and Muot, S. (2005). 'Prospect of CO_2 Sequestration in the Ishigari Coalfield, Japan,' in E.S. Rubin, D.W. Keith and C.F. Gilboy (eds), *Proceedings of 7th International Conference on Greenhouse Gas Control Technologies. Volume 1: Peer-Reviewed Papers and Overviews*, Elsevier Science, Oxford, pp. 423–30.

Zhang, Y. (1996), 'Dynamics of CO_2-driven Limnic Eruption', *Nature*, 379, pp. 57–59.

Chapter 3

Engineering Feasibility of Carbon Dioxide Capture and Storage

Jiri Klemeš, Tim Cockerill, Igor Bulatov, Simon Shackley and Clair Gough

3.1   Introduction

A CCS system may be viewed as being composed of four subsystems as illustrated in Figure 3.1, specifically: the electricity producing power plant, the CO_2 capture system, the CO_2 transport system and finally the CO_2 repository and associated injection equipment. This chapter starts with a critical state-of-the-art review and assessment of the major technological options that are available for the capture of CO_2 emissions from power plants and other process industries. The aim is to provide a user-friendly review which provides the reader with sufficient information to understand the basic technologies and their possible future evolution. In sections 3.2 and 3.3 we review the established techniques for capturing CO_2 from large industrial sources, followed by modelled results estimating the costs of CO_2 capture. In sections 3.4 to 3.6 we present a review of some of the novel approaches which have been proposed for capturing (and storing) CO_2. Once CO_2 has been captured it has to be collected and transported via a pipeline or other means to the geological storage site. A whole system techno-economic model was devised and constructed for characterising and costing the CCS system, depicted in Figure 3.1, and described in section 3.7. The main findings of the model are not discussed in section 3.7 but will be presented in future publications.

There are three major sources of CO_2 for capture:

1. Electricity generation: Fossil fuel based power generation from coal, gas and oil accounts for 29 per cent of the total CO_2 emissions at the global scale and 30 per cent of the UK's emissions. There are a number of key electricity generating technologies. Typically a coal power plant entails use of pulverised fuel (PF) which is used to heat a boiler thereby generating steam to power a turbine. Control of other pollutants from burning coal has become increasingly necessary in order to meet regulatory requirements for air quality. Particles arising from coal flue gases are removed in electrostatic precipitators, whilst flue gas desulphurisation (FGD) is an established technology for removal of sulphur. Natural gas is generally used within Combined Cycle Gas Turbines (CCGTs), in which gas combustion powers a

gas turbine; the hot flue gases are then passed through a heat exchanger which raises heat in a boiler to power a steam turbine. An emerging technology for use of coal is the Integrated Gasification Combined Cycle (IGCC). The coal fuel is reacted with steam and oxygen at pressure to produce a gas (syngas) which is then burnt and used to power a gas turbine. As in a CCGT, the heat is recovered and used to power a steam turbine, thereby improving the efficiency of the overall operation. There are three main ways in which fossil fuel electricity can be decarbonised:

- post-combustion decarbonisation – involving removal of CO_2 from flue gases;
- pre-combustion decarbonisation – involving removal of the CO_2 from fuel prior to its combustion;
- oxyfuel combustion – involving burning the fuel in an oxygen-enriched mixture which results in a concentrated CO_2 stream enabling capture.

2. Industrial processes: 23 per cent of world CO_2 emissions arise from the production of cement, iron, glass, chemicals and pulp;
3. Fuels production: refineries generate CO_2 emissions as do natural gas processing plants and other fuel production processes.

3.2   Technical Review of Existing CO_2 Capture Processes

Absorption processes are currently the most developed CO_2 removal technology. Absorption systems use continuous scrubbing to remove CO_2 from a gaseous stream. Three main absorption processes are available: chemical, physical and hybrid. CO_2 capture is already a commercial operation in industrial manufacturing, refining and gas processing. So far, all commercial CO_2 capture plants use chemical absorption with a monoethanolamine (MEA) solvent. The food and beverage industry, for example, commonly apply chemical absorption to recover CO_2 released during fermentation processes (Anderson and Newell, 2003). Flue gas streams from gas CCGT and PF coal power plants are characterised by low to moderate concentrations of CO_2 (3 to 4 per cent and 13 to 14 per cent respectively). At low CO_2 concentrations (below 10 per cent, as in flue gases from gas power plants) chemical absorption is the preferred approach because the energy required for the capture process is not very sensitive to the (low) CO_2 concentration and partial pressure (IEA, 2004). If the CO_2 concentration is higher, say over 15 per cent, then physical absorption is a better approach. In general, removal of CO_2 is easier and cheaper when the CO_2 concentration is higher and in section 3.4. we explore some of the alternative approaches to increasing CO_2 concentration for capture purposes.

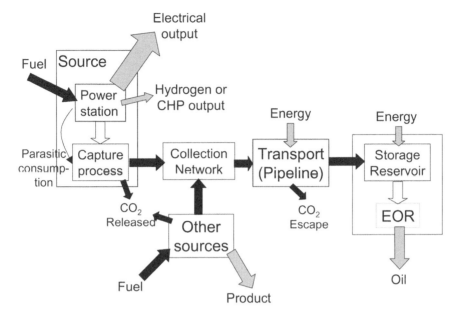

Figure 3.1   Overview of CCS system represented in the techno-economic model

Chemical Scrubbing Processes

The most effective way to capture CO_2 at low concentration currently is absorption using a chemical solvent such as monoethanolamine (MEA), diethanolamine (DEA) or potassium carbonate (Chinn *et al.*, 2005). Recent research shows that amino-acid salt solutions can be an alternative to amine based solutions (Feron *et al.*, 2004). CO_2 reacting with an MEA solvent forms an intermediate compound with weak bonds. By applying heat, this intermediate compound can be broken down, thus regenerating the original solvent and releasing the CO_2 (Figure 3.2). These processes can run at low CO_2 partial pressures, but the precondition is that the flue gas should be free of contaminants such as SO_2, O_2, hydrocarbons and particulates which may cause operating problems in the absorber. Because heat is required to break the bond between the solvent and CO_2, chemical absorption reduces energy efficiency. 1.5 tonnes of low pressure steam is required per tonne of CO_2 captured (3.2 GJ per tonne) for a boiler with a 90 per cent CO_2 recovery rate (IEA, 2004). The energy required for recovery decreases as the CO_2 concentration increases (from 3.4 GJ/tCO_2 to 2.9 GJ/tCO_2 as the CO_2 concentration in the flue stream rises from 3 to 14 per cent, representing the conditions for natural gas and coal-fired steam cycles respectively) (IEA, 2004). Whilst post combustion amine scrubbing is quite a mature technology it still has potential for development, with opportunities for much better performance through energy and process integration (Roberts *et al.*, 2005) and operation optimisation (Gibbins *et al.*, 2005). New chemical absorbents, which have a lower bonding strength between the solvent and CO_2 than does MEA, are being investigated, e.g. sterically-hindered amines (IEA, 2004).

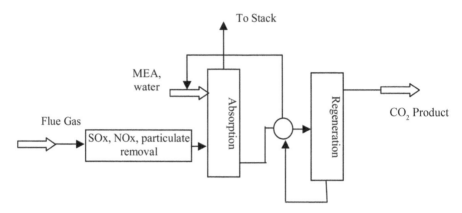

Figure 3.2   The chemical absorption process

Physical Absorption Processes

CO_2 can be physically absorbed in a solvent in accordance with Henry's Law. The binding of the solvent with CO_2 occurs at high pressure and a reduction in pressure releases the gas. The energy requirement is that needed to pressurise the gas and is proportional to the inverse of the CO_2 concentration, i.e. if the CO_2 concentration in the gas is doubled, the amount of energy required for capture is halved. At low CO_2 concentration, the cost of pressurising the gas is prohibitive. Where source streams have high concentrations of CO_2, as is the case for IGCC plant (Anderson and Newell, 2003), physical absorption using a solvent like Selexol (dimethylether of polyethylene glycol) or Rectisol (cold methanol) can be less costly than chemical absorption. One problem, however, is that C3+ hydrocarbons are soluble in the physical solvent.

Hybrid Solvents

Hybrid solvents combine the best characteristics of both chemical and physical solvents and are usually composed of several solvents that complement each other. Tailor-made solvents in which the proportions are varied to suit the application are a promising area for future innovation. Typical solvents are A-MDEA, Purisol, Sulfinol and UCARSOL. All such absorption processes operate in essentially the same manner by scrubbing the flue gas in towers to collect the CO_2 and regenerating the solvent and releasing the CO_2.

Adsorption

Adsorption is a process in which a gas fixes on to the surface of a solid substance by either chemical or physical attraction; the scheme for its application in CO_2 capture is illustrated in Figure 3.3. The nearest adsorption technology to a CCS application in current commercial use is separation of CO_2 from CO_2–H_2 gas mixtures during the production of hydrogen. Some porous solids with large surface areas are able to adsorb large quantities of gas per unit of volume. For separation of CO_2 from power plant flue gases, adsorbent beds of alumina, zeolite molecular sieves (natural or manufactured aluminosilicate) and activated carbon are at present considered to be most applicable. The trade-off between the stronger attraction of a gas to an adsorbent solid and the energy cost of regeneration (i.e. removal of the adsorbed gas) defines the economic performance of the process. After the gas has been adsorbed, the adsorbent bed may be regenerated using a variety of methods. Four methods are used commercially for regeneration (Anderson and Newell, 2003).

1. Pressure swing adsorption (PSA), in which the external pressure of the scrubber is lowered until trapped gases are released from the adsorbent bed. PSA is the most common method used in hydrogen production from steam-reformed natural gas.
2. Thermal (or temperature) swing adsorption (TSA) employs high temperature to drive off trapped gases from the adsorbent bed. The disadvantage of this process is that the regeneration cycles are quite slow (taking hours) and larger quantities of adsorbent are required than for PSA.
3. Washing away the trapped gases by running a stream of fluid over the adsorbent bed.
4. An electrically conductive bed of activated carbon selectively adsorbs CO_2, and then releases it when electricity is applied, allowing for regeneration without costly temperature and pressure changes (Riemer *et al.*, 1993).

The IEA notes that adsorption is not presently considered economically attractive for large-scale CO_2 capture because the capacity and CO_2 selectivity of the available adsorbents are low (IEA, 2004).

3.3   Technical Review of New and Emerging CO_2 Capture Processes

Whilst physical and chemical absorption represent the most developed technical options for CO_2 capture, significant research efforts are being made for more 'exotic' capture technologies. Most of these technologies have been developed for use in other applications and some are used in niche applications. Technologies such as cryogenic fractionation, membrane separation and adsorption using molecular sieves to capture the CO_2 from the flue gas of a power plant have been considered but they are presently even less energy efficient and more expensive than chemical absorption (Herzog, 2001).

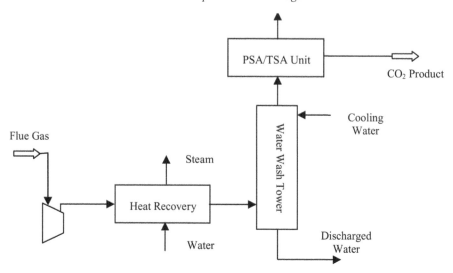

Figure 3.3   The adsorption process

Membrane Separation

Gas separation membranes make use of the differences in physical or chemical interaction between components of a gas mixture with the membrane material. This difference causes one component to permeate faster through the membrane than another (Figure 3.4 and Figure 3.5). The gas component dissolves into the membrane material and diffuses through it to the other side.

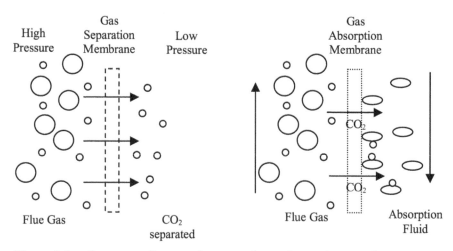

Figure 3.4   Gas separation membrane and gas absorption membrane

The separation process is governed by the *permeability* and *selectivity* of the membrane. The permeability of a gas through a membrane is its rate of flow through the membrane, given a pressure differential across the membrane. A limited reduction in pressure across the membrane is sufficient to achieve separation, hence this technology is potentially more energy efficient than absorption separation systems (IEA, 2004). The selectivity of a membrane is the ability of one gas to permeate faster than another. The ideal option would be the separation of CO_2 from a concentrated CO_2 source stream with some contaminant gases, combined with a permeable membrane that is highly selective with respect to CO_2. However, in practical terms, source streams generally have low pressures, low concentrations of CO_2, and many component gases (e.g. NO_x, SO_x, and water vapour). In addition, it is very difficult to produce membranes with high selectivity: increasing the permeability of one gas often increases the permeability of another. In practice, the selectivity of membranes is not sufficient to achieve the desired purity on the first pass and only a fraction of the CO_2 has been recovered in experimental trials (IEA, 2004). Hence multistage processes are required which imply increased compression and capital costs. Several gas separation membranes are available: polymer membranes, palladium membranes, facilitated transport membranes and molecular sieves – though some of these have been used only in laboratory settings (Riemer *et al.*, 1993). Polymer membranes are better developed, and can achieve higher levels of CO_2 recovery, but they cannot operate at high temperatures hence if used in an IGCC the fuel gases have to be cooled first. Ceramic membranes, whilst suitable for operation at high temperatures, are less efficient at recovering CO_2 (IEA, 2004).

Gas absorption membranes are another possible CO_2 capture technique (IEA, 1998). They serve as a way of putting gas mixtures (e.g. flue gases) and liquid absorbents (e.g. MEA solvents) into contact with one another, increasing the efficiency of physical or chemical absorption (Figure 3.5). By increasing the contact surface area, it is possible to reduce the size of the scrubbing plant, hence reducing the capital cost. The separation is caused by the presence of an absorption liquid on one side of the membrane which selectively removes certain components from a gas stream on the other side of the membrane. In contrast with gas separation membranes, it is not essential that the membrane has any selectivity at all. It is only necessary that there is an area of contact between the gas and absorption liquid flow, but without mixing. The membrane's function is to keep the gas and liquid flows separate, minimizing entrapment, flooding, channelling, and foaming. The selectivity of the process is determined by the absorption liquid.

Removal of flue gas components, such as SO_2 or CO_2, is achieved through the use of porous, hydrophobic membranes in combination with suitable absorption liquids, such as sulphite, carbonate or amine solutions. The equipment in a process using gas absorption membranes is more compact than for conventional membranes, thus reducing capital costs (Miesen and Shuai, 1997). The limitation of using membranes in this process is that the absorption liquid and gas stream need to have similar pressure levels.

Membrane technology is used commercially in hydrogen separation but still needs significant development before it can be used on a significant scale for the capture of CO_2. It is not clear to what extent the large capital costs can be reduced. One attraction of membranes is that they typically require less energy for operation than other methods of capture. Studies have shown that CO_2 removal using gas absorption membranes in conjunction with MEA is significantly better than using the membrane on its own (IEA, 1998). This is a promising finding as new improved gas absorption membranes are expected on the market shortly.

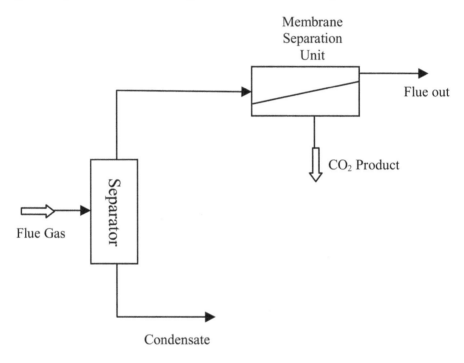

Figure 3.5   Membrane separation process

Hydrogen Production with CO_2 Capture Using Membrane Reactors

The combination of separation and reaction in membrane steam reforming offers higher conversion rates of the reforming and shift reactions at low temperatures due to the removal of products from these equilibrium reactions (Equations 3.1 and 3.2) and illustrated in Figures 3.6 and 3.7.

Equation 3.1 Steam Methane Reforming

$$CH_4 + H_2O \leftrightarrow CO + 3H_2$$

Equation 3.2 Water Gas Shift Reaction

$$CO + H_2O \leftrightarrow CO_2 + H_2$$

When applied in a membrane reformer or membrane water gas shift reactor, the retentate gas is predominantly CO_2 and excess steam which remains at the relatively high pressure of the reacting system. Condensation of the steam leaves a concentrated CO_2 stream at high pressure, avoiding the need for multiple shift reaction stages and eliminating the wet CO_2 scrubbing stage; the compression requirement is also minimised for the captured CO_2. Unlike conventional reforming, membrane reforming benefits from the high pressure of operation due to the increased H_2 partial pressure differential across the membrane which acts as the driving force for hydrogen permeation. Current proven membrane technology typically uses a separation layer of palladium (Pd) alloyed with 23 per cent silver. Development of a number of novel membranes is occurring, including Pd alloys, Pd Group V metal composites and microporous membranes including amorphous silica, zeolites and mixed conducting ceramic materials (Middleton *et al.*, 2004). For gasifier systems producing sour syngas, sulphur tolerant membranes offer significant operating benefits and reduction in cost, enabling the removal of the desulphurising unit from the shift reactor feed (Middleton *et al.*, 2004).

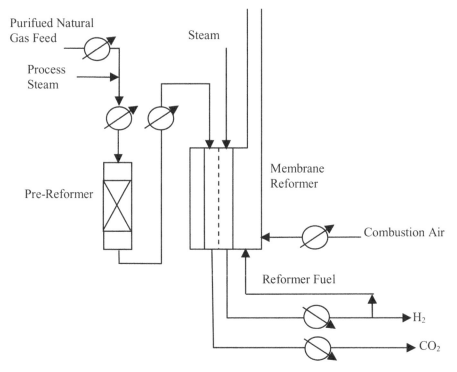

Figure 3.6   **Simplified flowsheet of natural gas decarbonisation by membrane reformer**

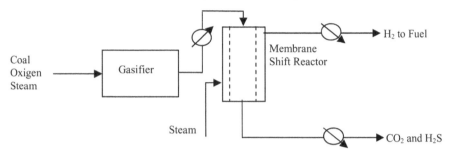

Figure 3.7   Simplified flowsheet of coal gasification with membrane water gas shift reactor

Cryogenic Separation

CO_2 can be physically separated from other gases by condensing the CO_2 out at cryogenic temperatures (Singh *et al.*, 2001). Cryogenic separation leads to phase changes in CO_2 and other gases induced by compression and cooling of gas mixtures in multiple stages. This process proves to be most effective when feed gases contain components with significant differences in boiling points (Herzog *et al.*, 1997). Cryogenic separation is widely used commercially for purification of CO_2 from streams that already have high CO_2 concentrations (typically >90 per cent) but it is not normally used for more dilute CO_2 streams. However, the process is often complicated by contaminants (SO_2 and NO_x) that can impede cryogenic processes. Water also presents problems in cryogenic systems, so the feed gas has to be dried before being cooled (Davison and Thambimuthu, 2005). Moreover, the phase transformation of CO_2 is also complicated and can lead to the formation of solids that plug equipment and reduce heat transfer rates. The need for pressurization and refrigeration makes cryogenic processes very energy intensive and hence expensive. Nonetheless, cryogenic separation could be effective for certain large, highly concentrated source streams of CO_2, e.g. in pre-combustion capture processes, or O_2/CO_2 combustion.

3.4   Alternative Power Station Designs to Improve CO_2 Capture

The low concentration of CO_2 in conventional flue gas streams (between 3 and 14 per cent) arises from the use of air for combustion, the large percentage of nitrogen in air (80 per cent) acting to dilute the CO_2. This low concentration has a large impact on the capital and operational costs of the capture process, since a very large volume of gas has to be handled for relatively low levels of CO_2 extraction. Only oxygen is required for combustion, hence one strategy is to modify the power plant and process by using pure oxygen rather than air, hence increasing the concentration of CO_2 in the flue gas. A second strategy is to remove the CO_2 from the fuel prior to combustion with air.

The O₂/CO₂ Recycling Process

The main idea behind an O_2/CO_2 recycling or 'oxyfueling' process is that oxygen (95 per cent purity or higher), rather than air, is fed into a boiler where the fossil fuel is combusted (Figure 3.8). Most of the flue gas (70–80 per cent), which is rich in CO_2, is recycled back to maintain an acceptable temperature in the gas turbine (Jordal *et al.*, 2005). This so-called MATIANT cycle would require the development of new turbines since retrofitting existing turbines would not be technically feasible (IEA, 2004). The remaining part of the flue gas, consisting mainly of CO_2 and water vapour and small quantities of Ar, N_2, NO_x, SO_x and other substances from air leakage and fuel, is cleaned, compressed and transported to storage or for another application. Since the nitrogen is removed from the combustion process, the flue gas contains much less NO_x, and the need for NO_x scrubbing is significantly reduced.

Unlike pre-combustion and post-combustion techniques for CO_2 removal, the oxyfuel process removes water and other non-condensable gases, purifying the CO_2 rich output stream. If it is not possible to store the CO_2 with traces of the remaining impurities, it will be necessary to remove the SO_x and NO_x though this should be less of a challenge, and less expensive, than in the case of other capture technologies because of the lower volumes of these impurities.

The present air separation technologies (cryogenic separation) are still rather costly and account for the higher costs of the oxyfuel process in comparison to other techniques (Gottlicher *et al.*, 1997). The electric power consumption of a Cryo-ASU (air separation unit) may account for roughly 20 per cent of the plant gross power output for the O_2/CO_2 recycle combustion power plant, which of course is very detrimental to plant efficiency (Jordal *et al.*, 2005). Some argue that this energy penalty means that oxyfueling is less efficient than post-combustion decarbonisation. However the latest developments in air separation technologies, such as ion transport membranes, oxygen transport membranes and mixed conducting membranes may prove to be promising solutions for lowering the high costs of the O_2/CO_2 process (Simmonds *et al.*, 2005). If such new membrane systems for oxygen generation can be made to work at the scale required for power plants, the energy requirement for air separation may be reduced by over 30 per cent compared to current cryogenic technology. The IEA concludes that oxygen-blown gasifiers with pre-combustion CO_2 removal will be essential for coal-fired IGCCs, though the case for oxyfueling of natural gas combined cycles and coal-fired steam cycles is less clear (IEA, 2004).

Integrated Gasification Combined Cycle (IGCC) with CO₂ Capture

An IGCC plant with CO_2 capture requires coal gasification, oxygen production, a shift reactor, hydrogen separation, hydrogen turbines and CO_2 separation (IEA, 2004). The synthesis gas (syngas) is a mixture of carbon monoxide (CO) and hydrogen (H_2) (Davison *et al.*, 2005). In an IGCC without capture, syngas is directly combusted in gas turbines. In an IGCC with capture, the syngas is reacted with steam in the presence of catalysts forming a mixture of H_2 and CO_2

(see Equations 3.1 and 3.2). The use of oxygen and steam at high pressure conditions makes the plant well suited for CO_2 removal, though this is counterbalanced by the high costs of IGCCs and their relative technological immaturity. The H_2 is separated and used in a combined-cycle gas turbine, leaving a pure stream of CO_2 that can be directly compressed and stored (Figure 3.9). Despite the fact that several IGCC demonstration projects operate (e.g. in Florida, Indiana and Tennessee) and several companies are planning to use coal gasification technologies in future power plants. So far, there have been no demonstration plants with CO_2 capture. The Futuregen project is expected to be the first IGCC plant with CO_2 storage; this 275 MW prototype plant, to be located in the US, is the result of an agreement between the US Government and a multinational industry alliance.

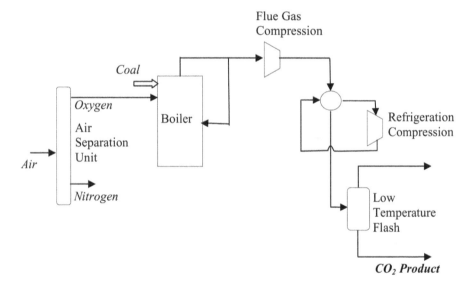

Figure 3.8   The O₂/CO₂ recycling process

There are currently three main gasifiers available: one-stage slurry fed such as the Texaco gasifier; two-stage slurry fed such as the E-Gas gasifier; and dry fed systems such as the Shell gasifier (IEA, 2004). Gas turbines will need to be modified to generate electricity from pure hydrogen, and NO_x emissions will have to be reduced to acceptable levels without excessive water/steam injection (IEA, 2004). Regeneration of physical solvents is less energy intensive than for chemical absorption and in terms of energy efficiency the pre-combustion removal of CO_2 from coal is generally considered superior to post-combustion removal, though the situation with respect to natural gas is less clear.

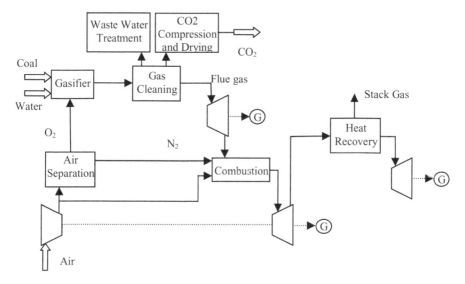

Figure 3.9 IGCC power plant with CO_2 removal by means of Selexol scrubbing (after IEA, 1998)

Chemical-looping Combustion

In chemical-looping combustion oxygen is transferred from the combustion air to the gaseous fuel by means of an oxygen carrier. The fuel and the combustion air are never mixed and the gases from the oxidation of the fuel, CO_2 and H_2O, leave the system as a separate stream. The H_2O can easily be removed by condensation and pure CO_2 is obtained without any loss of energy for separation. This makes chemical-looping combustion an interesting alternative to other CO_2 separation schemes, which have the drawback of considerable energy consumption. The system is composed of two reactors, an air and a fuel reactor, as shown in Figure 3.10. The fuel must be in a gaseous form and is introduced into the fuel reactor, which contains a metal oxide, MeO. The fuel and the metal oxide react according to Equation 3.3.

Equation 3.3

$$(2n+m)MeO + C_nH_{2m} \rightarrow (2n+m)Me + mH_2O + nCO_2$$

The exit gas stream from the fuel reactor contains CO_2 and H_2O and almost pure CO_2 is obtained when H_2O is condensed. The reduced metal oxide (Me) is transferred to the air reactor where the metal is oxidized according to Equation 3.4.

Equation 3.4

$$Me + \tfrac{1}{2} O_2 \rightarrow MeO$$

The advantage of chemical-looping combustion compared to normal combustion is that CO_2 is not diluted with N_2 but is obtained in a relatively pure form without any energy needed for separation (Lyngfelt *et al.*, 2001). At present there is insufficient experimental data for different oxygen carriers at high temperatures and high pressures to assess the commercial prospects for chemical looping. Similar processes involving particle transfer have, in the past, encountered plugging and abrasion problems. The metal oxide materials used need to be able to withstand chemical cycling and to be resistant to the physical and chemical degradation caused by impurities arising from fuel combustion (IEA, 2004). Some studies have suggested that a coal-fired circulating fluidised bed with chemical looping and CO_2 capture would generate electricity at lower costs than using IGCC with CO_2 capture. One study suggested a 54 per cent efficiency for a gas-fired power plant with chemical looping and CO_2 capture, including CO_2 pressurisation, which compares favourably with other capture technologies (IEA, 2004).

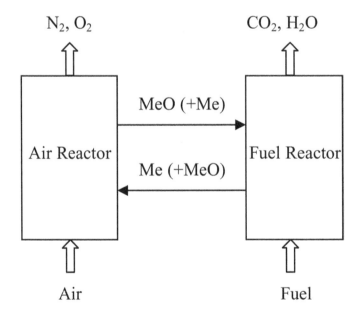

Figure 3.10 Chemical-looping combustion (after Lyngfelt *et al.*, 2001)

3.5 Novel Routes to CO_2 Capture and Utilisation

Fertiliser (NH_4HCO_3) Production with CO_2 Sequestration

The main idea behind this approach is to sequester CO_2 as ammonium bicarbonate using aqueous ammonia (West and Marland, 2002). Selective catalytic reduction (SCR), involving injection of gaseous NH_3 or aqueous NH_3 for removal of NO_x from the flue gas, is widely applied in power plants. It might also be economical to use NH_3 for capturing CO_2 from flue gases. Ammonia/water liquors can be used to scrub out CO_2 in the form of stable ammonium bicarbonate and carbonate compounds, by taking advantage of the acidic nature of CO_2 in aqueous media.

Equations 3.5 and 3.6

$$NH_3 + H_2O + CO_2 = NH_4HCO_3 \text{ (ammonium bicarbonate)}$$

$$2NH_3 + H_2O + CO_2 = (NH_4)_2CO_3 \text{ (ammonium carbonate)}$$

Ammonium bicarbonates (ABC) have been used as nitrogen fertilizers, e.g. in developing countries until the 1980s, and thereby contribute to storing organic carbon in the soil. Modern fertilizers, such as urea, ammonium nitrate and ammonium sulphate, have been replacing ABC because they contain more nitrogen and are more stable. However, in the 1990s a modified ABC, known as long-effect ABC, was developed. It is a combination of nanosized co-crystallized dicyanodiamide (DCD) and ABC. The hydrogen bonds between the DCD and the ABC affect the physical properties of ABC such as volatility, stability and ability to remain in the soil for as long as 100 days. There is less nitrite (NO_2 -) and nitrate (NO_3 -) run-off from the utilisation of ABC with DCD compared to fertilizers such as urea, ammonium nitrate and ammonium sulphate. Preliminary studies show that using ammonia to capture the CO_2 emitted from fossil-fuel combustion, with subsequent use of ABC as fertiliser, can be an effective and economical method to manage carbon; however the quantitative estimation remains to be done (Lal *et al.*, 1999).

Recovery and Sequestration of CO_2 by Photosynthesis of Micro-Organisms

Micro-algae and cyanobacteria are aquatic micro-organisms which undertake photosynthesis using water as the reducing agent. The productivity of micro-algae can be increased by artificially increasing the transfer rate of CO_2 from the atmosphere to the aqueous environment (provided that other essential minerals are not limiting factors) (Pedroni *et al.*, 2005; Brown and Zeiler, 1993; Benemann, 1997). Micro-algae have high growth rates and strain selection is assisted by their rapid reproduction rates. They can grow under conditions which higher plants find more difficult, such as in deserts, utilising salt water supplied from deep aquifers or from the sea. Cell suspensions of algae can be handled as liquids and for CO_2 capture purposes need to be tolerant to:

- high CO_2 and HCO_3 concentrations and consequently be able to withstand direct aeration by flue gases;
- low pH (down to pH=2) caused by the presence of SO_2 and NO_x, and higher than ambient temperatures;
- low concentrations of heavy metals.

Several strains of micro-algae and cyanobacteria having properties satisfying many of the above criteria are known. The highest CO_2 removal rate so far reported is 4.44 g CO_2/litre/day using a culture of marine *Synechococcus* sp. (a cyanobacterium) in a photobioreactor (IEA, 1998), whilst *Chlorella*- and *Synechocystis*-based systems absorb approximately 50 g CO_2/m²/day (Otsuki, 2001). Innovation in photo-bioreactor technology aims to improve the efficiency of biomass production (Figure 3.11). For example, the 'flashing light effect' entails repeatedly cycling cells from the dimly lit interior of the reactor to the higher illumination of the exterior (Degen *et al.,* 2001).

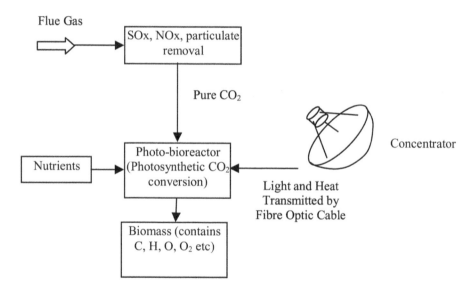

Figure 3.11   Conceptual design of a photo-bioreactor system for CO_2 conversion (after Stewart, 2005)

To capture the CO_2 from a 500 MW power plant, large open ponds of about 50-100 km² with algae suspension, into which power plant flue gas or pure CO_2 (captured from power plants) is introduced as small bubbles, would be required. The estimated mitigation costs for this type of scheme would be up to $100 per tonne CO_2 (with significant opportunities for further cost reduction) (Benneman, 1993). After harvesting, the biomass could be used as a feedstock or converted into a fossil fuel replacement. Strain selection perhaps with genetic modification could increase lipid production relative to carbohydrate production, so making better biofuels. Algal CO_2 capture systems would require considerable land, water and climatic conditions, the combination of which is seldom found in the vicinity of power plants. Furthermore, despite 50 years of development, closed photo-bioreactor systems have not yet been developed commercially (Benemann 1997).

Sequestration by Mineral Carbonation

A further possibility is to sequester CO_2 in stable solid carbonate minerals that can be returned to the environment. Calcium and magnesium carbonates are solid and could be stored in a suitable location as landfill. In principle, no energy inputs are required because carbonates exist at a lower energy level than CO_2. On the contrary, energy could be produced as these reactions are exothermic (Herzog, 2002) (Equations 3.7 and 3.8) though in practice implementing the option would require energy inputs.

Equations 3.7 and 3.8

$$CaO + CO_2 = CaCO_3 + 179 \text{ kJ/mol}$$

$$MgO + CO_2 = MgCO_3 + 118 \text{ kJ/mol}$$

Unfortunately calcium and magnesium are rarely naturally available as binary oxides, but mainly as calcium and magnesium silicates. The reaction of CO_2 with such common calcium and magnesium bearing minerals is still exothermic but the heat release is considerably reduced, as shown in Equations 3.9 and 3.10 for forsterite and serpentine respectively (Herzog, 2002):

Equations 3.9 and 3.10

$$\tfrac{1}{2}Mg_2SiO_4 + CO_2 = MgCO_3 + \tfrac{1}{2}SiO_2 + 95 \text{kJ/mol}$$

$$\tfrac{1}{3}Mg_3Si_2O_5(OH)_4 + CO_2 = MgCO_3 + \tfrac{2}{3}SiO_2 + \tfrac{2}{3}H_2O + 64 \text{ kJ/mol}$$

Commercially viable reaction pathways for mineral sequestration have not yet been identified. Among the methods currently being explored are the following:

1. Exposing calcium and magnesium silicates to CO_2 (Kojima, 1997). This is based on a natural process of CO_2 sequestration with pulverisation and dissolution of olivine sand and wollastonite, and their subsequent reaction with power plant CO_2 to form magnesium and calcium carbonates. The disadvantages of the approach include the large amounts of rock that have to be transported and handled (up to several times the weight of the CO_2 sequestered), the energy requirements to pulverise the rock and the use of significant quantities of hydrochloric acid.

2. Application of underground brines rich in chlorine and sulphate to produce carbonates (Dunsmore, 1992). The brines could be pumped to a CO_2 contacter and the precipitate slurry could be re-injected underground. An *in situ* processing option also exists. About 2.2 tonnes of precipitate would be formed per tonne of CO_2 sequestered. The drawback of the method is that the suitable brines are available in only a few locations and the environmental management of the acidic wastes generated would present a major problem. The extent of the waste streams, and the transport distances to bring power plant CO_2 to the disposal site, probably make this an impractical option for mitigation.

Chemicals Manufactured from CO_2

Approximately, 110 Mt CO_2 is used annually on a global scale as a raw material in the production of urea, methanol, acetic acid, polycarbonates, cyclic carbonates and speciality chemicals. The largest use is for urea production which reached about 90 Mt CO_2 per year in 1997 (Creutz and Fujita, 2000). A summary of the reactions using CO_2 is given in Table 3.1. The operating conditions, catalysts, selectivity and other parameters of these reactions are well known and documented (Xu *et al.*, 2003, 2005). Carbon dioxide is thermodynamically stable, so any use of CO_2 as a feedstock tends to require a significant amount of energy input. Some of the examples of reactions are briefly discussed below.

Reduction of CO_2 by alkanes The types of reactions shown in Equations 3.11 and 3.12 are well known with those based on the Mobil HZSM-5 series of catalysts being particularly effective.

Equations 3.11 and 3.12

$$CO_2 + C_3H_8 \rightarrow CO + C_3H_6 + H_2O \text{ (metal oxide based catalysts)}$$

$$CO_2 + C_3H_8 \rightarrow CO + BTX + H_2O \text{ (Zeolite based catalysts)}$$

The hydrogen form of the catalyst is normally cation exchanged with Ga, Zn or Pt. These cation exchanged catalysts can convert small chain alkanes (e.g. propane) into aromatics and, with the necessary temperature and pressure conditions, produce a reasonable yield of aromatic mixtures of Benzene, Toluene and Xylene (BTX). A series of tests over ZSM-5, and similar catalysts, showed that yields of BTX could be increased by the addition of CO_2 to the alkane feed. The best reported performer was Zn-ZSM-5 which had a conversion of 71.4 per cent of propane to over 43 per cent aromatic product. Without CO_2 additions the figures are 57 per cent and 37 per cent respectively (IEA, 1998). The use of CO_2 for this type of reaction scheme is limited by the demand for BTX and for the methanol by-product.

Table 3.1   Some catalytic reactions of CO_2 conversion into products

Hydrogenation		Hydrolysis and Photocatalytic Reduction
$CO_2 + 3H_2 \rightarrow CH_3OH + H_2O$	methanol	$CO_2 + 2H_2O \rightarrow CH_3OH + O_2$
$2CO_2 + 6H_2 \rightarrow C_2H_5OH + 3H_2O$	ethanol	$CO_2 + H_2O \rightarrow HC=O-OH + 1/2O_2$
$CO_2 + H_2 \rightarrow CH_3-O-CH_3$	dimethyl ether	$CO_2 + 2H_2O \rightarrow CH_4 + 2O_2$
Hydrocarbon Synthesis		
$CO_2 + 4H_2 \rightarrow CH_4 + 2H_2O$	methane and higher HC	
$2CO_2 + 6H_2 \rightarrow C2H_4 + 4H_2O$	ethylene and higher olefins	
Carboxylic Acid Synthesis		**Other Reactions**
$CO_2 + H_2 \rightarrow HC=O-OH$	formic acid	CO_2 + ethylbenzene → styrene
$CO_2 + CH_4 \rightarrow CH_3-C=O-OH$	acetic acid	dehydrogenation of propane $CO_2 + C_3H_8$ $\rightarrow C_3H_6 + H_2 + CO$
		reforming $CO_2 + CH_4 \rightarrow 2CO + H_2$
Graphite Synthesis		**Amine Synthesis**
$CO_2 + H_2 \rightarrow C + H_2O$		methyl amine and higher amines $CO_2 + 3H_2 + NH_3 \rightarrow CH_3-NH_2 + 2H_2O$
$CH_4 \rightarrow C + H_2$		
$CO_2 + 4H_2 \rightarrow CH_4 + 2H_2O$		

Source: After Xu et al., 2003

The oxidative coupling of methane with CO_2 In this reaction, a reverse water gas shift reaction uses methane as a reducing agent for CO_2 converting it to hydrogen and carbon monoxide. Though technically feasible, the economic prospects for this method are not promising. The product stream can be altered to produce methanol but the current demand for methanol is insufficient to make the process economically attractive, e.g. using the CO_2 from one 500MW power plant produces about twice the present demand for methanol of a large economy such as that of Japan (IEA, 1998).

CO_2 polymers The environmental impacts of plastic waste have raised considerable interest in the development and production of biodegradable plastics. Polyhydroxyalkanoates (PHAs) are polyesters that accumulate as inclusions in a wide variety of bacteria. These bacterial polymers have properties ranging from stiff and brittle plastics to rubber-like materials. Because of their inherent biodegradability, PHAs are regarded as an attractive source of non-polluting plastics and elastomers

that can be used for speciality and commodity products. The possibility of producing PHAs at a large scale and at a cost comparable to synthetic plastics has arisen from the demonstration of PHA accumulation in transgenic *Arabidopsis* plants expressing the bacterial PHA biosynthetic genes. There are three groups of organisms that can accumulate PHA from carbon dioxide: chemoautotrophic bacteria such as hydrogen-oxidizing bacteria, genetically engineered higher plants and cyanobacteria (Asada, 1999).

Although CO_2 has not generally been regarded as a promising monomer in chemical engineering it can feature in a number of reactions, particularly to form alkylene oxides and alkylene poly-carbonates. The reaction normally involves organometallics such as diethylzinc with a hydrogen donor (water, an amine or an aromatic dicarboxylic acid). The products of such reactions are currently used as binders in the electronics industry and are being further developed for film applications in the food and medical areas. The predicted market is around 100 tCO_2/y and these 'new' polymers may substitute for other more conventional oil based polymers. The cost of these polymers is high, mainly due to the cost of the catalyst.

Dimethyl Carbonate (DMC) The increased demand in DMC for a number of organic syntheses is caused by recent moves away from the use of phosgene and dimethyl sulphate. DMC is used as a solvent and in motor fuel as an octane booster. A process to produce DMC from CO_2 is now in commercial operation using cobalt based catalysts. In practice, methyl tertiarybutyl ether (MTBE) is already well established as a vehicle fuel additive, with a large capacity installed worldwide, so DMC would have problems in becoming competitive with MTBE in a large scale market. Assessments are that globally, 1 Mt of CO_2 per year might be used in DMC production.

Summary of Novel Approaches

Most of the technologies described in this section are relatively small scale solutions for CO_2 capture and sequestration compared to the total quantities of CO_2 emitted by the power and industry sectors. Furthermore, all of them have energy penalties, frequently considerable in size, in addition to other environmental impacts. The current demand for CO_2 in the chemical industry is also rather limited. Most of the processes which could increase the demand for CO_2 struggle to be economically viable, e.g. because of the need for expensive catalysts and process energy. In the longer term, basic and applied research and development on artificial photosynthesis and hydrogen bio-production (which could be used with CO_2 to produce methanol) is necessary to deliver economically promising and efficient CO_2 capture and sequestration technologies.

3.6 Cost Functions for CO_2 Capture Processes

There are a number of ways of expressing the costs of CO_2 capture and we will briefly discuss three approaches here: costs of CO_2 captured, costs of CO_2 avoided and costs per unit of output.

Costs of CO_2 Captured

The costs of pollution control are conventionally defined as the cost of removing a given quantity of pollutant from the waste streams. In the case of CO_2 this would be the costs arising from the construction, operation, maintenance and repair of the CO_2 capture technology and process. The total costs would then be expressed in terms of 'net present value' by converting the costs over the project life time into today's values using an appropriate discount rate. The capture cost is then calculated as \$ per tonne of CO_2 captured by dividing the annualised additional cost of the plant due to capture (M\$/yr) by the annualised quantity of CO_2 captured (tonnes CO_2/yr). This capture cost therefore takes into account the efficiency of the capture process (since it will influence the capital cost of the capture plant) and the energy penalty (since the costs of the additional fuel consumed will be included in the calculation).

Costs of CO_2 Avoided

If CO_2 capture is being compared to other carbon abatement technologies and options it is necessary to take account of the additional CO_2 which is generated through the use of energy from fossil fuels in the CO_2 capture process. Consider a 200 MW coal power station producing 1 million tonnes of CO_2 per year. CO_2 capture technology is then implemented which incurs, say, a 30 per cent energy penalty. This results in an increase in total CO_2 generated by combustion to 1.3 million tonnes per annum. If the capture process is 90 per cent efficient, then 1.17 million tonnes of CO_2 are captured and the capture costs per tonne CO_2 can be calculated as above by the additional plant and operational costs divided by 1.17 million. Now suppose that we are comparing costs with those for wind energy. Each one kWh generated by wind energy replaces one kWh of electricity generation by fossil fuels, hence if we replaced the above coal power station with wind turbines we would be avoiding the release of 1 million tonnes of CO_2 into the atmosphere. Notice, however, that the use of CO_2 capture has resulted in the generation of *more* than one million tonnes of CO_2, yet the quantity of electricity generated remains the same as the coal power plant with no CO_2 capture and as the wind farm. Hence somewhat less CO_2 production is avoided when using CO_2 capture than is implied by the amount of CO_2 captured, which in turn increases the capture costs. The costs of CO_2 avoided can be calculated by Equation 13:

Equation 3.13

$$\text{Cost (avoided)} = \text{Cost (captured)} \times CE \ / \ [\text{eff}_{new}/\text{eff}_{old} - (1\text{-}CE)]$$

Where: eff_{old} is the efficiency of the power plants without CO_2 capture, eff_{new} is the efficiency of the power plants with CO_2 capture and CE is the fraction of the CO_2 that is captured (IEA, 2004). The larger the reduction in the efficiency of the power plant, the higher the cost of CO_2 avoided, whilst a more efficient capture process limits the cost increase (though cost (avoided) is always greater than cost (captured)). For an eff_{old} of 43 per cent, an eff_{new} of 31 per cent and CE of 0.85, the correction factor is 1.48 and declines to 1.20 to 1.25 for the emerging, more efficient, CO_2 capture technologies (IEA, 2004).

Costs of CO_2-free Electricity

In order to allow for a full comparison, it is necessary to develop a common baseline against which alternative carbon mitigation options may be compared including indicators which reflect the costs of power output, such as cents or pence per kWh of CO_2-free electricity generation. The choice of the baseline is crucial and will have a great impact upon the perceived costs of alternative options. In the marginal costing approach the baseline is the supply technology selected by the market without any carbon constraints, hence the plant type which dictates the price of electricity in an ideal market (IEA, 2004). In most OECD countries this will frequently be a natural gas Combined Cycle Gas Turbine (CCGT). With no CO_2 controls, CCGTs typically produce half of the CO_2 per kWh of a PF coal power station. Imagine that a 200 MW coal power station with CO_2 capture is compared to one with no capture which generates 1 million tonnes CO_2 per annum; approximately 870,000 tCO_2 per annum might be avoided. If the comparison is instead made with a 200 MW gas CCGT then the CO_2 avoided is approximately 350,000 tCO_2/annum, i.e. less than half. If we imagine that the additional cost of the coal power plant with CO_2 capture is $50 million, then the cost per tonne of CO_2 avoided is $57 per tCO_2 if the comparison is with a coal power plant with no capture, but $140 per tCO_2 if compared with a gas CCGT (see Table 3.8).

Capture Cost Model Development

We developed a simple model to estimate the costs of CO_2 capture from different power plant configurations. The model is based on simple and reliable relationships developed from a more detailed model (IECM-CS) (version 3.5.5), developed by the Center for Energy and Environmental Studies at Carnegie Mellon University and which has been constructed and tested over a number of years to simulate CO_2 capture within power plants using amine scrubbing (Figure 3.12) (CEES, 2003). The ICEM-CS model was not suitable for application directly due to the large number of input and output parameters required and software compatibility issues. However, the detailed IECM-CS model has been used to provide data series for development of parametric Cost Estimation Relationships (CERs). The CERs are estimated using straightforward functions and the proscribed relationships do not need to be analytic; functions that rely on interpolation between data points are quite acceptable. The

CERs have been defined at three CO_2 removal efficiencies – 95 per cent, 90 per cent and 85 per cent – on plants ranging between 300 to 2000 MW in size.

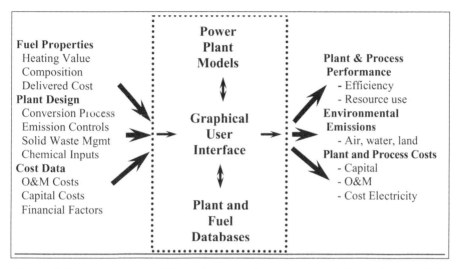

Figure 3.12   Amine scrubbing plant performance model implemented in IECM (version 3.5.5) software (after Center for Energy and Environmental Studies, 2003)

A separate set of functions is required for each combination of power station and capture technology, although some simplification may be possible. There are two types of information that have to be estimated. Firstly, for the design phase of the model, information about the overall costs and performance of the power station and capture equipment as a function of its maximum output power is required. Secondly, functions are required which allow the fuel consumption and carbon dioxide production arising from the operation of the plant to be calculated. For each fuel, power station and capture technology considered, the following must be estimated as a function of the parameter(s) in brackets:

- construction cost (station rated power);
- construction time (station rated power);
- overall conversion efficiency (instantaneous output power);
- annual plant O&M costs (rated power, average utilisation);
- CO_2 concentration in flue gas (instantaneous output power);
- flue gas flow rate (instantaneous output power);
- capture equipment costs (maximum CO_2 throughput, CO_2 output conditions);
- capture effectiveness (CO_2 concentration in stream, flue gas flow rate);
- capture O&M costs (maximum CO_2 throughput, average utilisation);
- onsite parasitic electrical power consumption (CO_2 concentration in stream, flue gas flow rate).

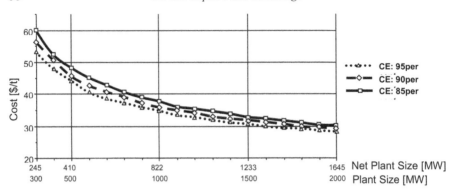

Figure 3.13   Cost of CO₂ avoided versus plant size and net plant size

Review of CO₂ Capture Cost Estimates

Because of the very large amounts of CO_2 handled the system inevitably generates large amounts of waste (MEA) and by-products. The model incorporates waste production, as far as it is possible to specify an MEA replacement rate and cost. At this stage, it also includes a by-product flow for hydrogen. The model is easily extendible and can include any other by-products and waste flows. The cost estimates produced by the model are provided in Table 3.2 expressed at 2004 $ levels for a new PF power plant with MEA capture at three power plant capacity ranges (2,000–1,500 MW, 1,500–900 MW and 900–300 MW). The cost of capturing a tonne of CO_2 is reduced significantly with plant size as illustrated in Figure 3.13. A higher capture efficiency (CE) reduces the cost of a tonne of CO_2 capture, also shown in Figure 3.13. The range of cost estimates is in line with other assessments as illustrated in Table 3.3.

Table 3.2   Model results for costs of CO₂ capture for new build PF coal fired power plants

Size of Plant	Efficiency of CO₂ removal	Costs (2004) in $t/CO₂	Costs (2004) in £t/CO₂
300 to 900 MW	85%	39–60	21–32
	90%	37–56	20–30
	95%	35–53	19–28
900 to 1,500 MW	85%	33–39	18–21
	90%	32–37	17–20
	95%	31–35	17–19
1,500 to 2,000 MW	85%	30–33	16–18
	90%	29–32	15–17
	95%	28–31	15–17

Note: Assumptions on base line efficiency are set out in Center for Energy and Environmental Studies (2003)

Table 3.3   Summary of new plant performance and CO_2 capture cost based on current technology (in 2002 prices)

Performance and cost measures	New natural gas CCGT Plant			New PF coal plant			New IGCC plant		
Range	Low	High	RV	Low	High	RV	Low	High	RV
Emission rate w/o capture (kg CO_2/MWh)	344	379	367	736	811	762	682	846	773
Emission rate with capture (kg CO_2/MWh)	40	66	52	92	145	112	65	152	108
Percent CO_2 reduction per kWh (%)	83	88	86	81	88	85	81	91	86
Plant efficiency with capture, LHV basis (%)	47	50	48	30	35	33	31	40	35
Capture energy requirement (% more input per MWh)	11	22	16	24	40	31	14	25	19
Total capital requirement w/o capture ($US per kWh)	515	724	568	1161	1486	1286	1169	1565	1326
Total capital requirement with capture ($US per kWh)	909	1261	998	1894	2578	2096	1414	2270	1825
Percent increase in capital cost with capture (%)	64	100	76	44	74	63	19	66	37
COE w/o capture ($US per MWh)	31	50	37	43	53	46	41	61	47
COE with capture ($US per MWh)	43	72	54	62	86	73	54	79	62
Increase in COE with capture ($US per MWh)	12	24	17	18	34	27	9	22	16
Percent increase in COE with capture (%)	37	69	46	42	66	57	20	55	33
Cost of CO_2 captured ($US/ tonne CO_2)	33	57	44	23	35	29	11	32	20
Cost of CO_2 captured (£/ tonne CO_2)	19	32	25	13	20	16	6	18	11
Cost of CO_2 avoided (($US/ tonne CO_2)	37	74	53	29	51	41	13	37	23
Cost of CO_2 avoided (£/ tonne CO_2)	21	42	30	16	29	23	7	21	13

RV= representative value; LHV: Lower heating value; COE: cost of electricity
Source: Adapted from Table 8.1, IPCC (2005), p. 343

CO_2 capture is an energy intensive process and increases overall energy consumption by up to 39 per cent for current power plant designs though this value can be reduced to 6 per cent for new designs (IEA, 2004). Table 3.3 is extracted from data provided by the IPCC (2005) and summarises state-of-the-art cost estimates for three technologies: new Natural Gas Combined Cycle Gas Turbine (NG-CCGT) plant, new PF coal plant and new IGCC plant. For a new high-efficiency coal-burning power station, CO_2 capture using an amine scrubber increases the cost of electricity generation (COE) by 40 to 70 per cent whilst reducing CO_2 emissions by 85 per cent per kWh (IPCC, 2005); there is a similar increase in the cost of natural gas CCGT with CO_2 capture. Because consumer prices for electricity are considerably higher (by 2 to 3 times) than production costs, this represents a 10 to 20 per cent increase in consumer price (IEA, 2004). A new coal-based Integrated Gasification Combined Cycle (IGCC) plant (with water gas shift reactor and physical absorption) increases the COE by 20 to 55 per cent (IPCC, 2005). The lower cost reflects the smaller volume of gas which is produced during the IGCC process from which CO_2 has to be removed. IGCC is more capital intensive than PF coal combustion, hence is slightly more expensive with no CO_2 capture. On the other hand, IGCC is slightly cheaper than PF coal plants in many studies when CO_2 capture is incorporated (IPCC, 2005). Natural gas CCGT produces electricity at a cheaper price than coal power plants, with or without capture, but the cost is highly sensitive to the gas price. This has risen markedly in the last few years, hence the historic COE from this fuel and technology may not reflect future market conditions. The IPCC (2005) note that the COE varies widely because of local and regional variations in fuel costs, plant utilisation, plant age, market circumstances, and many other factors. Therefore any meaningful comparison of alternative electricity generating technologies has to be conducted in quite a specific context; such a detailed assessment is beyond the scope of the current study.

Retrofitting of existing power plants for CO_2 capture has been widely discussed (e.g. IEA, 2004), though the IPCC (2005) notes that as yet there has not been any systematic comparison of the feasibility and costs of alternative retrofitting and repowering options for existing plant. The IEA (2004) has demonstrated that retrofitting for CO_2 capture from existing plants only makes sense when the efficiency has been enhanced. The net electrical efficiency of individual coal-fired power plants varies from 25 per cent to 48 per cent due to differences in steam conditions, coal quality, cooling water temperature and the installation of emissions mitigation equipment (ibid.). Investing in improving the efficiency of coal power plants is regarded by the IEA as the first step prior to considering CO_2 capture. This can be achieved by rebuilding the boiler and turbine so that they can operate as supercritical or ultra-supercritical units (with efficiencies of 44 to 48 per cent). Oxyfuelling may make sense for retrofitting of some PF coal-fired plants, whilst repowering with IGCC technology that includes CO_2 capture might be appropriate for other existing plants.

Summary

In sections 3.2 to 3.6 we have described the key technologies that are being investigated for the large-scale capture of CO_2 from point sources such as power stations, refineries and chemical works. We have also described some of the new and emerging technologies which are being evaluated in the hope that they can provide more efficient means for the removal of CO_2 from flue gases (post-combustion) or from fuels prior to combustion. Finally we have also evaluated the information on the costs of CO_2 capture, presenting data from our own model, as well as the state-of-the-art evaluation of the topic by the IPCC (2005).

Amine-based scrubbing is probably still the best available technology for post-combustion CO_2 capture. Yet, interest in oxyfuelling and pre-combustion removal of CO_2 through gasification is growing rapidly internationally. The costs of CO_2 capture are probably still somewhat too high for CCS to be widely considered as a desirable option for carbon mitigation (see Chapter 9). It is likely that new and improved CO_2 capture technologies, combined with advanced designs for power systems, can reduce the costs of CO_2 capture in the near- and medium-term future. The IPCC (2005) considers that a reduction of 20 to 30 per cent in the capture cost is feasible even with current commercially-available technologies. More substantial cost reductions are of course possible through innovation in new technologies.

3.7   A Whole System Techno-Economic Model of CCS

An intrinsic difficulty in the techno-economic assessment of entire CCS systems is the interdependence of the component subsystems. By way of example the conditions at which the capture equipment produces carbon dioxide will have an impact on the optimal pipeline design. The pipeline design in turn will impact on the conditions at which carbon dioxide is delivered to the storage repository and hence the specification of any injection equipment. In any study of the techno-economics of CCS, it is only sensible to consider optimal overall system configurations. The definition of optimum in this context is not crucial to the discussion here, but in general we will be interested in combinations that give the lowest energy cost. It is clearly possible to devise a wide range of perfectly functional CCS plant that are composed of non-optimal component combinations. However in view of the expense of construction, plant engineers would expend considerable effort in identifying optimal component configurations. Thus a realistic cost modelling study must make efforts to optimise system configurations, a task which is complicated by the interdependence of the subsystems.

To perform an entirely rigorous parameter study of costs requires that optimal CCS plant are designed and costed for the range of interesting parameter values. In principle the optimal parameters for any system can be identified through sensitivity studies. An engineer designing a CCS plant, may, for example, wish to investigate how the cost of energy varies with the rated output of the power plant. There is no exceptional difficulty in identifying local optimal parameters for each of the subsystems of a CCS plant. However, subsystem optimisation is unlikely to result in an optimal overall system.

Identification of whole system optima requires that the designer takes full account of the interactions between the subsystems. The complex, integrated nature of CCS makes this difficult to do. For this reason an integrated, computer based, technical model of the entire CCS system was developed, in order to facilitate rapid investigation of how system cost and performance varies as a function of several technical parameters.

Economic Considerations

The over-riding purpose of the model is to place a cost on carbon abatement for various technologies. Such abatement costs can only be considered in a comparative way, that is by citing the *differences* in energy cost and carbon dioxide production per unit of electricity between two or more technologies. By comparing two similar technologies, the additional expense required to produce less carbon dioxide per unit of energy can easily be seen. A commonly used measure is the cost of carbon avoided, that is:

Equation 3.14

$$(CC)_{avoided} = \frac{(COE)_2 - (COE)_1}{(CE)_2 - (CE)_1}$$

where:

CC	=	Cost of carbon (dioxide) avoided (£/kWh)
COE	=	Cost of energy (levelised production cost)
CE	=	Total carbon (dioxide) emissions per kWh (kg/kWh)

and the subscripts denote respective technologies.

To calculate the levelised energy production cost for a system, a methodology recommended by the IEA (Nitteberg *et al.*, 1983) is employed. With a test discount rate d and project economic lifetime L, the net present value of the whole project is given by:

Equation 3.15

$$NPV = -C_s - C_T - C_D - C_E - \sum_{i=1}^{i=L} \frac{c_{S,i} + c_{T,i} + c_{D,i} + c_{E,i}}{(1+d)^i} - \sum_{i=1}^{i=L+N} \frac{c_{D,MON,i}}{(1+d)^i} - \frac{D_S + D_T + D_D}{(1+d)^L}$$
$$+ \sum_{i=1}^{L} \frac{i_{S,i} + i_{E,i}}{(1+d)^i} - \sum_{i=1}^{L} \frac{c_{T,E,i} + c_{D,E,i}}{(1+d)^i}$$

where N is the number of years beyond project completion that reservoir monitoring must be maintained, and C_T, C_D etc. represent the sums of relevant capital costs listed in Table 3.4. The levelised energy cost is therefore:

Equation 3.16

$$COE = \frac{-NPV \times CRF}{E}$$

where the capital recovery factor is given by:

Equation 3.17

$$CRF = \frac{d(1+d)^n}{(1+d)^n - 1}$$

and E is the average annual energy production.

The user defines certain overall parameters for the CCS system, such as the rated power and type of the electricity generation plant, together with details of the pipeline route and storage reservoir. Model algorithms then 'size', in the sense that their major technical parameters are determined, the subsystems needed to produce a whole CCS system. Routines within the model also 'match' the various subsystems to produce, so far as possible, an overall optimal design. Finally the cost and performance of the overall system is predicted and output to the user.

Techno-economic models have been developed that predict the cost and performance, in terms of both energy output and carbon dioxide production, for each of the subsystems. So far as practicable the subsystem models are based upon rigorous technical analysis, but the scope of the problem means that resort has to be made to empiricism, especially for cost prediction. The model has been written as a computer code in Visual Basic.

Simplifying Assumptions

Attempting to model all possible options for each of the subsystems was beyond the scope of the project. To make the problem tractable, only a limited number of possibilities were treated in detail, these being listed in Table 3.5. The modelling approach and the computer code framework however are perfectly general and could be easily extended to treat a wider range of systems.

Table 3.4   Major costs in the techno-economic analysis of CCS

Major System	Cost	Type	Symbol	Note
Source (generator)	Plant	Capital	$C_{S,P}$	Only included for new-build plant, regarded as a sunk cost for re-fit cases.
	FGD	Capital	$C_{S,FGD}$	Flue gas desulphurisation equipment cost. Not included if already fitted to an existing plant.
	Fuel	On-going	$c_{S,FUEL,i}$	Fuel cost in year i from start of sequestration project.
	O&M	On-going	$c_{S,P}$	Plant O&M cost in year i from start of project.
	Income	On-going	$i_{S,i}$	e.g. Income from CHP or hydrogen in year i.
	Decommissioning	End of life	$D_{S,P}$	
Source (capture process)	Purchase and construction	Capital	$C_{S,C}$	
	O&M	On-going	$c_{S,C,i}$	
	Energy	On-going	$c_{S,E,i}$	Cost of energy required for capture plant operation – where possible will be treated implicitly as parasitic consumption of generated electricity.
	Decommissioning	End of life	$D_{S,C}$	
Transport	Collection network construction	Capital	$C_{T,NET}$	Only required if multiple sources used.
	Pipeline construction	Capital	$C_{T,1}$	
	Pumping station construction	Capital	$C_{T,2}$	
	Other construction costs…	Capital	$C_{T,3}…$	Other large capital costs (case specific).
	O&M	On-going	$c_{T,OM,i}$	
	Energy	On-going	$c_{T,E,i}$	Cost of energy required for pumping carbon dioxide through pipeline.
	Decommissioning	End of life	D_T	
Disposal	Well drilling	Capital	$C_{D,1}$	
	Well head equipment	Capital	$C_{D,2}$	
	Other construction costs…	Capital	$C_{D,3}…$	Other large capital costs (case specific).
	O&M	On-going	$c_{D,OM,i}$	
	Completion	End of life	D_D	
	Monitoring	On-going	$c_{D,MON,i}$	This cost is unique in that it continues after project completion.
EOR	Equipment	Capital	$C_{E,1}$	
	Transport	On-going	$c_{E,T,i}$	Cost of transporting oil to shore.
	O&M	On-going	$c_{E,OM,i}$	
	Income	On-going	$i_{E,i}$	Benefits from oil sales.

Note: Not all of the costs listed are considered in detail in the techno-economic model

Table 3.5   Overview of technical options considered by the model

Systems modelled	Limitations and comments
Power plants and capture system	
Coal fired pulverised fuel with flue gas desulphurisation and chemical (MEA) scrubbing (PFFGD+MEA)	In principle can model any rated capacity station but data availability limits capabilities
Integrated coal gasification cycle with physical scrubbing (IGCC)	In principle can model any rated capacity station but data availability limits capabilities
Combined cycle gas turbine with chemical (MEA) scrubbing (CCGT)	In principle can model any rated capacity station but data availability limits capabilities
Carbon dioxide transport system	
Pipeline transport, including repressurisation stations where needed	Only deals with single source to single sink transport. Routes must be determined using a GIS that is currently independent of the CCS model
Storage reservoirs	
Offshore oil and gas reservoirs	Data describing the reservoirs must be provided by the model user

Major Technical Parameters Treated by the Model

Conceptually there are two parts to the model. Firstly, there is a modelling framework provided by the model computer code. This provides a mechanism by which the techno-economics of CCS may be investigated; built on rigorous analysis so far as practicable. Secondly, there is a set of component performance and cost data that inform instances of the model calculations. This is essentially empirical in origin, having been derived from literature sources, and in certain cases, from parameter studies with other models. As will become apparent, the greatest constraint is due to the limited cost and performance data available. The modelling framework has been designed to allow study of the following main parameters:

- power station rated capacity;
- power station type;
- power plant utilisation, including variations over time;
- capture plant CO_2 capture efficiency;
- rate of solvent consumption (for MEA plant, due to sulphur products not removed by FGD);
- fuel composition (in so far as it has an impact on CO_2 produced per unit of energy);

- pipeline route;
- storage reservoir capacity and (limited) geological parameters;
- 'consumables' costs, i.e. energy supplied, fuel, solvent, etc.

The extent to which it is practical to investigate the performance of any system configuration depends on the availability of component cost and performance data. As will become apparent, the greatest constraint on model use is due to the limited component cost and performance data available

Design and Capital Cost Estimation for Power Plant and Capture System

The calculations rely on a look-up table approach for characterisation of the power plant and capture system, with tables populated by data produced either from specific simulation tools or from literature derived data such as that provided in section 3.6. For each power plant type and capture system type, tables have been formulated that give the plant capital cost and performance data as a function of rated capacity. Slightly different table formats are required to deal with retro fit and integrated cases. For the non-integrated power plant, the performance data comprises the overall efficiency of the energy conversion process from fossil fuel to electricity. Simple combustion calculations allow the peak rate of flue gas production and concentration of carbon dioxide in the flue gases to be estimated. This is then matched with data from look-up tables describing the capture plant, and the cost of the capture plant thereby estimated. The modelling framework allows the capture efficiency to be specified as an input, so that the effect of having more or less effective capture equipment can be investigated. Cases that do not match data within the look-up tables are dealt with via linear interpolation.

For integrated plant, a similar methodology is used. However the procedure is simpler as, by definition, there is no requirement to be able to 'mix and match' production plant and capture plant. Thus, the look-up tables directly provide data on the proportion of carbon dioxide from fuel combustion that is captured and fed to the storage reservoirs. In practice, it was found that insufficient reliable data was available in the open literature to make full use of the model capabilities. For example, there is little data on the variation of the capital cost of capture equipment with overall capture efficiency. The poor data availability has constrained the extent of the study, rather than any limitations in the model itself.

On-going Cost Estimation

The major on-going costs associated with CCS power plant operation are:

- fuel costs;
- solvent replacement costs, where a chemical process is used to capture carbon dioxide;
- operation and maintenance costs.

Fuel costs are calculated directly from the plant load factor, the per unit calorific value of the fuel and the overall efficiency of the electricity production process. The model has a time dependent mode which allows the effect of varying the plant load factor over time to be investigated, for example to represent the gradual run down of a plant as it approached the end of its life. The calculations reported here, however, all assume a constant load factor.

Solvent replacement costs arise in MEA based carbon dioxide capture plant because the MEA is gradually poisoned by combustion products in the flue gases. Sulphur dioxide is a particular problem in this case. The calculations reported herein assume that the flue gas from coal fired plant pass through FGD before the capture plant, but clean-up is far from perfect and quantities of sulphur dioxide will reach the MEA. Only limited data on the impact of sulphur dioxide on MEA consumption is available in the open literature; this constrains the validity of the calculations reported here. Operation and maintenance costs are treated in a simple manner. The annual expenditure is assumed to be a small percentage of the total capital cost of the plant, with values taken from the literature.

Pipeline Design Model

Realistic options for transport of carbon dioxide from commercial scale UK CCS plant to offshore reservoirs are pipeline, or in some cases bulk ship. The work reported here considered only pipeline transport as this is the most general solution, and from the limited studies in the literature, appears to offer the most economic solution. In general, previous studies have found that the pipeline infrastructure accounts for a relatively small proportion of the capital costs of CCS. Nevertheless, it is important to expend some effort in modelling pipelines as they have the potential to dominate disposal site selection decisions. The key point here is that the costs associated with the CO_2 disposal site are largely independent of the site itself, at least to the extent that they can currently be modelled. The only economic basis for reservoir selection that is currently justifiable, for an overview study at least, is the relative cost of pipeline construction compared to the capacity of the storage reservoir.

There is some discussion in the literature of the conditions under which captured carbon dioxide can be most economically transported. Outline calculations demonstrate that, for CCS on a commercial scale, unfeasibly large diameter pipelines will be required unless transport takes place under supercritical conditions, and these have been assumed here. The choice of transport conditions also has implications for the requirement for re-pressurisation stations, needed to overcome the frictional losses associated with the flow of the carbon dioxide through the pipe and to maintain the pressure and temperature of the carbon dioxide. Over the relatively short distances involved in implementing CCS in the UK, parameter studies have demonstrated that the exact transport conditions have a negligible impact on the need for re-pressurisation stations so long as supercritical conditions are adopted.

Design and Capital Costs for Pipeline Model

Pipeline capital costs are divided into material costs, and those associated with construction operations. In principle, both classes of cost depend on the same parameters, specifically:

- the chosen route, defined here as the distance the pipeline extends over different types of terrain. Costs are calculated from a cost per unit distance associated with each type of terrain;
- the pipeline diameter, which is assumed to be constant over the entire route;
- the flow conditions, which in combination with the above effects the specification of the pipeline and the need for re-pressurisation stations along the length of the route.

The cost model as written contains a facility to model complex interactions between the parameters. For example, it is possible to identify optimum diameters that minimise the overall cost of the pipeline. Again the utility of this has been rather constrained by the limited cost data available in the open literature. In particular, we have not accounted for the impact of pipeline diameter and specification on construction cost. The cost model requires that details of the route be used as inputs describing distances travelled across a limited number of terrain types. For the results reported here, unit costs for crossing each type of terrain have been obtained via a literature survey. However there is a facility for the model user to input revised costs.

The need for re-pressurisation stations is determined by a physical simulation of the flow through the pipeline. Pressures and temperatures, accounting for frictional effects, heat transfer from the pipe surface, and changes in elevation are calculated at stations along the length of the pipeline. If the pressure falls below a user specifiable value, then a re-pressurisation station is inserted into the pipeline, which restores the pressure and temperature to specifiable conditions. The model framework can cost stations based on the throughput of carbon dioxide and the effort of re-pressurisation (which influences the size of the plant required), but the limited cost data available means that the numerical results use a fixed cost. Construction of re-pressurisation stations offshore would be prohibitively expensive. At pipeline landing points therefore, the model checks to see if any re-pressurisation will be required before the pipeline reached the storage site. If so, then a re-pressurisation station is inserted at the landing point.

On-going costs associated with pipeline transport are those of maintenance and the energy consumed in any re-pressurisation equipment. Annual maintenance costs are taken to be a fixed fraction of the construction costs, which can be specified by the model user and for the work reported here was based on data in the literature. A physical calculation estimates the power required to re-pressurise the carbon dioxide in each pumping station. Taken with an assumed re-pressurisation plant efficiency, this allows the energy consumed by any pumping station to be calculated. The user may specify the per unit carbon dioxide release and monetary cost associated with this energy. For the studies reported here it has been assumed that this energy has been supplied electrically, with the current UK average per unit cost and associated carbon dioxide emission.

Pipeline Routing

In order to cost pipelines on the basis of their routes, a route description is required. Producing a detailed pipeline route for each case considered was beyond the scope of this work, so a simplified semi-automated methodology for devising representative pipeline routes was devised. It is important to note that the objective of this portion of the study was only to produce routing data that would reflect the general characteristics of the location considered, and did not consider detailed pipeline proposals.

After a review of the readily available literature and data it was concluded that attempting to custom-design the offshore portion of the pipeline routes required for the study cases was impractical. There are many factors, including seabed obstacles, that can constrain offshore pipelines, and accounting for them all was beyond the scope of the work. To simplify the analysis, therefore, it was assumed that the offshore portions of pipelines follow the routes of existing offshore oil and gas pipelines, as described in the IDEAL database (Davison, 2003). Informal consultations revealed that it is unlikely that owners of existing pipelines would be prepared for CCS pipelines to be constructed in close proximity, but again, the intention here is merely to produce representative data.

It turns out that pipelines originating from the UK already connect to all the storage sites of interest. Thus, by re-using these existing routes, the on-shore pipeline routing problem is 'reduced' to one of devising routes from carbon dioxide sources to the five existing pipeline landing points, shown in Figure 6.7. To this end a Geographical Information System (GIS), using the ArcView software, has been established that can calculate least cost routes from carbon dioxide sources to landing points (Davison, 2003). Sample output from the GIS is shown in Figure 6.7. Details of the GIS implementation are not covered here, but it is pertinent to outline the major points. Based on a literature survey, a cost is allocated to the construction of a representative pipeline across several types of terrain. Drawing on terrain use information derived from the UK portion of the 'Digital Chart of the World', a map can be produced that shows the per kilometre cost of building the representative pipeline at any point in the UK (Davison, 2003). A least cost routing routine can then be used to identify the cheapest route from any carbon dioxide source to pipeline landing points. The following features are amongst those accounted for in developing the cost maps:

- urban areas;
- major roads;
- major rail lines;

- large rivers and other water features;
- the effect of elevation on accessibility and construction costs.

For safety reasons, it is likely that there are constraints on the routes over which it is practicable to build CCS pipelines. A survey of the literature revealed no guidance on the construction of supercritical carbon dioxide carrying pipelines. A British Standard (BSI, 1992) does however refer to the construction of high-pressure carbon dioxide pipes and, although the range of conditions considered do not approach those proposed here, the most extreme constraint was adopted. The major constraint is that pipelines are not permitted to pass within a kilometre of populated areas.

The GIS system has been used to conduct studies investigating the relative cost of constructing pipelines from candidate CCS sites to landing points (Davison, 2003). In order to assess the impact of wider considerations on the construction costs, routes were developed with varying degrees of environmental sensitivity. Table 3.6 compares the relative cost of constructing pipelines from several large UK carbon dioxide sources to six offshore pipeline landing points that serve the UK.

Table 3.6   Relative costs of onshore CCS pipelines between some UK sources and landing points

Landing Point	Coal-fired Power Station		
	Fiddlers Ferry	Ratcliffe	Teesside
Point of Ayr	1.55	3.52	5.87
Barrow	4.81	5.10	3.57
Teesside	5.01	4.80	1.00
Easington	4.78	3.76	2.84
Theddlethorp	4.46	2.54	3.93
Bacton	6.61	4.06	6.87

It must be pointed out that the 'Digital Chart of the World' data is rather approximate. As a result, the generated routes should be regarded as indicative rather than definitive. Several parameter studies have, however, demonstrated that in practice the uncertainty in the geographical and land use data does not have a great influence on the calculated routes.

Disposal Site Model

The disposal site model considers primarily the cost of constructing injection wells. Calculations follow closely the approach reported in Holloway (1996) as no more rigorous approach could be identified from the literature. On-going costs to be considered by the model are the income from enhanced oil recovery (EOR) and the cost of long-term monitoring of stored carbon dioxide. It was originally intended to include a facility to model use of a distribution hub that could convey carbon dioxide to one of a number of reservoirs. However once again cost data availability rendered this impractical and the model only treats single source to single sink CCS schemes. The code has been written in such a way that this facility could easily be added.

Capital Cost of Injection Wells

The cost of each injection well required is calculated using the methodology reported in Holloway (1996) but with costs updated to 2005 values. The number of wells required is calculated from the maximum carbon dioxide flow rate and an estimate of the maximum rate at which carbon dioxide can be injected into a well. This latter value is estimated using a simplified version of a methodology due to Hendriks *et al*. (2000), which relies on estimates of the reservoir permeability and thickness. It is assumed that no additional compression takes place at the injection site, so that injection is driven only by the pressure at the well head and gravity. As such, there is no energy consumption at the injection site.

In addition to each well, there is also a cost associated with the mechanical equipment needed to complete each well head, so called well-head completion costs. These are assumed to be fixed for each well head and are again updated versions of previous estimates (Holloway, 1996).

Enhanced Oil Recovery

The model includes a very simple facility to include income streams from enhanced oil recovery (EOR), but this has not been used for the current studies.

Monitoring

It is likely that storage sites will require long term monitoring in order to satisfy regulatory requirements. There is currently much uncertainty around the form and extent of any monitoring that will be required, and the associated costs. As such, developing a cost model of monitoring processes proved to be prohibitively difficult. Monitoring is not therefore included in the model explicitly, although the user can enter an annual monitoring cost derived from other sources.

Integrated System Design

The overall system design is driven by the power station with all components being sized to deal with peak carbon dioxide flow. Thus the capture plant, the pipeline, the recompression stations and the reservoir injection equipment are all 'designed' in order after the power station. There may be cases where an improved energy cost could be achieved by sizing the components in a different way, but these have not been considered here.

Validation of the Techno-Economic Model

There are three areas of concern in the validation of cost models such as the one described here, and in particular:

- validation of the overall model and its cost predictions;
- validation that the model code performs correctly;
- validation of the theory underlying the model calculations.

Validation of the entire model is not practical, as it is predictive in nature and models systems for which no prototypes exist. Moreover many of the cost calculations depend on data taken from the literature which almost by definition cannot be validated. The validity of the theory underlying the model is also difficult to assess rigorously. The calculation methods are derived from published sources, and it has been assumed that they are correct. It is possible to assess whether the model code functions correctly. This has been achieved through a series of test cases run on the separate elements of the model and these have been used to test the validity of the implementation.

Table 3.7   **Range of total costs for CO_2 capture, transport and geological storage based on current technology for new power plants (in 2002 prices)**

Power plant with capture and storage	Pulverised coal power plant		Natural gas combined cycle plant		Integrated coal gasification combined cycle plant	
	US$	UK£	US$	UK£	US$	UK£
Cost of electricity without CCS per MWh	43–52	25–30	31–50	18–29	41–61	23–35
Cost of electricity per MWh	63–99	36–57	43–77	25–44	55–91	31–52
Electricity cost increase per MWh	19–47	11–27	12–29	7–17	10–32	6–18
% increase	43–91		37–85		21–78	
Cost per tonne CO_2 avoided	30–71	17–41	38–91	22–52	14–53	8–30
Cost per tonne C avoided	110–260	63–149	140–330	80–189	51–200	29–114

Source: Adapted from Table 8.3a, page 347, IPCC (2005)

Whole System Costs

Due to unavoidable personal circumstances it has not been possible to include the actual model results here. Table 3.7 presents data on the costs of CO_2 capture, transportation and storage from PF coal, natural gas CCGT and IGCC plant. Table 3.8 presents the same data but using natural gas CCGT as the reference plant against which cost comparisons are made. Estimates of the costs for transporting CO_2 are sensitive to the volume being moved. The costs more or less stabilise for volumes larger than approximately 10 Mt CO_2 per year at 1 to 1.5 US$ (0.5 to 1 £) per tCO_2 per 250km onshore, and 1 to 2 US$ (0.75 to 1.25 £) per tCO_2 per 250km offshore (IPCC, 2005). Storage costs vary widely depending upon the specific site selected, and a wide range from 0.5 to 8 US$ (0.3 to 4.6 £) per tCO_2 stored in saline formations and disused oil and gas fields has been quoted (IPCC, 2005). Monitoring is estimated to cost approximately 0.1 to 0.3 US$ (0.05-0.2 £) per tCO_2 stored. As an approximate estimate in the UK context, transport, storage and monitoring accounts for between 20 and 33% of the total costs of CCS, with the remainder being made up of the capture costs.

Table 3.8 Costs of CO_2 avoided through CCS for three power plants relative to natural gas CCGT reference case plant (in 2002 prices)

Power plant with CO_2 capture and geological storage	Natural Gas Combined Cycle Gas Turbine reference	
	US$ tCO_2 avoided	£ tCO_2 avoided
NG-CCTG	40–90	25–50
Pulverised Coal	70–270	40–155
Integrated Gasification	40–220	25–125

Source: Adapted from Table 8.4, page 348, IPCC (2005)

3.8 Conclusions

The major outcomes from this work are as follows:

1. A whole system cost model for the implementation of CCS plant in the UK has been developed and demonstrated, though the results have still to be presented.
2. The model has been used to investigate the techno-economics of building CCS systems in the UK, as a function of plant type and geography. Some consideration was also given to the impact of energy supply scenarios. However as the model has been formulated only for single source to single sink analysis, this was not treated in detail in the current work.
3. The extent of the study was largely limited by poor data availability on the capital cost of carbon dioxide capture plant as a function of the overall technical parameters. For techno-economic assessment purposes, a detailed parameter of the costs of capture equipment would be of great value. This

however would only be practical with significant industry involvement and issues of commercial confidentiality would need to be tackled.

4. Due to the data uncertainties, the modelling framework and computer code developed are the primary outputs from the study. The capabilities of the model significantly exceed the validity of the input data available from the open literature. Thus the model, and its underlying methodology, provide a basis for future studies of CCS in the UK as more reliable data becomes available.

5. A GIS for the techno-economic investigation of candidate pipeline routes from carbon dioxide sources to offshore pipeline landing points has been developed. The GIS has been used to investigate the impact of routing constraints, and in particular environmental impact considerations, on the relative pipeline construction cost.

Recommendations for Further Work

Data availability, rather than the capabilities of the developed model, has constrained the study and thus the main suggestions for further work are oriented around the collation of better input data. Specific recommendations are as follows:

1. The key recommendation from this work is that the scope for assessment and overall design optimisation of CCS systems is severely limited by the availability of cost data in the open literature. There is an urgent need for collation of cost data associated with the fossil fuel supply chain. It is unlikely that reliable decisions regarding the viability of CCS in the UK can be taken until such data is collated.

2. The issue of monitoring of storage sites needs to be considered in more detail. In particular a detailed model for the cost of monitoring, which is currently almost impossible to predict, should be developed. There is considerable debate over the extent of the monitoring required and the linked question of the time period over which it is desirable that carbon dioxide is retained in storage reservoirs (see Chapter 9). This debate requires as an input an economically oriented risk analysis, which in turn requires a cost model for the monitoring processes.

3. There is a need for design guidance on the construction and routing of CCS pipelines. Existing UK guidance, perhaps unsurprisingly, makes no reference to the construction of long pipelines carrying supercritical carbon dioxide.

3.9   References

Anderson, S., Newell, R. (2003), *Prospects for Carbon Capture and Storage Technologies*, Resources for the Future, Discussion Paper 02–68, Washington DC.

Asada, Y., Miyake, M., Miyake, J., Kurane, R., and Tokiwa, Y. (1999), 'Photosynthetic Accumulation of Poly-(Hydroxybutyrate) by Cyanobacteria: The Metabolism and

Potential for CO_2 Recycling', *International Journal of Biological Macromolecules*, **25**(1–3), pp. 37–42.

Benemann J. (1997), 'CO_2 Mitigation with Microalgae Systems', *Energy Conversion and Management*, **38**, Supplement 1, pp. S475–S479.

Benneman, J. (1993), 'Utilisation of Carbon Dioxide from Fossil Fuel-Burning Power Plants with Biological Systems', *Energy Conversion and Management*, **34**, pp. 999–1004.

Brown, L.M, and Zeiler, K.G. (1993), 'Aquatic Biomass and Carbon Dioxide Trapping', *Energy Conversion and Management*, **34**, pp. 1005–13.

BSI (1992), *Code of Practice for Pipelines, Part 2: Pipelines on Land: Design, Construction and Installation*, BS 8010–2.8:1992, British Standards Institute.

Center for Energy and Environmental Studies (2003), *Integrated Environmental Control Model User Documentation*, Carnegie Mellon University, Pittsburgh, PA. http://www.iecm-online.com.

Chinn, D., Choi, G. N., Chu, R. and Degen, B. (2005), 'Cost Efficient Amine Plant Design for Post Combustion CO_2 Capture from Power Plant Flue Gas', in M. Wilson, T. Morris, J. Gale, and K. Thambimuthu (eds), *Proceedings of 7th International Conference on Greenhouse Gas Control Technologies, Volume 2: Contributed Papers and Panel Discussion*, Elsevier Science, Oxford, pp. 1133–38.

Creutz, C. and Fujita, E. (2000), 'Carbon Dioxide as a Feedstock', in *Carbon Management: Implications for R&D in the Chemical Sciences and Technology: A Workshop Report to the Chemical Sciences Roundtable*, National Academies Press, Washington, DC.

Davison, J, Bressan, L. and Domenichini, R. (2005), 'CO_2 Capture in Coal-Based IGCC Power Plants', in E.S. Rubin, D.W. Keith and C.F. Gilboy (eds), *Proceedings of 7th International Conference on Greenhouse Gas Control Technologies. Volume 1: Peer-Reviewed Papers and Overviews*, Elsevier Science, Oxford, pp. 167–75.

Davison, J. and Thambimuthu, K. (2005), 'Technologies for Capture of Carbon Dioxide', in E.S. Rubin, D.W. Keith and C.F. Gilboy (eds), *Proceedings of 7th International Conference on Greenhouse Gas Control Technologies. Volume 1: Peer-Reviewed Papers and Overviews*, Elsevier Science, Oxford, pp. 3–13.

Degen, A., Uebele, A., Retze, U., Schmid-Steger and Trosch, W. (2001), 'A Novel Airlift Photobioreactor with Baffles for Improved Light Utilization through the Flashing Light Effect', *Journal of Biotechnology*, **92**, pp. 89–94.

Davison, T. (2003), *Investigating the Environmental Disruption Values and Financial Construction Costs of Possible Pipelines Routes for the Purpose of Carbon Sequestration Projects*, MSc Dissertation, University of Sunderland, Sunderland.

Dunsmore H.E. (1992), 'A Geological Perspective on Global Warming and the Possibility of Carbon Dioxide Removal as Calcium Carbonate Mineral', *Energy Conversion and Management*, **33**(5–8), pp. 565–72.

Gibbins, J., Crane, R., Lambropoulos, D., Booth, C., Roberts, C.A. and Lord, M. (2005), 'Maximising the Effectiveness of Post Combustion CO_2 Capture System', in E.S. Rubin, D.W. Keith and C.F. Gilboy (eds), *Proceedings of 7th International Conference on Greenhouse Gas Control Technologies. Volume 1: Peer-Reviewed Papers and Overviews*, Elsevier Science, Oxford, pp. 139–46.

Gottlicher, G. and Pruschek, R. (1997), 'Comparison of CO_2 Removal Systems for Fossil-Fueled Power Plant Processes', *Energy Conversion Management*, **38**, pp. S173–S178.

Hendriks, C.A., Wildenborg, A.F.B., Blok, K., Floris, F. and van Wees, J.D. (2000), 'Costs of Carbon Removal by Underground Storage', *Proceedings of the 5th International Conference on Greenhouse Gas Control Technologies*, CSIRO Publishing, Collingwood.

Herzog, H., Drake, E. and Adams, E. (1997), *CO_2 Capture, Reuse, and Storage Technologies for Mitigating Global Climate Change*, A White Paper Final Report, DOE Order No. DE-AF22-96PC01257, Washington DC.

Herzog, H. (2001), 'What Future for Carbon Capture and Sequestration?' *Environmental Science and Technology*, **35**(7), pp. 148 A–153 A.

Herzog, H. (2002), *Carbon Sequestration via Mineral Carbonation: Overview and Assessment*, http://sequestration.mit.edu/pdf/carbonates.pdf.

Holloway, S. (1996) (ed.), *The Underground Disposal of Carbon Dioxide, Final Report of the Joule II Project*, No. CT92-0031, British Geological Survey, Nottingham.

IEA (1998), *Carbon Dioxide Capture from Power Stations*, International Energy Agency, Paris.

IEA (2004), *Prospects for CO_2 Capture and Storage*, International Energy Agency, Paris.

IPCC (2005), *IPCC Special Report on Carbon Dioxide Capture and Storage*, prepared by Working Group III of the Intergovernmental Panel on Climate Change, B. Metz, O. Davidson, H.C. de Coninck, M. Loos and L.A. Meyer (eds), Cambridge University Press, Cambridge.

Jordal, K., Anheden, M., Yan, J. and Strömberg, L. (2005), 'Oxyfuel Combustion for Coal-fired Power Generation with CO_2 Capture: Opportunities and Challenges', in E.S. Rubin, D.W. Keith and C.F. Gilboy (eds), *Proceedings of 7th International Conference on Greenhouse Gas Control Technologies, Volume 1: Peer-Reviewed Papers and Overviews*, Elsevier Science, Oxford, pp. 201–209.

Kojima, T., Nagamine, A., Ueno, N. and Uemiya, S. (1997), 'Absorption and Fixation of Carbon Dioxide by Rock Weathering', *Energy Conversion and Management*, **38**, pp. S461–S466.

Lal, R., Kimble, J.M. and Follett, R.F. (1999), *The Potential of U.S. Cropland to Sequester Carbon and Mitigate the Greenhouse Effect*, Lewis Publishers, Boca Raton, Florida.

Lyngfelt, A., Leckner, B. and Mattisson, T. (2001), 'A Fluidized-bed Combustion Process with Inherent CO_2 Separation: Application of Chemical-Looping Combustion', *Chemical Engineering Science*, **56**(10), pp. 3101–3113.

Middleton, P., Solgaard-Anderson, H. and Rostrup-Nielsen, H.T. (2004), *Hydrogen Production with CO_2 Capture using Membrane Reactors*, Phase 1 Report of CO_2 Capture Project, www.CO2captureproject.org/reports/reports.htm.

Miesen, A. and Shuai, X. (1997), 'Research and Development Issues in CO_2 Capture', *Energy Conversion Management*, **38(S)**, pp. S37–S42.

Nitteberg, J., de Boer, A.A. and Simpson, P.B. (1983), *Recommended Practices for Wind Turbine Testing 2: Estimation of the Cost of Energy from Wind Energy Conversion Systems*, International Energy Agency, Paris.

Otsuki, T. (2001), 'A Study for the Biological CO_2 Fixation and Utilization System', *The Science of The Total Environment*, **277**(1–3), pp. 21–25.

Pedroni, P.M., Lamenti, G., Prosperi, G., Ritorto, L., Scolla, G., Capuano, F. and Valdiserri, M. (2005), 'ENITECNOLOGIE R&D Project on Microalgae Biofixation of CO_2: Outdoor Comparative Tests of Biomass Productivity Using Flue Gas CO_2 from a NGCC Power Plant', in M. Wilson, T. Morris, J. Gale, and K. Thambimuthu (eds), *Proceedings of 7th International Conference on Greenhouse Gas Control Technologies, Volume 2: Contributed Papers and Panel Discussion*, Elsevier Science, Oxford, pp. 1037–42.

Riemer, P., Audus, H. and Smith, A. (1993), *Carbon Dioxide Capture from Power Stations*, IEA Greenhouse Gas R&D Program, Cheltenham.

Roberts, C.A., Gibbins, J., Panesar, R. and Kelsall, G. (2005), 'Potential for Improvement in Power Generation with Post-Combustion Capture of CO_2', in E.S. Rubin, D.W. Keith and C.F. Gilboy (eds), *Proceedings of 7th International Conference on Greenhouse Gas Control Technologies, Volume 1: Peer-Reviewed Papers and Overviews*, Elsevier Science, Oxford, pp. 1555–63.

Simmonds, M., Miracca, I. and Gerdes, K. (2005), 'Oxyfuel Technologies for CO_2 Capture: A Techno-Economic Overview', in M. Wilson, T. Morris, J. Gale, and K. Thambimuthu (eds.), *Proceedings of 7th International Conference on Greenhouse Gas Control Technologies, Volume 2: Contributed Papers and Panel Discussion*, Elsevier Science, Oxford, pp. 1125–30.

Singh, D.J., Croiset, E, Douglas, P.I. and Douglas, M.A. (2001), 'CO_2 Capture Options for an Existing Coal Fired Power Plant: O_2/CO_2 Recycle Combustion vs. Amine Scrubbing', *Proceedings of the First National Conference on Carbon Sequestration*, Washington, D.C.

Stewart, C. and Hessami., M.-A. (2005), 'A Study of Methods of Carbon Dioxide Capture and Sequestration: The Sustainability of a Photosynthetic Bioreactor Approach', *Energy Conversion and Management*, **46**(3), pp. 403–20.

West, T.O. and Marland, G. (2002), 'A Synthesis of Carbon Sequestration, Carbon Emissions, and Net Carbon Flux in Agriculture: Comparing Tillage Practices in the United States', *Agriculture Ecosystems & Environment*, **91**, pp. 217–32.

www.peer.caltech.edu/projects/cat_chem_2.htm, accessed 10 August 2005.

Xu, A., Indala, S., Hertwig, T., Pike, R., Knopf, F., Yaws, C., and Hopper, J. (2003), 'Identifying and Developing New Carbon Dioxide Consuming Processes', *Proceedings of the Sustainable Development Topical Conference*, American Institute of Chemical Engineers 2003 Annual Meeting, Paper No. 408b, San Francisco, CA.

Xu, A., Indala, S., Hertwig, T., Pike, R., F. Knopf, F., Yaws, C. and Hopper, J. (2005), 'Development and Integration of New Processes Consuming Carbon Dioxide in Multi-Plant Chemical Production Complexes', *Clean Technologies and Environmental Policy*, **7**, pp. 97–115.

Chapter 4

Geological Carbon Dioxide Storage and the Law

Ray Purdy

4.1 Introduction

Whilst geological carbon dioxide capture and storage (CCS) remains a potentially attractive climate change mitigation option, there are uncertainties and complexities surrounding the legality of such projects. The importance of the legal position when considering geological CCS cannot be understated. Even if suitable storage sites for CO_2 have been identified, the technologies for transportation and injection are feasible and available, and there is broad support from within government, industry, and the public for such projects taking place, the significance of these is limited if the current applicable laws prevent or restrict such projects from taking place. This chapter identifies the relevant international and European legislation that potentially impinges on offshore geological carbon dioxide storage projects taking place and considers the key legal questions of ambiguity concerned with geological CCS beneath the waters surrounding the United Kingdom (UK). Although this chapter focuses on the legal position in relation to the UK, any conclusions that are reached will be relevant to others in the international community. It is clear that the international community would benefit from greater legal clarification of existing relevant legislation. The chapter cannot, and its aim is not to, always offer definitive answers as to whether certain geological CCS projects are legal or not. Where there is uncertainty as to the legal position, comment is provided on why this is the case and, where appropriate, the authors' opinion as to what might be the correct legal interpretation.

Any geological carbon dioxide storage project taking place under the seas surrounding the UK will most likely fall within the remit of a number of overlapping legal regimes. In other words, they will be covered by international, European Community (EC) and national legal regimes. This is because the UK is a signatory to many of the international laws which are potentially relevant to geological carbon dioxide storage projects, and is also a member of the EC, and as such is bound by Community law. It is important to note that there is currently no legislation, either national, European or international, which specifically covers the legal issues surrounding carbon dioxide storage. The laws that could apply to geological CCS were not designed with this is mind. The storage of CO_2 is a relatively new concept and was not envisaged until relatively recently, whereas for example, some of the international marine conventions are thirty years old. Although carbon dioxide storage is not specifically mentioned in legislation it can of course still fall under the legislation's remit. Many laws in particular are still developing and evolving and can react to such changes in society. Laws can also be amended or replaced to reflect the changing objectives of national governments, members of the EC or the international community. However, amending and replacing laws can be time consuming and difficult in practice because they often require unanimous or majority agreement. Although this chapter only comments on the legislation as it currently stands, the procedures for amending individual laws are discussed, where appropriate.

The question whether the legal principles that are currently in place in existing legislation will apply to the storage of CO_2 is also untested water in the courts. Comparative jurisprudence concerning the storage of other materials in the seabed provides some clues as to the direction the courts might be willing to take but this is not certain. There has also been very little sustained legal analysis in academic books and journals on geological carbon dioxide storage. Much of the published legal analysis is also only partly relevant, because it has primarily focused on the legal issues associated with the storage of CO_2 directly in the oceans, which no longer seems to be a well backed mainstream mitigation option. Research conducted by the author early on in this project revealed a dearth of in-depth research studies on geological carbon dioxide storage, combining international, European and national legal dimensions, in the UK, and other European and developed countries (Purdy and Macrory, 2004).

The importance of a systematic analysis of the legal dimension of the long term sub-strata storage of CO_2 is increasingly being recognized in the UK, as are some of the uncertainties and complexities involved. A report by the Royal Commission on Environmental Pollution[1] (RCEP) in 2000 concluded that it was 'open to interpretation whether disposal of carbon dioxide into the ocean or under the seabed would be permissible under current international law' (RCEP, 2000). The UK Government started to consider the legal implications of CCS after the Royal Commission's report. In 2001 a Department of Trade and Industry (DTI) report acknowledged the uncertainties of the current legal status of sub-seabed storage, concluding that 'there is a strong case for assessing in a systematic way the legal, scientific, engineering and economic aspects of both EOR [enhanced oil recovery] and geological CO_2 capture and storage. Such an assessment needs to precede any further analysis of the policy case for support for steps to use CO_2 in this way' (DTI, 2001). The Cabinet Office's Performance and Innovation Unit's energy review in 2002 also recognized these legal uncertainties and supported the recommendations of the DTI in stressing the need for a more detailed assessment (Cabinet Office, 2002).

---

1   The Royal Commission on Environmental Pollution is an independent standing body established in the UK in 1970 to advise the Queen, the Government, Parliament and the public on environmental issues.

In early 2002, the UK Government concluded that it was premature to finance a carbon capture demonstration project because, amongst other things, the uncertainty over the legal status of disposal in sub-seabed strata. The DTI, however, continued to examine the feasibility of carbon dioxide capture and storage in the UK. The legal services section in the Department of the Environment Food and Rural Affairs (DEFRA) produced a short report in August 2002 for the Sustainable Energy Policy Unit of the DTI, advising them on the main legal issues arising from CO_2 storage (DEFRA, 2002). Towards the end of the year the DTI organized a CO_2 capture and storage stakeholder meeting in London. A conclusion of this meeting was that one of the main economic and commercial barriers to capturing and storing CO_2 was its legality; one of the recommended steps was for the Government to determine the legality of CCS. Delegates pointed out that there appeared to be a lack of a framework and disagreement between the international laws as to whether geological dioxide storage could take place. The UK Government concluded in its formal response to the RCEP that any disposal of carbon dioxide into submarine strata would also depend on the resolution of legal issues under the international marine conventions, such as the London and OSPAR Conventions (UK Government, 2003).

A reading of the UK Government response to the RCEP report suggests that they appear to have adopted a wait and see approach until the legal issues are resolved at international level (UK Government, 2003). However, it is clear that in practice the UK Government has been particularly active in promoting international discussion as to the legal issues surrounding CO_2 capture and storage, and the possible need for changes to legislation. The UK Government organized a seminar on CO_2 storage in London for OSPAR Contracting Parties in October 2003 and coordinated a legal review for the Contracting Parties to the London Convention (IMO, 2005).

Similarly, at an international level there has been increasing interest and discussion in the last few years as to whether the storage of CO_2 in the seabed is consistent with international law. It should be noted that CO_2 storage is not a new concept and some actors in the international community anticipated it becoming a potentially thorny legal issue for many years. CO_2 storage in the seas was first brought to the attention of the London Convention at the fifteenth meeting of its Scientific Group as far back as 1992. In 1997 GESAMP[2] conducted a study that noted that 'dumping from vessels and platforms of both liquid and solid CO_2 is prohibited by the LC (London Convention) and the 1996 Protocol and unless these instruments can be amended to permit such dumping, it seems unlikely that any of the current parties could give approval to such a practice' (GESAMP, 1997). This conclusion was significant because GESAMP comprises some of its members from the London Convention Secretariat and several delegates from London Convention Contracting Parties.

The next most significant review into the legal issues raised by CO_2 storage then took place during 1999, when the Scientific Group to the London Convention examined whether CO_2 fell within the definition of industrial waste (IMO, 1999).

---

2 GESAMP is an advisory body consisting of specialized experts nominated by the Sponsoring Agencies (IMO, FAO, UNESCO-IOC, WMO, WHO, IAEA, UN, UNEP).

They concluded at their twenty-second meeting that fossil fuel derived CO_2 was an industrial waste, and that delegations at the twenty first consultative meeting should be consulted concerning the priority to be accorded to consideration of these issues. The consultative meeting was presented with the Scientific Groups report and a further report on Ocean Storage of CO_2 put together by the International Energy Agency (IEA) (Brubaker and Christiansen, 2001). The conclusion of the Consultative Parties at the twenty-first meeting was that the Scientific Group should continue to keep a watching brief on the relevant research being carried out and that they would consider the legal, political and institutional dimensions of a potential proposal to amend the London Convention or the 1996 Protocol at a later stage (IMO, 2000).

The initiative to examine the legal implications of CO_2 storage was again lost for a number of years, before Norwegian CO_2 trials in the North Sea brought it back into the political spotlight.[3] These trials fell under the direct geographical remit of the OSPAR Convention, a regional marine environmental protection law. In June 2002, the Secretariat of the OSPAR Commission, who manages the OSPAR Convention, asked its legal experts in the Group of Jurists and Linguists to provide advice on the compatibility with the Convention of possible placements of carbon dioxide in the sea and the seabed. In a press release the OSPAR Commission stated that it was desirable to establish as soon as possible an agreed position on whether such placing of CO_2 in the sea or the seabed was consistent with the OSPAR Convention (OSPAR Commission, 2002). The OSPAR Commissions Group of Jurists and Linguists completed a preliminary legal paper in May 2003 (OSPAR Group of Jurists and Linguists, 2003), and this was discussed at the meeting of the parties to the OSPAR Convention in Bremen in June 2003 (OSPAR Commission, 2003). The Group of Jurists and Linguists final report was accepted by the OSPAR Contracting Parties in May 2004 and was published shortly afterwards (OSPAR Commission, 2004). The Group of Jurists and Linguists main conclusion was that compatibility with the Convention depended on the method by which the carbon dioxide was placed in the maritime environment. They decided that further consideration was needed on the interrelations between the current legal position, the possible physical impacts of the placement of CO_2 on the marine environment, and the appropriate regulatory approach. The current legal framework does not appear to have been the subject of any further discussion by OSPAR Contracting Parties since the Group of Jurists and Linguists report was agreed, although the OSPAR Commission organized a workshop in October 2004 looking at the environmental impact of CO_2 placement in the marine environment and the technical options for placement.

Early legal interest in carbon dioxide storage was dominated by bodies associated with marine environmental protection. Since then other international actors in the fields of energy and climate change have also been increasingly drawn into the debate. The IEA[4] developed an interest in CO_2 placement and held a workshop in July 2004 looking specifically at the legal issues surrounding CO_2 storage, in both international

---

3   The Norwegian Institute for Water Research wanted to release 5.4 tonnes of CO_2 into the sea off the coast of Norway at a depth of 800 metres.

4   The IEA acts as energy policy advisor to 26 member countries.

law, and some national laws. In 2005 they published an extensive report examining the legal issues of CO_2 storage, which concluded that existing national and international regulations were not fitted to large-scale experiments in CO_2, and that urgent legislative work was needed to keep pace with technical progress (IEA, 2005). The IEA noted that in their opinion 'the contracting parties to these [marine] agreements need to interpret, clarify or, as the case may be, amend these treaties with a view to accounting for some form of controlled carbon dioxide storage. There is significant room for such interpretation and clarification under these treaties' (IEA, 2005).

The Intergovernmental Panel on Climate Change[5] (IPCC) published a report on CO_2 capture and storage in September 2005 (IPCC, 2005). The report concluded that the 'actual use of CCS is likely to be lower than the estimates for economic potential indicated because of barriers including perceived environmental impacts, risks of leakage, lack of a clear legal framework and public acceptance' (IPCC, 2005). The report, which only included a limited analysis of the legal framework, arrived at the same conclusion as the IEA, that 'generally, it is unclear whether cases of offshore CO_2 injection into the geological sub-seabed or the ocean are compatible with international law' (IPCC, 2005).

In late 2004 the Contracting Parties to the London Convention revived their interest in the issue of CO_2 placement. They agreed at their 26th Consultative Meeting that this issue should be formally included in their work programme, and that they would initially focus on storage of CO_2 in geological structures in the marine environment (IMO, 2004). A correspondence group was established under the lead of the UK to consider the legal issues associated with CO_2 placement in geological structures under the London Convention and Protocol. The UK was charged with preparing and communicating a list of legal questions, which were delivered to the Contracting Parties in March 2005 (IMO, 2005). The UK Government prepared a consolidated paper on the legal views of Contracting Parties for submission to the 27th Consultative Meeting in October 2005. The document was discussed by the Contracting Parties at this meeting and a work plan formulated to help establish a consensus.

Although there has been increasing discussion in the international community on CO_2 storage offshore, there is, however, probably no international consensus on the interpretation of key provisions in these marine conventions at present. Some of the reviews that have already taken place have reached similar conclusions as to the existing legal position. However, many of the bodies undertaking reviews have their own agendas, ranging from environmental protection to economic growth. The IEA and the IPCC, for example, are concerned with energy and climate change frameworks respectively, rather than marine environmental protection – where existing legislation creates potentially serious impediments to some CO_2 storage projects taking place.

---

5   The IPCC was established by the WMO and UNEP to assess scientific, technical and socio-economic information relevant for the understanding of climate change, its potential impacts and options for adaptation and mitigation.

Whilst it is convenient to blame any existing legal uncertainties on the current drafting of marine conventions and call for these to be revised, this might be considered to be something of a distraction. In many ways the current marine conventions are drafted reasonably clearly. Any CO_2 projects that can take place at the current time, are arguably only because of loopholes in the marine legislation, as CO_2 storage was not envisaged at the time that these laws were adopted. The current marine legislation therefore only needs amending if there is a need (e.g. states want to go ahead or restrict such projects) and there is political will to amend legislation. One might argue that the existing legal framework concerning climate change is actually less clear in relation to geological CO_2 storage in marine waters. If the climate change legal regime, expressly allows and creates binding rules concerning geological CO_2 storage in marine waters to be classed as an emission reduction then there could be a potential need to change other legislation. Additionally, if a consensus of international states sees CO_2 storage in geological formations in marine waters as more important than greenhouse gases entering the atmosphere, then this provides the political will to amend the marine legislation. If the climate change regime restricts, or alternatively is ambiguous as to whether geological CO_2 storage in marine waters does count as an emission reduction, then there is arguably less incentive and need to amend the marine conventions. This chapter will therefore consider all of the major legislation affecting CO_2 storage, because their objectives and how they operate together are, in practice, extremely important. Although the importance of the law cannot be understated, it should also be considered alongside the objectives and political will of countries, as laws can be amended to reflect the changing objectives of the European and international community.

4.2   Applicable International and European Laws

International Laws

There are a number of international laws that could be relevant to geological CO_2 storage. International conventions normally become operational in two stages. The first stage is when a certain number of countries sign the convention, signalling their intention to become parties to it. The UK is a signatory to all of the international laws discussed below. The second stage in becoming legally operational is when a sufficient number of states have acceded to the convention and it becomes international law. Details of the legal status of each international convention are summarized in Table 4.1.

Four of the international laws contained in Table 4.1 are marine conventions. In practice these marine conventions operate together, not separately, in strengthening environmental protection. The most important convention concerning marine waters is the UNCLOS. This Conventions aim is to regulate all uses of the sea and it establishes basic legal rules for all aspects of the use and protection of the sea, including shipping, scientific research, exploration of natural resources, disaster prevention, avoidance of pollution and protection of the marine environment. The

Convention is constructed in a framework nature, leaving the elaboration of precise rules to other bodies, such as national Governments and international Organizations. All the states which are Contracting Parties in UNCLOS are obliged to issue laws and to take other measures to regulate pollution by dumping, and they must not be less effective than the 'global rules and standards'.[6] It is generally accepted that these 'global rules and standards', in relation to marine pollution, are found in the rules and standards of the Convention on the Prevention of Marine Pollution by Dumping of Wastes and Other Matter, 1972 (the London Convention) and its 1996 Protocol. At the current time the London Convention (and then the Protocol when it enters into force) is the most significant global convention to protect the marine environment and conserve its species and ecosystems. States which have ratified both UNCLOS and these conventions are obliged to uphold the laws and regulations in accordance with UNCLOS.

In order to enhance protection of the marine environment, parties to both the London Convention and the 1996 Protocol are encouraged to create regional agreements which further their objectives.[7] The regional agreements can provide a greater degree of environmental protection in the regions that they cover but they must endeavour to be consistent with the Convention and Protocol. For the purposes of this chapter, the OSPAR Convention is a relevant regional agreement, because the UK falls within its geographical coverage and is a signatory. Whilst the OSPAR Convention is similar to the Protocol and London Convention, it is stricter in scope as it is more modern and takes into account other sources of pollution.

Table 4.1   International laws relevant to geological CO_2 storage in marine waters

Convention	Purpose	In force
UNCLOS (1982)	Marine – Regulates Uses of the Sea	Yes
London Convention (1972)	Marine Environmental Protection	Yes
OSPAR Convention (1992)	Marine Environmental Protection	Yes
Protocol to the London Convention (1996)	Marine Environmental Protection	No
Framework Convention on Climate Change (1992)	Climate Change	Yes
Kyoto Protocol to the Climate Change Convention (1997)	Climate Change	Yes
ESPOO Convention (1991)	Environmental Assessment	Yes
Kiev Strategic Environmental Assessment (SEA) Protocol (2003)	Environmental Assessment	No
Convention on Biological Diversity (1992)	Habitat Protection	Yes

---

6   UNCLOS, Article 210.
7   London Convention Article VIII; 1996 Protocol, Article 12.

As noted above, the UNCLOS encourages the adoption of further international legislation on marine pollution, which was created through the London Convention and Protocol, which in turn anticipates the creation of regional agreements. This means that in practice, each convention envisages compliance with the other more specific or regional convention, and indeed, encourages it. In practice, the conventions contain provisions making sure the objectives and provisions of their convention are followed and establish procedures for co-operation between them in order to develop harmonized procedures to be followed by Contracting Parties to the different conventions concerned.[8] This means that every effort is made to make the provisions of prior treaties consistent with the provisions of more recent treaties.

In general, states are only bound to conventions to which they ratify or accede. As the UK is a party to all of the conventions mentioned above, it is obliged to follow the laws of them all. In practice it is obvious that if there are overlapping conventions, then it is possible that one will contain more stringent provisions than another. If consistent reading is not possible between two conventions and neither contains a provision expressly stating which will prevail, the legal position under international law is that the provision of the most recent treaty controls over an older treaty, and a specific treaty controls over a general treaty. In the context of this chapter, in such cases of inconsistency, the Convention, 1996 Protocol, and OSPAR regimes should be viewed as being distinct, although, in order to adhere to international obligations, a state would need to apply the standard of the most specific and stringent treaty. As the OSPAR Convention often goes further than the London Convention and Protocol, Contracting Parties to all of these marine conventions will most likely be obliged to take OSPAR as setting the standard in such cases. The UK Government's view seems to be that if there were conflicting provisions then they would follow OSPAR over the other conventions (DEFRA, 2002). Where there is inconsistency between the London Convention and the 1996 Protocol, the UK Government consider that the latter will prevail in terms of the obligations they impose, but this would not mean that there would not be a breach of OSPAR obligations (DEFRA, 2002).

The position, in relation to a conflict between the marine conventions and the international laws concerning climate change, namely the United Nations Framework Convention on Climate Change (UNFCCC) and Kyoto Protocol, is not so clear. It is not known whether there was some discussion in the drafting of the latter conventions over the impact that CO_2 storage could have with the compatibility with other international obligations under the marine conventions. If one follows the rule that a specific treaty controls over a general treaty, it seems that marine conventions are the more specific laws relevant to controlling CO_2 storage under marine waters. In the absence of any reference in these conventions to one another, the guidance contained in the Vienna Convention on the Law of Treaties will also have to be considered.[9]

---

8   E.g. London Convention, Article VIII.
9   See Article 30 on the application on of successive treaties relating to the same subject matter.

Applicable European Laws

There are also a number of European laws that could be relevant to geological carbon storage below marine waters. The UK is a member of the EC and must adhere to the laws of the Community. Most European environmental legislation is in the form of Directives, which are addressed to Member States and require implementation into national law. If the European law is not implemented fully or applied correctly the EC can take enforcement action against the Member State. In addition, national courts are obliged to as far as possible to interpret national law consistently with EC Directives, and where public authorities are involved, precise and certain provisions of Directives may in certain circumstances still be legally binding on those authorities even where national legislation has not yet been implemented. Details of the purpose and legal status of the relevant directives are summarized in Table 4.2.

The three EC Directives which seem most relevant to carbon dioxide storage projects, and which are examined in this chapter, include: the EIA Directive, the Strategic Environmental Assessment (SEA) Directive and the Habitats Directive. An important consideration is whether these Directives have legal application in marine waters. It seems that the EC Treaty itself does not explicitly deal with the territorial application of treaty rules, it simply states that it shall apply to the UK of Great Britain and Northern Ireland.[10] It is generally agreed that this implies that the Treaty can extend to areas within the jurisdiction of Member States. Although Community law can be applied outside the land and territorial waters of Member States, and this has indeed been the practice in many areas of Community law, not all Community legislation has such a widened territorial scope. Each Community law has to be examined individually.

Table 4.2   European legislation relevant to geological CO$_2$ storage in marine waters

Directive	Purpose	Date coming into force
Habitats Directive	Habitat Protection	June 1994
Environmental Impact Assessment Directive	Environmental Assessment	March 1999 (in its amended form)
Strategic Environmental Assessment (SEA) Directive	Environmental Assessment	July 2004

It is clear that the EIA Directive extends to certain marine waters because the UK Government has implemented regulations applying the Directive to offshore installations. The Department of Trade and Industry has also carried out an SEA in the UK's continental shelf. The UK Government has also conceded the point in relation to the Habitats Directive, after a judicial review case brought by Greenpeace challenging the nineteenth licensing rounds for offshore exploration and drilling

---

10 EC Treaty.

initiated by the Secretary of State in 1997.[11] The UK Government tried to claim that by its very nature the Habitats Directive was restricted to land and the territorial waters of EC Member States. The High Court held in 1999 that the UK Government was wrong in its view that the Habitats Directive did not extend beyond the UK's twelve mile territorial limit and accepted Greenpeace's argument that the Habitats Directive applies to the continental shelf and the two hundred mile fishing limit. Therefore it is highly likely that CO_2 projects taking place in the territorial waters or continental shelf of the UK will be subject to the legal conditions and principles laid out in the EIA Directive, SEA Directive, and the Habitats Directive. These European Directives will not apply to projects taking place in the high seas.

As EC environmental law is drafted by the same body there are usually no concerns with Directives being in conflict with one another. With Directives dealing with a similar subject area there is usually a provision referring to the other Directive and how it should be considered.[12] Whilst there are unlikely to be conflicts between European laws, there are questions of competency between EC laws and international laws. Most international environmental agreements are for the purposes of EC law, mixed agreements, in that both the EC and Member States have the power to ratify. The scope of their respective competences is determined in part by Treaty provisions and in part by the extent to which Community measures have been made covering particular areas. The precise boundaries though are often unclear, especially when international agreements are ratified, and within the EC this is ultimately a matter for the European Court. In a number of areas the European Court of Justice has been called to rule on the question of competence, and in the past has generally favoured a Community approach at the expense of Member States in order to preserve the unity of the Community legal order.

The UNCLOS is an example of a mixed agreement. In 2003 there was an illustration of how questions of competence could be raised in an international dispute between Ireland and the UK, in relation to radioactive discharges from the Sellafield nuclear engineering centre into the Irish Sea.[13] Only five days before the hearing in this dispute before the Arbitral Tribunal established under UNCLOS, the EC had given a written answer to the European Parliament indicating that it was examining the question of whether to commence proceedings before the European Court of Justice against Ireland to prevent it taking unilateral action before the Law of the Sea Tribunal, on the grounds that under Community law it no longer had the competence to do so. Neither Ireland nor the UK argued that all the provisions of UNCLOS fell within the exclusive competence of the European Court, but the Tribunal accepted that it could not be certain that the European Court of Justice would not adopt such a view. The Tribunal considered that some provisions of UNCLOS could be exclusive

---

11 R v Secretary of State ex parte Greenpeace, Queens Bench Division, 5 November 1999, CO/1336/99.

12 E.g. in the Environmental Impact Assessment Amendment Directive 1999, there is a provision in Schedule 3 (2) (v) concerning habitats.

13 Ireland v UK. The Mox Plant Case. Law of the Sea Arbitral Tribunal Order no.3, 24 June 2003.

to the EC, but not all of them. As it was not clear at this stage of the hearing who had competence in this case, the Tribunal suspended the proceedings for six months until the matter was resolved at EC level. In doing so it recognized the need to avoid conflicts between different international judicial bodies.

As with the UNCLOS, the EC is also a party to the OSPAR Convention. Ireland also bought a case in 2003 before the Arbitration Tribunal established under the OSPAR Convention, concerning access to information about the MOX reprocessing plant at Sellafield.[14] In contrast to the parallel Irish action before the Law of the Sea Tribunal, the Commission does not appear to have raised concerns about Ireland's competence to take unilateral action under OSPAR, nor did the Tribunal appear so concerned about possible conflict between these two regimes. For now, the substantive legal issues concerning the dispute have shifted to the EC stage. If proceedings are brought by the Commission before the European Court of Justice, the Court may decide to use the opportunity to give a significant and broad ruling on the complex constitutional issues involved in the relationship of the Community and the Member States concerning the marine environment.

4.3 CO_2 Storage and Marine Legislation

UNCLOS

Background and Objectives of UNCLOS The most significant international marine convention is the 1982 United Nations Convention on the Law of the Sea (UNCLOS), which came into force in 1994. The UK acceded to the Convention and became a Contracting Party in 1997. The aim of the Convention is to regulate all uses of the sea so it contains provisions governing all aspects of ocean space, such as delimitation, environmental control, marine scientific research, economic and commercial activities, transfer of technology, and the settlement of disputes relating to ocean matters. UNCLOS is therefore considerable in length and comprises 320 articles and nine annexes. Although the text of the Convention is extremely large it provides little in the way of substantive regulations itself and contains few detailed rules of substance. The Convention is more of a framework document, leaving any precise rules to be elaborated further in other more specific international conventions.

Zones of the Sea From a legal perspective, different locations of the ocean are subject to different prescriptions under the UNCLOS. This is important in relation to the location storage sites of CO_2, because there are different rights and duties for each zone. The zones divided by UNCLOS are repeated and adhered to in all other international marine laws, such as the London Convention and 1996 Protocol. Nations have the greatest amount of coastal jurisdiction and control over the waters closest to shore with increasing responsibility to accommodate uses by other nationals

---

14 Ireland v UK. Dispute concerning access to information under Article 9 of the OSPAR Convention, Final Award, Permanent Court of Arbitration, The Hague, 2 July 2003.

as the distance from shore increases. The primary zones in increasing distance from shore are the internal waters, the territorial sea, the exclusive economic zone, the continental shelf and the high seas. These can be seen in Figure 4.1.

It is important to consider the implications of each of these different zones in the sea, as this could prove crucial in selecting CO_2 disposal sites. Whilst the UK has full sovereignty over its territorial sea, its rights over the waters beyond this boundary are more limited, and the role that other states could have in objecting to such projects varies significantly.

The territorial sea is the region of ocean that extends up to a rough limit about twelve miles from the coastline. The territorial sea of the UK as prescribed by UNCLOS is implemented by the Territorial Sea Act 1987, and the baselines are contained in the Territorial Waters Order in Council 1964.[15] Within this zone, coastal state sovereignty over activities is limited only by the freedom of navigation.[16] In relation to CO_2 storage, it seems that the territorial sea could be of limited relevance. In the unlikely event that suitable disposal sites were available in the 12 mile territorial sea of a state then this would only require the consent of that state. The same applies if a pipeline is to be placed across the territorial waters.

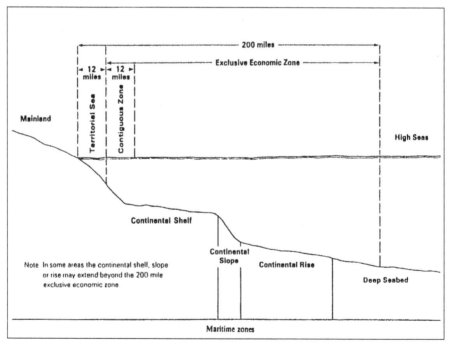

Figure 4.1   The zones of the sea
Source: Churchill (1996)

---

15 Territorial Waters Order in Council 1964, SI 1965, Part III, p. 6452A.
16 UNCLOS, Article 2(3).

The exclusive economic zone (EEZ) extends from the end of the territorial sea out to a maximum of two-hundred nautical miles from the baselines of the coast. Within this zone the coastal state has sovereign rights of exploration, exploitation and management of the natural resources of the EEZ in both the seabed and waters above it.[17] It is possible that a state could claim that their right to exploit the EEZ extends to exploiting the empty spaces in the geological formations for storage purposes. It is regarded as more likely that the CO_2 is considered to be dumped. Dumping can be carried out in the EEZ by a coastal state as long as they have due regard to the rights and duties of other states, they respect their obligations under other international marine pollution legislation, and they are placed under a duty not to cause damage by pollution to the territory of other states or areas beyond national jurisdiction.[18] Dumping and the construction of platforms or pipelines by other states cannot be undertaken in a coastal state's EEZ without the approval of the coastal state, who can permit it if they wish.[19] Coastal states have the power to regulate pollution arising from or in connection with seabed activities.[20]

If CO_2 is planned to be dumped/stored then the coastal state has the option to approve or prohibit the activity, after consideration of the relevant provisions of related international and national legislation. A coastal state also has jurisdiction to authorize pipelines within their exclusive economic zone (EEZ). If a nearby coastal state complained that any storage project affected their rights or caused damage in some form then presumably the burden of proof of proving this would be on them. The jurisdiction of a coastal state also extends to controlling research and development in their EEZ, so trials of CO_2 could in theory take place. Under UNCLOS, research and development projects can also take place which may introduce harmful substances into the marine environment.[21] An EEZ can be claimed by a coastal state around its territory, if they want to exercise their rights to explore and exploit natural resources in the two-hundred mile radius from their coastline. Jurisdiction over the EEZ can only be claimed in so much as international law is acceptable and before a state can exercise EEZ rights conferred by UNCLOS, there has to be legislation at national level which vests such rights with an authority competent to exercise them. States are not under any obligation to claim an EEZ, but if a coastal state does not claim jurisdiction to the extent international law provides for, jurisdiction remains limited. The UK has not declared an EEZ, choosing only to register an established Exclusive Fisheries Zone in which it exercises EEZ fisheries rights only.

The continental shelf extends from the natural prolongation of the land territory to the outer edge of the continental margin or a minimum distance of two-hundred nautical miles from the territorial sea baselines, subject to a maximum of three-hundred and fifty miles from the baselines or one hundred miles beyond the two-

---

17 Ibid., Articles 55–57.
18 Ibid., Article 194(2).
19 Ibid., Article, 210(5).
20 Ibid., Article 194(3).
21 Ibid., Articles 56 and 246.

thousand and five-hundred metre isobath.[22] Within two-hundred miles of the coast the continental shelf and an exclusive economic zone (EEZ) can overlap. Coastal states have sovereign rights to explore and exploit the natural resources of the seabed and subsoil of the continental shelf.[23] Their control over an area is limited to the regulation of interference with minerals and other non-living resources, and sedentary species of living organisms only. The Law of the Sea Convention (UNCLOS) also confers jurisdiction over dumping on the continental shelf, whereby the coastal state has the same rights and obligations to control dumping of matter such as CO_2 onto the continental shelf, as it has in the EEZ.[24]

In practice, the continental shelf of some states can be co-existent with an EEZ, which provides much wider jurisdiction, but the concept of the continental shelf remains significant where no EEZ has been declared or the continental shelf extends beyond such a zone. The continental shelf can extend beyond two-hundred miles, as it does in the UK, so the UK has not declared an EEZ, and instead has relied on its continental shelf rights under the Geneva Convention on the Continental Shelf 1958, not UNCLOS. The continental shelf was defined in the Geneva Convention on the Continental Shelf 1958 as 'the seabed and subsoil of the submarine areas adjacent to the coast but outside the territorial sea to a depth of two-hundred metres or, beyond that limit, to where the depth of the super adjacent waters admits of the exploitation of their natural resources'.[25] The UK has implemented this Convention through the Continental Shelf Act 1964 and other secondary legislation.[26] This provides the legal basis in international law for the UK to prospect for and to extract oil and gas from the continental shelf. The reference to the seabed and subsoil and their natural resources could be construed to cover things such as CO_2 storage. The UK's rights to explore and exploit the Continental Shelf are vested in the Crown Estate (the Queen). The Crown Estate can grant leases or licences as appropriate to permit such activities. In the case of oil and gas, the power to licence exploration and extraction on the Continental Shelf, is vested in the Secretary of State for Trade and Industry. Development consents and regulatory control of marine activities are matters for the appropriate Government Department.

The waters beyond the 200 mile limit of the EEZ are known as the high seas. The high seas are open to all states, but fall under what is known as 'the common heritage of mankind'. The body empowered to administer the common heritage of mankind and to regulate its exploration and exploitation is the International Seabed

---

22 Ibid., Article 76 (note – this is subject to a maximum of 350 miles from the baselines or 100 miles beyond the 2,500 metre isobath).

23 Ibid., Articles 76 and 77.

24 Ibid., Articles 210–216.

25 Geneva Convention on the Continental Shelf 1958, Article 1.

26 Section 1(1) of the Continental Shelf Act 1964 states that 'any rights exercisable by the UK outside territorial waters with respect to the sea bed and subsoil and their natural resources, except so far as they are exercisable in relation to coal, are hereby vested in Her Majesty'.

Authority.[27] All states enjoy the freedom to act within this zone but are required to give due regard to the interests of other states[28] and also due regard to the rights under UNCLOS with respect to activities in the international seabed area.[29] No specific requirements over duties to protect the marine environment exist within the articles that specifically address the high seas and a separate part of UNCLOS is concerned with protecting the marine environment. CO_2 projects will not be prohibited in the high seas under international law, although states where the CO_2 has originated from and who are storing it must have due regard for any states whose interests would be affected,[30] and observe any legal obligations under international marine laws. The due regard obligation may well impose restrictions on the storage of CO_2 in the high seas (Churchill, 1996). Other states may complain that the CO_2 storage affects their interests, for activities like fishing, and may require international arbitration. It may be that access to storage sites in the high seas would prove too costly to transport the CO_2 and build new platforms anyway. It is clear it would also generate a significant amount of international opposition and would not be a particularly politically sensitive decision.

Protection of the marine environment Although the text of UNCLOS is often framed in general terms there are a number of relevant provisions that control activities impacting marine environments. Article 192 of UNCLOS imposes a general obligation on states to protect and preserve the marine environment in all of the territorial zones of the seas. Article 194 requires states to take individually or jointly all measures necessary to prevent, reduce, or control pollution using the best practicable means at their disposal and in accordance with their capabilities. This duty increases under this article where the activity threatens to damage the territory of another state, whereby states must take all measures necessary to ensure that the activity does not cause damage to other states. Article 194 states the measures taken pursuant to this part shall deal with all sources of pollution of the marine environment, including dumping. UNCLOS defines dumping to be 'any deliberate disposal of sea of wastes or other matter from vessels, aircraft, platforms or other man-made structures at sea', but does not include 'placement of matter for a purpose other than mere disposal'.[31] If CO_2 is transported by ship or by a pipeline to a disposal site and then injected from a platform or a ship then it might be considered to be dumping under the purposes of the Convention. The definition of dumping in UNCLOS is the same as that in both the London Convention and the 1996 Protocol and this will be considered in more detail below.

---

27 The International Seabed Authority is an autonomous international organization established under UNCLOS.

28 UNCLOS, Article 86.

29 Ibid., Article 87(2).

30 Ibid., Article 86.

31 Ibid., Article 1(5) (a+b).

In effect, UNCLOS sees the open sea as open-access commons, where any use can be regulated, unless it causes harm to other states or is prohibited by international law. UNCLOS does not specifically prohibit or even refer to the legality of CO_2 storage offshore, but it seems the provisions in Article 194 will apply if the proposed activity is determined to be pollution. Pollution is defined as 'the introduction by man, directly or indirectly, of substances or energy into the marine environment, including estuaries, which results or is likely to result in such deleterious effects as harm to living resources and marine life, hazards to human health, hindrance to marine activities, including fishing and other legitimate uses of the sea, impairment of quality for use of sea water and reduction of amenities'.[32] It is not clear from this definition whether CO_2 is pollution. Some commentators have argued that it probably is not a pollutant, although if large quantities of CO_2 are stored then this could cause pollution if it resulted in harm to living marine resources (McCullagh, 1996). It should be noted that UNCLOS makes no explicit reference to the precautionary principle in determining whether some activity might cause harm to others.

As UNCLOS is very broad based, it obliges other international Organizations and states to introduce more specific laws, or as it call it 'global rules and standards'.[33] In the case of marine pollution and dumping these are widely accepted to be contained in the London Convention and its 1996 Protocol. Contracting Parties to both UNCLOS and the London Convention should follow the requirements under the London Convention first, as it is the more stringent treaty, and they will in practice not refer to the general requirements imposed by UNCLOS.

The London Convention and the 1996 Protocol

Background and objectives The Convention on the Prevention of Marine Pollution by Dumping of Wastes and Other Matter 1972 (known more commonly as the London Convention) was the first truly global convention to control and regulate the deliberate disposal at sea of wastes and other material in the seas. In the 1990s there was recognition that a more modern approach to waste management at sea was needed, to enhance the level of environmental protection. The Contracting Parties to the 1972 London Convention adopted a Protocol in 1996 to revise the London Convention. This 1996 Protocol (hereafter the Protocol) is in fact an entirely new Convention, modifying and adding to virtually every aspect of the London Convention. The Protocol has not yet entered into force, but when it does it will supersede the London Convention, for those parties to the Convention which have subsequently become parties to the Protocol.[34] The requirements under the London Convention and Protocol are of global application to all signatories. The provisions contained in the Convention and Protocol are not always the same and will be dealt with separately where appropriate.

---

32 Ibid., Article 1(4).
33 Ibid., Article 210.
34 1996 Protocol, Article 23.

The London Convention controls ship and platform based dumping activities. The principle objective of the London Convention is to prevent, reduce and where practicable, eliminate pollution caused by disposal or incineration at sea. It does not define pollution, but recognizes that dumping is one of the many sources of marine pollution and seeks to control pollution by managing dumping of wastes and other matter that is liable to create hazards to human health, to harm living resources and marine life, to damage amenities or to interfere with other legitimate uses of the sea. Therefore, on a basic level if CO_2 injection and storage into geological formations under the sea could cause pollution then it could be prohibited. The Protocol embodies a more simplified, modern and comprehensive regulatory framework than the London Convention, and is intended to provide greater protection to the marine environment. It is based far more on precaution and prevention. Instead of regulating dumping like the Convention its objective is to prevent, reduce and where practicable eliminate pollution. Unlike the Convention it does define pollution – as meaning the 'introduction, directly or indirectly, by human activity, of wastes or other matter into the sea which results or is likely to result in such deleterious effects as harm to living resources and marine ecosystems, hazards to human health, hindrance to marine activities, including fishing and other legitimate uses of the sea, impairment of quality for use of sea water and reduction of amenities'.[35]

The UK is a Contracting Party to the London Convention 1972 and is bound by its provisions. The 1996 Protocol is not yet in force as not enough states have ratified it. This delay in implementation is not unusual in international law, where conventions can take many years to receive ratification by the required number of countries before they come into force. The UK ratified the Protocol in December 1998. As the Protocol is not yet in force the London Convention continues to apply. The UK Government have commented that it is current UK policy to apply the requirements under the Protocol where possible (DEFRA, 2002), but they are under no legal obligation to apply the latter Protocol, though under general principles of international law they are, as a signatory party, obliged not to frustrate its objectives.

Geographical coverage and application to the seabed The London Convention and Protocol applies to all marine waters world-wide other than the internal waters of states.[36] The London Convention does not refer to the seabed anywhere in its text and only concerns dumping in the 'sea', which is defined as meaning 'all marine waters other than the internal waters of states'.[37] This would probably not be enough to cover the storage of CO_2 in the seabed or subsoil of the seabed. The only way that the seabed can be included in the Conventions remit is if a purposive approach is

---

35 Ibid., Article 1(5)(10).

36 London Convention, Article III (3); 1996 Protocol, Article 1(7). Note – the term 'internal waters' carries a specific meaning in international parlance and does not include certain bays, estuaries, navigable rivers and other inland waters as defined by international law.

37 Ibid., Article III (3).

adopted when interpreting this provision, so that it could be argued that the purpose of the Convention was not just to protect the sea but also activities in the seabed that have the potential to harm the sea as well. The Protocol goes further in its scope than the Convention and applies to the 'sea, seabed and subsoil'.[38] However, it expressly excludes 'sub-seabed repositories accessed only from land'.[39] This inclusion of 'seabed and subsoil' would appear at first sight to cover the storage of CO_2 in geological formations. The IEA commented that the 'Protocol will therefore prohibit without distinction the storage of CO_2 both in the water column and in sub-seabed repositories' (IEA, 2005). It could be that the Protocol possibly prohibits the storage of CO_2 in geological formations, but this is dependant on what exactly subsoil is legally interpreted as meaning. For example, it is arguable that the 'subsoil' could just be the layer of rock and soil immediately under the seabed, not the geological formations underneath.

The UK Government have commented on this point and concluded that the Protocol, has been drafted with the purpose of covering activities in areas below the sea column and that the Convention should be interpreted in that light (DEFRA, 2002). They considered that if there was a possibility that the storage of CO_2 in sub-seabed reservoirs could result in pollution to that environment, with an effect on such life or ecosystems, which is possible through leakage, the Protocol should be interpreted in such a way as to give effect to this wider purpose, that is protection of the marine environment from pollution,[40] including the prevention of pollution liable to harm living resources and marine life.[41] The UK Government's position is that their express policy of adhering to the more stringent requirements of the Protocol, and that the limitation of the London Convention in this area should not be taken as denying its application to sub-seabed CO_2 storage; rather it should be read in the light of the current standards set by the Protocol (DEFRA, 2002). However, there is no obligation in international law to interpret conventions in light of current developments,[42] i.e. in this case, before the Protocol enters into force.

Legality of CO_2 storage There are several important considerations in determining the legality of CO_2 storage under the London Convention and 1996 Protocol. The first legal question to consider is how the CO_2 gets into the storage site. The operational framework for the London Convention is based on controlling the input of substances into the sea. It provides a framework for general considerations which determine the acceptability of dumping in the sea. The Protocol only allows approved wastes to be dumped. The Convention and the Protocol both define dumping to be 'any deliberate disposal of sea of wastes or other matter from vessels, aircraft, platforms or other

---

38 1996 Protocol, Article 1 (5) (7).
39 Ibid.
40 London Convention, Article 1, 1996 Protocol, Article 2.
41 Ibid., Article 1.
42 E.g. Ireland v UK. Dispute concerning access to information under Article 9 of the OSPAR Convention, Final Award, Permanent Court of Arbitration, The Hague, 2 July 2003.

man-made structures at sea'.[43] If CO_2 is transported by ship or by a pipeline to a disposal site and then injected from a platform or a ship then it will be covered by the provisions of both the Convention and the Protocol and considered to be dumping, as the purpose is to dispose of CO_2 – subject to the exemptions discussed later in this chapter. However, the Convention and Protocol only applies to activities using ships or platforms to inject CO_2 into the marine environment and there are no controls governing pipeline discharges from land based sources. This can be supported by the provision in the Protocol stating that its remit does not extend to sub-seabed repositories accessed only from land.[44] This was confirmed at the thirteenth meeting of the consultative parties to the London Convention in 1990 (Snelders, 2002). The use of pipelines from land based sources to transport CO_2 direct to off-shore repositories is therefore a legitimate activity under the Convention and Protocol.

The second key legal question is whether CO_2 is a waste or not? The London Convention prohibits the disposal of all wastes or other matter specified in Annex I (known as the black list).[45] This is because these are known to cause harm to aquatic organisms, even in low concentrations. Wastes or other matter listed in Annex 2 (known as the grey list) requires special consideration if the quantity exceeds 'significant amounts', and a permit can be issued under certain circumstances.[46] CO_2 is not specifically referred to in any of the lists that are prohibited for disposal in Annex 1. Similarly it is not listed in Annex 2, which covers wastes requiring a permit. What is relevant in considering whether CO_2 is a waste or other matter, is whether it is classed as an 'industrial waste', which was added to the Annex 1 list with effect from 1 January 1996. 'Industrial waste' means 'waste materials generated by manufacturing or processing operations' and the Convention lists a number of substances that this does not apply to.[47] These exclusions are probably not relevant to CO_2. The OSPAR Secretariat examined the definition of 'industrial waste' in the London Convention as part of its CO_2 review (OSPAR Group of Jurists and Linguists, 2003). They considered it was far more unclear whether the generation of electricity is covered by either 'manufacturing or processing', and thought that electricity generation might be claimed to be an activity of a different kind. The UK Government have commented that they consider that CO_2 would fall within this definition of 'industrial waste' (DEFRA, 2002).

Consultative parties to the London Convention have recognized that the text of Annex I(II) to the Convention is ambiguous and open to varying interpretations with regard to the definition of 'industrial waste'. They have considered providing policy interpretations of 'industrial waste', but no consensus amongst the Contracting Parties could be reached. The Consultative Parties agreed to discuss this in the future with the aim of achieving consensus on the interpretation of industrial waste. The

---

43 London Convention, Article III; 1996 Protocol, Article 1(4).
44 1996 Protocol, Article 1(5) (7).
45 London Convention Article IV.1 (a).
46 Ibid., Article IV.1(b).
47 Ibid., Annex I Paragraph 11.

operating procedures of the London Convention have placed a strong reliance on high quality scientific advice and there is a permanently constituted scientific group which reviews existing provisions. The scientific group were asked to advise on the issue of whether CO_2 derived from fossils fuels was considered to be an 'industrial waste' at their twenty-second meeting in 1999 (IMO, 1999). They concluded that CO_2 was an industrial waste and that the twenty-first consultative meeting of all the parties to the Convention should be consulted concerning the priority to be accorded to these issues. The scientific group can only provide advice, and it is up to the Consultative Parties if there should be any amendment to the Convention or introduction of guidance. At the consultative meeting there was no consensus about whether or not storage at sea of CO_2 derived from fossil fuels should be seen as industrial waste. Some state delegations supported the conclusion that fossil fuel derived CO_2 falls within the definition of 'industrial waste', some were opposed and other states decided that it was premature to be decided at this time. The chairman to the consultative meeting decided that no consensus existed on whether CO_2 disposal would be considered an 'industrial waste' (Brubaker and Christiansen, 2001). The debate whether CO_2 is a waste is still ongoing, and was one of the key legal questions in the London Convention review in 2005 of CO_2 storage in geological formations (IMO, 2005).

In conclusion, if it can be shown that the CO_2 that is captured derives from manufacturing or processing operations, it will fall under the 'industrial waste' category and be prohibited under the London Convention. If it is considered not to be an 'industrial waste', it will not be prohibited by Annex I and will instead be subject to the permit procedure contained elsewhere in the Convention. Permits are issued by the state where the waste originates, who must take into account the provisions contained in Annex III of the London Convention. These include such general factors as the possible effects on marine life, or other uses of the sea, and the practical availability of alternative land-based methods of treatment or disposal.

The position under the 1996 Protocol is a little more straightforward because there is a general prohibition on the dumping of wastes or other matter with the exception of those wastes or other matter listed in Annex I.[48] It is extremely unlikely that CO_2 would fall under any of the categories approved for dumping in Annex I. The UK Government are also of the opinion that none of the seven categories listed in Annex 1 could be read as including CO_2 (DEFRA, 2002). The OSPAR Secretariat have commented that they too believe that there is considerable doubt whether CO_2 could be said to be included in any of the categories in the Protocol, but thought that

---

48 1996 Protocol, Article 4.1.1. Waste includes: '1. dredged material; 2. sewage sludge; 3. fish waste, or material resulting from industrial fish processing operations; 4. vessels and platforms or other man made structures at sea; 5. inert, inorganic geological material; 6. organic material of natural origin; and 7. bulky items primarily comprising iron, steel, concrete and similarly unharmful materials for which the concern is physical impact, and limited to those circumstances where such wastes are generated at locations, such as small islands with isolated communities, having no practicable access to disposal options other than dumping'.

there could be arguments where CO_2 deriving from combustion could fall under the category concerning 'organic material of natural origin' (OSPAR Group of Jurists and Linguists, 2003). This was because they considered that there was a respectable body of opinion that 'organic chemistry' and 'organic compound' covered any carbon compound. They also considered that CO_2 produced as a burning of fossil fuel could probably be regarded to be 'of natural origin' as burning is a natural process. However, they concluded that CO_2 storage offshore would probably be in conflict with the 1996 Protocol. At the current time it seems likely that CO_2 would be covered as a waste under the Protocol and the dumping of CO_2 from man-made structures and ships would be prohibited. In the unlikely event that it is considered to fall under one of the seven categories in Annex 1 then it may be authorized for disposal dependent on a permit being required from the Contracting Party.[49]

The third key legal question is whether the Convention and Protocol contains any exclusions that might possibly be relied upon in storing CO_2 in sub-seabed storage sites. The first exclusion considered here is the storage of CO_2 derived from off-shore platforms. Both the Convention and the Protocol specifically excludes from the definition of 'dumping' the 'disposal of wastes or other matter incidental to, or derived from, the normal operations of vessels or aircraft, platforms or other man made structures at sea and their equipment, other than wastes or other matter transported by or to vessels, aircraft, platforms or other man made structures at sea, operating for the purpose of disposal of such matter or derived from the treatment of such wastes or other matter on such vessels, aircraft, platforms or other'.[50] Therefore, if CO_2 was transported to such a site to store it to prevent it from entering the atmosphere, rather than being part of the operation of the installation, this exclusion could not be relied. If however, the CO_2 ends up in the sub-seabed during the operations of the installation and it is stored there because it has nowhere else to go this might be allowed. It was decided at the seventeenth consultative meeting of the London Convention that 're-injection' of produced water and other matter associated with offshore oil and gas operations does not fall within the Conventions definition of dumping (Brubaker and Christiansen, 2001). CO_2 operations involving enhanced oil recovery are permissible under the Convention.

The second possible exclusion covers placement in the maritime area. The London Convention and Protocol both specifically exclude from the definition of 'dumping' the 'placement of matter for a purpose other than the mere disposal thereof, provided that such placement is not contrary to the aims of this Convention [Protocol]'.[51] It could be argued that the CO_2 is not in fact disposed of, but placed, arguably until the deteriorating climate change situation is bought under control with new clean technologies. It is unclear what 'placement' is intended to constitute and its scope. One could guess it is intended to cover things such as the placement of artificial reefs. 'Placement' could be read in terms of the purpose behind the original placement

---

49 1996 Protocol, Article 4(1)(2).
50 London Convention, Article III 1(b)(i); 1996 Protocol, Article 1(4)(2)(1).
51 Ibid., Article III(1)(b); 1996 Protocol, Article 1 (4)(2)(2).

of the installation or structure (if it was for the purposes of undertaking offshore activities, it will be an offshore source, if not, then a land-based source) regardless of any subsequent change of use, or, alternatively, placement in the light of its present use (i.e. the placement is deemed to have occurred at the commencement of the new use for it). The former more literal interpretation would give the benefit of the doubt to the object and purpose of the Convention, to protect the marine environment, in the case of a conversion of a redundant offshore installation. But the latter interpretation has a degree of logic to it, and would avoid an inconsistency in the permissibility of CO_2 storage depending on the original purpose of the installation or structure used. On balance, the former interpretation is more likely because of its consistency with the aims of the Convention.

The term 'placement of matter for a purpose other than mere disposal thereof' was considered at the twenty-second Meeting of the Consultative Parties in 2000 (IMO, 2000). The Contracting Parties considered having guidance on this point, but the UK delegation expressed the view that guidance on placement would be undesirable, because it should be allowed provided that such placement is not contrary to the aims of the Convention. The meeting agreed that any guidance to be developed on this issue should include that placement should not be used as an excuse for disposing of waste, that placement should not be contrary to the aims of the convention and that information of the placement activities be provided to the secretariat, as available. The meeting report noted that no consensus could be reached on whether or not 'placement' was covered by the Convention (IMO, 2000).

The Precautionary Principle Disposing of CO_2 in sub-seabed storage could bring into play the precautionary principle. In general the precautionary principle recognizes that it is often advantageous to prohibit or limit an activity despite the absence of scientific certainty that the activity will result in a detrimental result (McCullagh, 1996). Both the Convention and the Protocol embrace the precautionary principle approach. Although the precautionary principle is not mentioned in the Convention, the Contracting Parties agreed to apply the precautionary approach in environmental protection within the framework of the London Convention in a Resolution to the Convention.[52] Article 3 of the 1996 Protocol also states that in implementing this Protocol 'Contracting Parties shall apply a precautionary approach to environmental protection from dumping of wastes or other matter whereby appropriate preventative measures are taken where there is reason to believe that wastes or other matter introduced into the marine environment are likely to cause harm even where there is no conclusive evidence to prove a causal relation between inputs and their harm effects'. It would seem if the precautionary principle is correctly applied that the weight would seem overwhelmingly to fall on the side of caution, unless there is compelling scientific evidence and opinion that the CO_2 will remain in the seabed repository throughout the whole of the storage period or for a very significant length of time.

---

52 Resolution LDC.44(14)) 1991.

Potential for future storage of CO₂ under the London Convention and Protocol If a Contracting Party to the London Convention went ahead with a carbon dioxide storage project at the current time and this was against the wishes of one or more of the Contracting Parties, this could result in some form of action being taken against them. This would normally take the form of resolution in the first instance by negotiation or conciliation, and then possibly arbitration in the second instance. Contracting Parties can agree to sidestep arbitration and use one of the procedures for court action listed in the UNCLOS. As noted above, the Contracting Parties to the London Convention are currently considering the legal questions associated with CO_2 storage in geological formations. This could identify the Contracting Parties which are in favour or against such projects and their opinions as to the current legal position. If CO_2 disposal is found to be in conflict with the Convention in certain circumstances, the only means by which Contracting Parties could still pursue this as an option would be to amend the Convention. The London Convention allows for Contracting Parties to review and adopt amendments to the Convention and its Annexes.[53] Amendments to the Convention may be passed by a two-thirds majority of those present at consultative or special meetings. Any amendment will only come into force for those Parties accepting it on the sixtieth day after two-thirds of all the parties (i.e. not just those present at the meeting) have deposited an instrument of acceptance for the amendment.[54] It is expected that this could be time consuming as there is no limit set out as to when states have to accept the amendment by.

Contracting Parties may also make amendments to the Annexes of the Convention. In relation to CO_2 it is most likely that the definition of 'industrial waste' in the Annexes will be amended. This procedure is easier than changing the Convention because all that is required all is a two-thirds agreement of those present at the meeting. The amendment will then enter into force immediately for any party agreeing to it, and for all other parties (whether they agreed to it or not) after a period of a hundred days following the relevant meeting, unless a declaration against acceptance is made by a party within that period.[55] It is unclear whether a two-thirds majority of the Contracting Parties would support CO_2 storage proposals, particularly in light of the fact that there was no consensus on the issue of whether it was an 'industrial waste' at an earlier consultative party meeting (IMO, 2000). Amendments to the Annexes also have to be based on scientific and technical considerations and previous research by the Scientific Committee to the Convention have suggested that it is probably an industrial waste (IMO, 1999).

After the Protocol has entered into force there are similar provisions within it that allow for meetings or special meetings to be held to review and amend the Protocol.[56] There is a requirement that any amendment to an article or annex proposed by a

---

53 London Convention, Article XIV (4)(a).
54 Ibid., Article XV1(a).
55 Ibid., Article XV1(b).
56 1996 Protocol, Article 18(1)(1).

Contracting Party must be notified by the International Maritime Organization to all Parties at least six months prior to its consideration at such a meeting.[57] The position on voting and entry into force of amendments to articles and of annexes is the same as for the London Convention.[58] Similarly, it would seem easier for an amendment to be made to the permitted list in Annex 1 under the Convention. The IEA have commented that 'necessary amendments might include putting CO_2 on the "reverse list" of the 1996 protocol of the London Convention, after its entry into force' (IEA, 2005).

In the unlikely event that the UK would seek to withdraw from the Convention and Protocol there are procedures allowing for this. The London Convention allows Contracting Parties to withdraw from the Convention by giving six months notice to the depositary.[59] After the Protocol enters into force Contracting Parties are not permitted to withdraw from the Convention for two years.[60] After this date, withdrawal takes effect one year after receipt of the notice to withdraw from the Protocol.

If a Contracting Party to the London Convention proceeds with a geological CO_2 storage project they could be liable for any damage caused in the event of an escape. Both the Convention[61] and Protocol[62] contain provisions stating that liability is in accordance with the principles of international law regarding state responsibility for damage caused to the environment of other states or to any other area of the environment. The Contracting Parties to the Convention and Protocol must also undertake to develop procedures regarding liability arising from the dumping of wastes or other matter.

OSPAR Convention

Background and objectives of OSPAR The OSPAR Convention is a framework document which sets out the overall principles of the Convention. The main text contains legal obligations, provisions on definitions, and the managerial aspects in the implementing and application of the Convention. An integral part of the Convention is a number of annexes and appendices which contain more detailed provisions than in the main text. The five separate annexes to the Convention cover pollution from land-based sources (Annex I); pollution by dumping and incineration at sea (Annex II); pollution from offshore installations and structures (Annex III); monitoring and assessment of the marine environment (Annex IV); and the protection of ecosystems and biological diversity (Annex V).

The primary objective of the OSPAR Convention is to protect the marine environment against the adverse effects of human activities, so as to safeguard

---

57 Ibid., Articles 21(1) and 22(1).
58 Ibid., Protocol, Article 21(2) + (3), Article 22(2) + (4).
59 London Convention, Article XXI London Convention.
60 1996 Protocol, Article 27 1996 Protocol.
61 London Convention, Article X.
62 1996 Protocol, Article 15.

human health and to conserve marine ecosystems and, when practicable, restore marine areas which have been adversely affected.[63] This is significant because the older Conventions, such as the London Convention, refer to the prevention of pollution of the sea, whereas the OSPAR Convention refers to the protection of the marine environment. The Contracting Parties to OSPAR must adopt measures to achieve the protection of the marine environment – this objective is the minimum legal obligation placed on them. The only discretion a Contracting Party has is the decision whether to adopt even more stringent measures than the main objective to protect the maritime area.[64] The main objective of the Convention is extremely important because in questions of interpretation of the legal text the court or tribunal will look too the underlying purpose of the Convention.

It is clear from the OSPAR Convention that its general purpose is to stop adverse activities and subsequently the risk of pollution taking place in the marine environment. The Convention is concerned with 'the introduction by man, directly or indirectly, of a substance into the marine area which results, or is likely to result, in hazards to human health, harm to living resources and marine ecosystems, damage to amenities or interferences with other legitimate uses of the sea'.[65] As compared to its predecessors, such as the London Convention, the OSPAR Convention also has increased scope of coverage and is legally tighter. It gives legally binding status to the precautionary principle and the polluter pays principle and Contracting Parties must also take into account best available techniques and best environmental practice in any measures they adopt.

Geographical coverage and application to the seabed The OSPAR Convention is a regional agreement and it applies to the waters of the Contracting Parties in the geographical maritime area around the North Sea and parts of the Atlantic and Artic oceans. The Convention applies to pollution in the 'maritime area', which includes in its interpretation 'the bed of all those waters and its subsoil'.[66] This would appear at first sight to cover geological CO_2 storage projects which would store CO_2 under the sea. There might well be argument over the legal interpretation of what is the 'the bed of all those waters' and its 'subsoil' and in geological terms whether the storage grounds for the CO_2 fall under this definition. The UK Government have already considered this point in an internal document and commented that it could be argued that if the CO_2 was placed in the sub-seabed this would not be caught by the Convention because 'subsoil' would only refer to the layer of broken rock immediately under the seabed (DEFRA, 2002). However, they went on to adopt a purposive approach in interpreting this legal point and after considering that the overriding objective of the Convention was the protection of the marine environment from pollution and the conservation of ecosystems, they concluded that seabed was

---

63 OSPAR Convention, Article 2(1)(a).
64 Ibid., Article 2(5).
65 Ibid., Article 1(d).
66 Ibid., Article 1(a).

intended to be given a broad interpretation, encompassing oil and gas reservoirs. They considered the Convention had been drafted with the purpose of covering activities in areas below the sea column and if there was a possibility that the storage of CO_2 in sub-seabed reservoirs could result in pollution to that environment, with an effect on such life or ecosystems, which was possible through leakage, the OSPAR Convention should be interpreted in such a way as to give effect to this wider purpose. The Group of Jurists/Linguists to the OSPAR Commission concluded that the definition of maritime area covers placements onto or into the seabed and the underground strata beneath it (OSPAR Group of Jurists and Linguists, 2003).

There is no case-law to guide us as to a complete definition of seabed and subsoil. A strong counter argument is that geological formations are not intended to be protected under the OSPAR Convention. If a purposive approach is adopted it seems that the OSPAR Convention was drafted to protect marine ecosystems against pollution and the risk of pollution. The OSPAR Convention has provisions concerning pollution from land-based sources (Article 3); pollution by dumping and incineration (Article 4); and pollution from offshore sources (Article 5). These do not specifically cover underground geological formations, but rather particular forms of pollution, and when drafted they plainly did not have CO_2 in mind. If OSPAR's purpose is to protect the maritime area, this itself might not include oil and gas reservoirs as an area for protection. If consideration is given as to what activities under OSPAR might affect the areas protected, this might encompass pollution from underground reservoirs, because geological storage of CO_2 could pose a risk to the maritime area. Article 2 of the OSPAR Convention sets out the general obligations under the Convention and Contracting Parties must 'take all possible steps to prevent and eliminate pollution and shall take the necessary measures to protect the maritime area'. Therefore, even if geological formations are not classed as the subsoil and are not covered in articles 3–5, states are still under a general duty to protect the marine environment under Article 2 of the Convention.

Legality of CO_2 storage There are several important considerations in determining the legality of CO_2 storage under the OSPAR Convention. The first key legal question to consider is how the CO_2 gets to the storage site. There are three methods by which CO_2 could reach an offshore location for subsequent disposal: transportation from a land-based source by a pipeline directly to the sub-seabed storage site; by ship for direct injection from the ship; or transportation by pipeline or ship to an installation such as a rig and then injection into the sub-seabed storage site. For the purposes of this study only the third method seems to apply to geological storage. The OSPAR Convention institutes three separate regimes to control pollution. It covers (i) pollution from land-based sources;[67] (ii) pollution from dumping and incineration;[68] and (iii) pollution from offshore activities.[69] These regimes are mutually exclusive. Annex I

---

67 OSPAR Convention, Article 3 and Annex I.
68 Ibid., Article 4 and Annex II.
69 Ibid., Article 5 and Annex III.

covers pollution from land-based sources and for the purposes of this study relates to offshore pipelines. The OSPAR Convention introduces a general prohibition against incineration and the dumping of all waste from ships in Annex II, except for certain wastes that are listed in the Convention. Annex III applies to offshore installations, and this states that any dumping of wastes or other matter from offshore installations is also prohibited.

Under the OSPAR Convention the Contracting Parties must take all possible steps to prevent and eliminate pollution from land-based sources.[70] 'Land-based sources' is defined in the Convention as including 'point and diffuse sources on land from which substances or energy reach the maritime area by water, through the air, or directly from the coast'.[71] It includes 'sources associated with any deliberate disposal under the seabed made accessible from land by tunnel, pipeline or other means and sources associated with man-made structures placed, in the maritime area under the jurisdiction of a Contracting Party, other than for the purpose of offshore activities'.[72] It is clear from the above definition that the transportation of CO_2 by pipeline from land *directly* (emphasis added) to sub-seabed storage sites falls within the scope of the Convention. To be clear, it should be pointed out that the situation is quite different if the CO_2 is transported by pipeline to an installation before being injected into the seabed. This part of the Convention is covered by Annex I and anyone could in practice release CO_2 into storage sites direct from a pipeline, subject to the authorisation of the Contracting Party.[73] The disposal of CO_2 in this way could be authorized by a Contracting Party as it could be classed as pollution from a land-based source. However, this discretion is limited by the fact that Contracting Parties are obliged to use the best available techniques for point sources and follow best environmental practice for point and diffuse sources.[74]

It is worth noting that although pollution from land-based sources may be permissible in some situations (i.e. if best practice and best techniques are adopted), Contracting Parties to the Convention are also under a duty to prevent and eliminate pollution from land-based sources in accordance *with the provisions of the Convention* (emphasis added).[75] This suggests that if the general obligation of the Convention is to 'prevent and eliminate pollution',[76] the storing of a substance under the sea which risks causing pollution to the maritime area may not be compatible with the discretion of the Contracting Parties. Contracting Parties must also take into account decisions made by the OSPAR Commission, which bind the Contracting Party – i.e. the contracting parties to OSPAR could decide to pass a decision prohibiting the disposal of CO_2 in this manner. The OSPAR Jurists/Linguists Group also concluded

---

70 Ibid., Article 3.
71 Ibid., Article 1(c).
72 Ibid., Article 1(e).
73 Ibid., Article 4(1); and Annex I, Article 2(1).
74 Ibid., Annex I Article 1(1).
75 Ibid., Article 3.
76 Ibid., Article 2(1)(a).

that there was no prohibition on introducing substances from a land-based source into the marine environment and such means are therefore permissible for placing CO_2 in the marine environment, irrespective of the purpose of their placement (OSPAR Group of Jurists and Linguists, 2003). We merely take note of this here as it may apply to proposals for direct ocean storage since geological CO_2 storage entails the use of an intermediate installation between transport and disposal.

CO_2 will not be disposed from aircraft so this is not relevant to this study. It is possible that CO_2 could be disposed of directly from a ship. It appears that the dumping of CO_2 directly from ships will be prohibited by the Convention, unless it can be said to be one of the wastes contained in Annex II. This is because Annex II prohibits dumping of waste from ships unless it is an approved waste under Article 3(2). Whether it is a waste will be discussed further later. The OSPAR Jurists/Linguists group also concluded that placements of waste from ships, constituted dumping and were prohibited (OSPAR Group of Jurists and Linguists, 2003).

The most likely way that CO_2 will be disposed of is from some form of offshore installation after being transported by ship or pipeline before being injected into the storage site. Annex II covers offshore installations and states that any dumping of waste from these is prohibited, but this prohibition does not apply to discharges or emissions from offshore sources.[77] If the CO_2 is piped or shipped to an offshore installation before going into the storage site then the less stringent requirements contained under Annex I (pollution from land-based source) could be superseded by other provisions in the Convention. Anyone wishing to dispose of CO_2 will try and show that the pollution comes from a land-based source, so the disposal operation will not be prohibited under the Convention.

In the case of ships transporting CO_2 to an installation it seems that although the pollution comes from an onshore source, the transport of the CO_2 cannot be treated as an offshore source. This is because the definition of land-based source states that 'it includes sources associated with any deliberate disposal under the seabed made accessible from land by tunnel, pipeline or other means'. This would seem to not include ships, and it would be classed as dumping of wastes under Annex III. The OSPAR Commission also reached the conclusion that ships could not be treated as other means for the purposes of the definition of 'land-based sources', and after consideration of the *ejusdem generis* rule (OSPAR Group of Jurists and Linguists, 2003). This rule says that 'where general words follow an enumeration of person or things, by words of a particular and specific meaning, such general words are not to be construed in their widest extent, but are to be held as applying only to persons or things of the same general kind or class as those specifically mentioned' (Garner, 1999).

In the case of pipelines transporting CO_2 to an offshore installation the definitions in the Convention are crucial. 'Offshore Installations' are defined as including man-made structures and vessels placed for the purpose of offshore activities. 'Offshore activities' is defined as 'activities carried out in the maritime area for the purposes of

---

77 Ibid., Annex III, Article 3.

the exploration, appraisal or exploitation of liquid and gaseous hydrocarbons'.[78] It is likely that the CO_2 that is captured will be transported to an offshore installation that already exists. If an offshore platform already exists it would have been used for oil and gas drilling and would fall under the hydrocarbon provision. The prohibition on disposal from offshore installations would therefore catch CO_2 disposal from such installations.

However, the definition of 'offshore activities' seems to offer a potential exemption in the Convention as it only applies to activities in the maritime area concerning liquid and gaseous hydrocarbons. Hydrocarbons cover a wide range of carbons containing hydrogen and carbon molecules, but probably do not include CO_2. Therefore a Contracting Party could in theory build a platform or other form of fixed structure, or even use an installation that has not been used for previous offshore activities for the specific purpose of disposing of CO_2. Activities such as these would therefore not fall within the more restrictive provisions of the Convention and would be regulated by the provisions contained in Annex I as they would be a land-based source. The requirements in Annex 1, relating to land-based sources were discussed above.

It is worth noting that although pollution from offshore sources may be permissible in some situations (i.e. if it is transported to the installation by pipeline and the installation has not been used for the exploitation of hydrocarbons), Contracting Parties to the Convention are also under a duty to take all possible steps to prevent and eliminate pollution from offshore sources in accordance *with the provisions of the Convention*[79] (emphasis added). This suggests that if the general obligation of the Convention is to 'prevent and eliminate pollution',[80] the storing of a substance under the sea which risks causing pollution to the maritime area may not be compatible with the discretion of the Contracting Parties.

The second key legal question is determining whether CO_2 is a waste or not. In general terms it is open to interpretation whether CO_2 can be classified as a waste. 'Wastes or other matter' is left undefined in the Convention, except to give a number of exclusions to this definition,[81] which have no relevance to CO_2. The concept of waste in the Convention is drafted in very wide terms and the inclusion of the phrase 'other matter' is designed to provide a catch all situation to respond to evolving threats. Under the Convention the dumping of all wastes or other matter is prohibited, unless these are specifically listed.[82] This means that if CO_2 is not expressly listed in the Convention its disposal will not be authorized. The categories of waste that are currently authorized for dumping include: dredged material; inert materials of natural origin, that is solid, chemically unprocessed geological material the chemical constituents of which are unlikely to be released into the marine environment; fish

---

78 Ibid., Article 1(j).

79 Ibid., Article 3.

80 Ibid., Article 2(1)(a)

81 Ibid., Article 1(o).

82 Ibid., Annex II and Article 3(1).

waste from industrial fish processing operations; and vessels or aircraft.[83] This corresponds closely with the list in the Protocol to the London Convention. It is clear that none of the above categories of waste could be said to include CO_2. The Group of Jurists/Linguists of the OSPAR Commission also considered that the disposal of CO_2 would not fall under any of the exceptions in Annex II (OSPAR Group of Jurists and Linguists, 2003).

Dumping is defined in the OSPAR Convention as (i) any deliberate disposal of wastes or other matter from vessels or aircraft or from offshore installations, and (ii) any deliberate disposal in the maritime area of vessels or aircraft or offshore installations and offshore pipelines.[84] The second provision in the Convention relating to dumping is obviously not relevant to CO_2 storage. The key words in the definition of dumping are 'deliberate disposal', which are undefined. The notion of disposing is inherent in most definitions of waste in most countries around the world. The use of this term is the decisive factor in determining whether a material is a waste and whether it falls within the remit of the Convention. Some jurisdictions such as the EU have indicated in case law that in certain circumstances discarding can take place even when arguably the opposite occurs. Examples of this include, keeping a substance rather than transferring it to another owner, or being able to use or treat it in an environmentally sound manner. It seems clear that this approach is often adopted to achieve the objectives of the legislation – i.e. environmental protection, even if this distorts the ordinary sense of the word. In the case of CO_2 there can be little argument that the CO_2 still has a use, as the intention of the person wishing to store the CO_2 would be to put it into another location to get rid of it. For the purposes of the Convention it seems that CO_2 is dumped if it is stored in geological formations in marine waters.

The third key legal question is whether the OSPAR Convention contains any exclusions that might possibly be relied upon in storing CO_2 in sub-seabed storage sites. The first exclusion considered here is the storage of CO_2 derived from an off-shore platform. The Convention specifically excludes from the definition of 'dumping' any 'disposal of wastes or other matter incidental to, or derived from, the normal operations of vessels or aircraft or offshore installations other than wastes or other matter transported by or to vessels or aircraft or offshore installations for the purpose of disposal of such wastes or other matter or derived from the treatment of such wastes or other matter on such vessels or aircraft or offshore installations'.[85] This exclusion is also contained in the London Convention. This provision clearly does not exempt the storage of CO_2 that is collected on land and transported for injection. Some commentators have argued that if the CO_2 is generated on an offshore platform during normal operations, then this falls outside the definition of dumping and is therefore permitted under the Convention (Snelders, 2002). It would seem on closer

---

83 Ibid., Annex II and Article 3(2) (a, b, d, e).
84 Ibid., Article 1(f).
85 Ibid., Article 1(g)(i).

analysis that the sub-seabed storage of CO_2 produced in this manner would more likely to still be prohibited under the OSPAR Convention.

The reason why this exemption would probably not come into play for CO_2 storage is because Article 3 of the Convention states that this exemption applies to *discharges* and *emissions* (emphasis added) from offshore sources.[86] It seems that the injection of CO_2 into a sub-seabed storage site would not constitute a discharge or emission through the normal operations of an offshore installation. CO_2 capture and storage projects will involve CO_2 going to such a site to store it to prevent it from entering the atmosphere, rather than being part of the operation of the installation. This would suggest that locating power stations offshore with the intention of collecting the CO_2 and disposing of it at source would not fall under this exemption. If however, the CO_2 ends up in the sub-seabed during the operations of the installation and it is stored there because it has nowhere else to go this might be allowed. This would suggest that operations involving enhanced oil recovery are only permissible under this part of the Convention.

The second possible exclusion covers placement in the maritime area. The OSPAR Convention also specifically excludes from the definition of 'dumping' the 'placement of matter for a purpose other than the mere disposal thereof, provided that, if the placement is for a purpose other than that for which the matter was originally designed or constructed, it is in accordance with the relevant provisions of the Convention'. This exclusion is also contained in the London Convention. It could be argued that the CO_2 is not in fact disposed of, but placed in the seabed, arguably until the deteriorating climate change situation is brought under control with new clean technologies e.g. the CO_2 is basically stored for a period of time before it is released back into the atmosphere. Although this does provide a form of an exemption the last part of this exclusion is crucial because it says it must be in accordance with the relevant provisions of the Convention. In Annex II, which concerns the prevention of pollution by dumping, this says that although placement of matter may take place in certain situations, this provision shall not be taken to permit the dumping of wastes or other matter otherwise prohibited under this Annex. It seems that the competent authority of the relevant Contracting Party gives the authorisation for this.[87] In the UK, this will be the Government's responsibility to decide whether matter can be placed in the sub-seabed. This authority is however curtailed by the fact that their authorisation must in accordance with the relevant applicable criteria, guidelines and procedures adopted by the OSPAR Commission – who are under a legal duty to issue such guidance.[88] It is not known whether such guidance exists.

It is unclear from the 'placement' provision in the Convention what this is intended to constitute and its scope. It could determine whether dumping from an installation or structure constitutes pollution from an offshore or land-based source,

---

86 Ibid., Article 3(2).
87 Ibid., Annex II Article 5.
88 Ibid., Annex II Article 6.

and thus whether it is permissible or not. 'Placement' could be read in terms of the purpose behind the original placement of the installation or structure (if it was for the purposes of undertaking offshore activities, it will be an offshore source, if not, then a land-based source) regardless of any subsequent change of use, or, alternatively, placement in the light of its present use (i.e. the placement is deemed to have occurred at the commencement of the new use for it). The former more literal interpretation would give the benefit of the doubt to the object and purpose of the Convention, to protect the marine environment, in the case of a conversion of a redundant offshore installation. But the latter interpretation has a degree of logic to it, and would avoid an inconsistency in the permissibility of CO_2 storage depending on the original purpose of the installation or structure used. On balance, the former appears to be most consistent with the aims of the Convention. The Group of Jurists/ Linguists to the OSPAR Commission thought that CO_2 was a substance that was not wanted in its present form or location (OSPAR Group of Jurists and Linguists, 2003). They considered that the placement of it in the maritime area was a deliberate action to dispose of it by putting it somewhere else and this was not an action to achieve another purpose. They concluded that the dumping regime would therefore seem to exclude it from the cases to which that regime applied.

The Precautionary Principle   An important legally binding provision in the Convention is the incorporation of the precautionary principle both in the definition of pollution[89] and the general obligations of the Convention.[90] Whilst this does not provide an exemption it will be very important in relying on any exemption. The Contracting Parties must apply the precautionary principle, by virtue of which preventive measures are due to be taken where there are reasonable grounds for concern that substances or energy introduced, directly or indirectly, into the marine environment may bring about hazards to human health, harm to living resources and marine ecosystems, damage to amenities or interfere with other legitimate resources of the sea, even where there is no conclusive evidence of a causal relationship between the input and the effects.

Unlike some definitions of the precautionary principle in other Conventions, which merely state that scientific uncertainty should not delay the taking of preventative measures (e.g. Climate Change Convention), the formulation in the OSPAR Convention is more proactive and positively requires preventative measures to be taken when there is a reasonable apprehension of a hazard.[91] Secondly, it does not require the potential damage to be serious or irreversible before action is taken. This is essential for a truly precautionary approach, for if there is uncertainty in other respects it may not be certain how serious the potential damage may be. Thirdly, the formulation does not even require 'damage', but only the possibility of a hazard, which is the mere risk that damage might occur. If it can be shown the CO_2 storage

---

89 Ibid., Article 1(d).
90 Ibid., Article 2(2) (a).
91 Ibid.

projects could cause a hazard the precautionary principle may be relied upon to prohibit them.

Potential for the future storage of CO₂ under the OSPAR Convention   It is not known whether the legal review that was undertaken by the OSPAR Commission in 2003/4 will result in any amendments to the Convention. The OSPAR Commission may choose instead to give a decision or a recommendation taken under the procedures of the Convention.[92] Any recommendation given by the OSPAR Commission is not legally binding.[93] Although the Committee can adopt a decision as to the correct interpretation of the relevant provisions in the Convention it will not become binding until adopted by voting of the Contracting Parties. The Commission needs a three-quarters majority of the Contracting Parties to obtain this.[94] It seems however that the decision is only binding on the Contracting Parties which have notified that they accept the decision.[95]

There is provision in the Convention for dispute resolution.[96] If any Contracting Parties to the Convention have a dispute relating to the interpretation or application of the Convention, they may request arbitration. If both parties agree to arbitration the dispute will go before a tribunal consisting of three appointed members. The rules concerning the running of this tribunal are contained in Article 32 of the Convention. The cost of going to the tribunal will usually be borne by the parties in dispute in equal shares. The OSPAR Commission will in practice notify the other Contracting Parties to the Convention, who can intervene in the proceedings if they have an interest in the legal nature in the subject matter of the dispute which may be affected by the decision in the case. The tribunal will reach a decision according to the rules of international law, and in particular, those of the Convention. The decision made by the tribunal is final and binding upon the parties to the dispute.

A further option for Contracting Parties pursuing the geological storage option is to propose amendments to the Convention. Under Articles 15 and 17, any Contracting Party may propose an amendment to either the main text or the Annexes of the OSPAR Convention. A Contracting Party proposing the amendment must do so at least six months before the meeting of the Commission at which it is proposed for adoption. Any changes to the Convention could in practice prove time consuming in practice. An amendment to the main text of the Convention itself requires the unanimous vote of all the Contracting Parties. If the amendment is accepted it will enter into force on the thirtieth day after deposit of the instrument of ratification, acceptance or approval by at least seven of the Contracting Parties. Amending the Convention may prove difficult in practice, as it is likely that one Contracting Party will object, particularly as this is such a controversial topic.

---

92 Ibid., Article 13.
93 Ibid., Article 13(5).
94 Ibid., Article 13(1).
95 Ibid., Article 13(2).
96 Ibid., Article 32.

It seems that a Contracting Party would be more likely to consider proposing amendments to the Annexes under Article 17 of the Convention. This only requires a three-quarters majority vote by the Contracting Parties bound by the Annex concerned. However, the entry into force procedures are the same as for changes to an article under Article 15. It would seem sensible for the UK to try and forge consensus for such an amendment amongst Contracting Parties from the EC. European countries negotiate climate change as a group at the Conference of Parties for the Climate Change Convention. Amongst the Contracting Parties to the OSPAR Convention are twelve European Member States[97] as well as the Commission of the European Communities. Only three Contracting Parties to the OSPAR Convention are not members of the EC.[98] As the EC countries make up the three-quarters majority they could come to some agreement as to the benefits in terms of sharing emissions reduction credits and cost.

If geological CO_2 storage is incompatible with the OSPAR Convention, any proposed amendment to the Convention is dismissed, and the UK Government is still determined to press ahead with using such technologies, then the only option would be for them to withdraw from the Convention. In practice it is unlikely that the UK Government would withdraw from the OSPAR Convention, but if they do chose to do so they may withdraw from the Convention by notification in writing.[99] The withdrawal takes effect one year after the notification is received. If they go ahead with CO_2 storage projects before this year is reached then the OSPAR Commission can call for steps to bring about full compliance with the Convention.[100]

If a Contracting Party to the OSPAR Convention went ahead with a geological CO_2 storage project there could also be financial consequences in case of an escape. As with many other international and regional agreements, the polluter pays principle has been added to the OSPAR regime.[101] It refers to the legal obligation of polluters to pay for damage caused by their operations to human health and the environment. It is assumed that this will cover the cost of clean up after an accident, and by paying compensation for any consequences of harm to human health and the environment. It is debateable whether CO_2 would cause much damage to human health or the environment, even if released in large quantities, but a Contracting Party could be under a legal duty to clean up any CO_2 that has escaped. A key question will be whether the polluter pays principle exempts excluded activities or not.

---

97   Belgium, Denmark, Finland, France, Germany, Ireland, Luxembourg, the Netherlands, Portugal, Spain, Sweden and the UK.

98   Iceland, Norway, Switzerland.

99   OSPAR Convention, Article 30.

100   Ibid., Article 23.

101   Ibid., Article 2(b).

4.4   CO₂ Storage and Climate Change Legislation

The UNFCCC was concluded in 1992 with the purpose of stabilising concentrations of greenhouse gases in the atmosphere at a level that prevents dangerous disruption to the climate. All greenhouse gases, including CO_2, fall within the scope of the UNFCCC, which came into force in 1994. The commitments under the Convention depend on which Annex to the Convention a state falls under. The industrialised countries, including the UK are Annex 1 countries and have to take a greater burden in reducing greenhouse gas (GHG) emissions. The UNFCCC is a framework Convention which requires contracting parties to adopt policies aimed at the stabilisation of concentrations of GHG in the atmosphere at 'levels that prevent dangerous (...) interference in the climate system'. It provides for a review of the adequacy of commitments and for annexes and protocols to be attached to the framework document, as more information becomes available.[102]

The UNFCCC provides a general obligation to adopt policies to limit emissions, and the Kyoto Protocol (the Protocol), which was agreed in 1997, provides for actual targets – quantified emission limitation and reduction commitments. The Protocol must be read in conjunction with the UNFCCC, because the latter is the parent law and the definitions contained in Article 1 of the UNFCCC apply in the Protocol. The Kyoto Protocol entered into force in 2005.

The underlying objectives of the UNFCCC is to both prevent greenhouse gas emissions from entering into the atmosphere, as well as to remove greenhouse gases (GHG) once they have been emitted. Under the UNFCCC emissions and removals are recognized as contributing to stabilisation of GHG concentrations in the atmosphere. There is an important legal distinction between what is an 'emission' and what is an 'emission reduction'. An emission is defined in the UNFCCC as 'the release of GHG and/or their precursors into the atmosphere'.[103] If CO_2 is captured at source and stored, it does not find its way into the atmosphere and therefore does not become an emission for the purposes of the Convention. Therefore, if a GHG does not find its way into the atmosphere, there is no emission, but an emission reduction.

The significance of the distinction between emissions and emission reduction is that parties to the Convention are more constrained in how they can deal with emissions. If the GHG is released into the atmosphere, it can be the subject of a storage project which removes the GHG from the atmosphere by storing it in what is known as 'sinks'.[104] The location where the greenhouse gas is stored is known as a 'reservoir'.[105] Oceans or forests can both be used as reservoirs to capture released

---

102   UNFCCC, Article 4(2)(d).

103   Ibid., Article 1(4).

104   The definition of a 'sink' is given in Article 1(8) UNFCCC: '"Sink" means any process or activity which removes a greenhouse gas, an aerosol or a precursor of a greenhouse gas from the atmosphere.'

105   The definition of reservoir is given in Article 1(7), UNFCCC: 'Reservoir' means a component or components of the climate system where a greenhouse gas or a precursor of a greenhouse gas is stored.'

emissions. The UNFCCC and Protocol encourages the protection and increase of natural CO_2 'sinks' and 'reservoirs', and allows Annex 1 parties to implement projects which reduce greenhouse gases at source, or to increase their removal by obtaining sinks and to credit the resultant emission reduction units against their own emission targets.[106] CCS is therefore certainly not incompatible with the UNFCCC or Protocol because it is an active use of a sink and reservoir. Neither the UNFCCC nor the Protocol specifically mentions offshore CO_2 storage, and Contracting Parties may only offset their emissions from land based sources by afforestation, reforestation and deforestation sinks.[107] The Independent World Commission on the Oceans (IWCO) commented in 1998 that in relation to using the oceans as sinks, 'the Framework Convention on Climate Change and its Kyoto Protocol do not provide for parties to dump or store CO_2 in international waters and thereby to offset their emissions' (quoted in Johnston *et al.*, 1999). It seems that the IWCO is correct, but additional sink activities may be agreed at a later date by the Conference of Parties to the UNFCCC, although under the current drafting of the Protocol these are also limited to land-use activities, not offshore activities.[108]

The term 'sequestration' is therefore often used in connection with the mitigation of greenhouse gases, but it is clear that the type of geological sequestration considered here is in principle concerned with emission avoidance by capturing CO_2 at sources such as industrial power stations. The Protocol allows Annex 1 parties to implement projects which reduce greenhouse gases at source,[109] and these can be counted as an emission reduction. It would seem if the CO_2 is captured at an industrial power station or similar facility then it cannot be released into the atmosphere and will not be counted as an 'anthropogenic emission' under the Protocol.[110] The UNFCCC and Protocol provides a clear option for the use of emission reductions, but are silent on how the emissions could be reduced at source and make no reference to storage sites. There is therefore nothing in the UNFCCC or Protocol which expressly prohibits captured CO_2 from being stored in geological formations under the sea. In fact the Protocol requires Annex 1 parties 'to implement policies and research, on the promotion, development and increased use of carbon dioxide sequestration technologies'.[111]

The IEA are also of the opinion that 'neither the UNFCCC nor the Protocol include or exclude CCS as an encouraged or permitted emission reduction device giving rise to emission credits (IEA, 2005). The IPCC report also found that both the UNFCCC and Protocol allows for projects that reduce greenhouse gases at their sources. The UK Government are similarly under the opinion that geological

---

106   Kyoto Protocol, Article 6.

107   Ibid., Article 3(3).

108   Ibid., Article 3(4): 'additional human-induced activities related to changed in greenhouse gas emissions by sources and removals by sinks in the agricultural soils and the land-use change and forestry categories'.

109   Ibid., Article 6.

110   Ibid., Article 5(2).

111   Ibid., Article 2(1) (a)(iv).

storage of greenhouse gases is not incompatible with the Convention, and have even commented that any Annex 1 Party that does not deploy CCS as provided for in Article 2(1)(a)(iv) might be presumed to be failing to develop the full range of policies and measures provided for in the Protocol (DEFRA, 2002).

Whilst it would appear that emission reductions by CCS are not prohibited, the UNFCCC appears to endorse using a precautionary approach, where policies and measures taken to deal with climate change are cost-effective, to ensure global benefits at the lowest possible cost.[112] Environmentalists could argue that it might be more cost effective to promote energy efficiency and demand reduction. If it is determined that the presently considered options for using the sub-seabed as a storage site for CO_2 do not ensure global benefits at the lowest possible costs, a higher level of scientific certainty may be required before allowing these activities (McCullagh, 1996).

Although sequestration projects appear to be encouraged under international climate change legislation, it should be noted that the Kyoto Protocol also provides that greenhouse gas emission reductions from sources and removals by sinks shall be reported in a transparent and verifiable manner. This is because there are concerns over the permanence of the CO_2 storage (i.e. that it is not spontaneously released to the atmosphere by fire or escape). There is an obvious problem about counting these as reductions because then there is potentially less or no incentive to minimize the risk of their escape after the credit has been given. With regards to geological sequestration where CO_2 could be stored for a scientifically uncertain period of time, legitimate concerns can be raised regarding leakage and security. The use of sequestration projects generally to count towards emission reductions and removals has also already attracted much controversy amongst Contracting Parties to the UNFCCC. Most of the focus of debate to date has been on terrestrial biosphere sequestration projects which are notably different from long-term CO_2 storage in geological reservoirs. The result of this is that methodologies for inventories and accounting of greenhouse gas reductions through sequestration still need to be developed and approved by Contracting Parties to the UNFCCC and Kyoto Protocol (IPCC, 2005).

As the UNFCCC stands, it appears to distinguish between two types of methodologies for measuring greenhouse gas reductions. These are inventories for the sake of yearly national inventories of greenhouse gas emissions, and accounting of greenhouse gas reductions in flexible mechanisms in the Kyoto Protocol; i.e. carbon trading. At the current time, IPCC Guidelines and Good Practice Guidance Reports are used in preparing inventories under the UNFCCC framework. The IPCC guidelines do not yet specifically include CO_2 capture and storage options, though revised Guidelines providing guidance on incorporating CCS in greenhouse gas inventories are planned to be published in 2006 (IPCC, 2005). The recent IPCC report commented that capture and storage of CO_2 could then be incorporated in emission factors of industry and the power sector, meaning that carbon capture and

---

112   UNFCCC, Article 3(3).

storage would be a mitigation option reducing CO_2 from the source, rather than a sink enhancement option (IPCC, 2005).

It is clear that key greenhouse gas accounting and inventory issues must be addressed before CO_2 capture and storage activities can be included in the portfolio of climate change mitigation mechanisms. There are however potential problems in addressing these issues, such as the uncertainty regarding leakage rates, fraction retained and chances of accidental release of CO_2. It is unclear whether IPCC guidelines will suffice and maybe binding rules will be required to be agreed by Contracting Parties to the Convention and Kyoto Protocol to resolve this. One solution under consideration is the introduction of compulsory insurance, and/or parties only being allocated temporary renewable credits, which will expire unless monitoring demonstrates that the stored greenhouse gases remain in place. The value of the temporary storage of CO_2 will ultimately have to be decided through national and international political processes (IPCC, 2005).

4.5 CO_2 Storage and Environmental Assessment Legislation

Environmental Impact Assessment (EIA)

EIA is a procedure that seeks to ensure the acquisition of adequate and early information on likely environmental consequences of development projects or activities, on possible alternatives, and on measures to mitigate harm (Kiss and Shelton, 2000). The person who seeks to undertake the project or activity may be required to complete an environmental assessment before the project can receive authorisation. An EIA could be required under international, European or domestic legislation.

The Law of the Sea Convention (UNCLOS) contains provisions concerning the environmental assessment of potentially damaging activities that take place in the oceans. UNCLOS provides that 'when states have reasonable grounds for believing that planned activities under their jurisdiction or control may cause substantial pollution of or significant and harmful changes to the marine environment, they shall, as far as practicable, assess the potential effects of such activities on the marine environment'.[113] This obligation applies to any part of the marine environment including marine waters under national jurisdiction. After the state undertaking the project has completed an assessment of the environmental effects they must communicate the results of the assessment to competent international organizations, which will make this available to all states.[114]

The environmental assessment provision in UNCLOS is loosely drafted and it appears it will only be triggered if a state considers that their project or activity could cause pollution or significant and harmful changes to the marine environment. The use of the word 'practicable' suggests that a full assessment of the environmental

---

113   UNCLOS, Article 206.
114   Ibid., Article 205.

impacts could in practice be somewhat limited, as states could provide the barest information to other states if they considered that this was 'practicable'. The state where the activity originates is also placed under a separate duty to monitor the risk or effects of pollution after the environmental assessment is complete and the activity is underway.[115] UNCLOS requires continuous environmental monitoring to determine whether the activities taking place are likely to pollute the marine environment, although the state is again only obliged to go as far as is 'practicable'. A Contracting Party to UNCLOS could challenge another Contracting Party under the procedures laid down in the Convention, if they considered that the environmental assessment provisions in UNCLOS had not satisfactorily been complied with.

The Convention on EIA in a Transboundary Context (hereafter ESPOO Convention) is another international agreement that stipulates the obligations of parties to assess the environmental impact of certain activities at an early stage of planning. The ESPOO Convention entered into force in 1997 and has been ratified by the EU and the UK. The ESPOO Convention is more specific and detailed than UNCLOS in setting out the procedures and substantive requirements as to the EIA. It applies to the area under the Contracting Parties jurisdiction, which can include offshore projects on the continental shelf because certain projects at sea are included in the Annexes.[116] Contracting Parties must establish procedures with regard to listed activities that are likely to cause significant adverse transboundary impact. Impact is defined to mean any effect caused by a proposed activity on the environment, including on human health and safety, flora, fauna, soil, air, water, climate, landscape and historical monuments or other physical structures, or the interaction among these factors.[117]

The ESPOO Convention lists specific activities in Appendix I that are subject to the EIA requirements. Among the activities listed are major storage facilities for chemical products.[118] It is questionable whether geological formations under the oceans can be classed as storage facilities. Appendix I also include oil and gas pipelines,[119] which could be relevant if the CO_2 is transported by pipeline to the storage site. Non-listed activities may be subject to the Convention requirements if the party of origin and the affected party or parties agree.

The most significant aspect of the Convention is that it lays down the general obligation of states to notify and consult each other on all major projects under consideration that are likely to have a significant adverse environmental impact across boundaries.[120] This limits its application to situations where impacts from activities of one state affect another state's territory. It is arguable whether geological carbon dioxide storage might have an impact on another state and this is most likely

---

115   Ibid., Article 204.
116   ESPOO Convention, e.g. Annex I (15).
117   Ibid., Article 1(vii).
118   Ibid., Appendix 1 (16).
119   Ibid., Appendix 1 (8).
120   Ibid., Article 2(4).

to be dependant upon where the activity takes place. If a state considers that another party could be affected they must notify the affected party who then have the right to participate in the EIA process. The public in the territories affected also have the right to be informed of, and participate in the assessment procedure.

A Directive on EIA was also introduced into European law in 1985 (and was later amended in 1997). Member States had to transpose the directive in its amended form by 1999. The EIA procedure under the Directive is more significant in its impact than the ESPOO Convention because it ensures that environmental consequences of projects are identified and assessed *before authorisation is given* (emphasis added). The process of EIA under the Directive can be broken down into a number of discrete stages. The first is to determine whether or not the project falls within the criteria for the requirement of EIA. The EIA Directive outlines which project categories shall be made subject to an EIA, and this determines which procedure shall be followed. Projects are categorized into Annex I, where EIA is compulsory; and Annex II, where an EIA is only needed if there are such significant effects on the environment by virtue of their nature, size or location.

There are no clear project categories in Annex I of the Directive that CO_2 storage projects could expressly fall under. One category covers waste disposal installations for the incineration or chemical treatment of waste.[121] Although there is a strong argument that CO_2 is a waste it is not incinerated and probably not treated. Another possible category that CO_2 geological storage might fall under is installations for the storage of petroleum, petrochemical, or chemical products with a capacity of 200,000 tonnes or more.[122] This might catch CO_2 storage and require an assessment of whether geological formations can be interpreted to be installations, CO_2 can be determined to be a chemical product, and whether more than 200,000 tonnes is stored. If the CO_2 is transported by pipeline an EIA will also be required if the chemicals pipeline is more than 800 mm diameter and a length of more than 40 km.[123] Where there is any degree of uncertainty over whether or not a project falls within Annex I of the Directive, a ruling on the need for an assessment can be obtained in the UK from either the Secretary of State or the local planning authority.

There are similarly no clear project categories contained in Annex II that cover CO_2 storage projects. Some project categories are similar to those in Annex I, and include installations for the storage of petroleum, petrochemical, or chemical products (without the minimum tonne requirement)[124] and oil and gas pipeline installations (without the size and length requirement).[125] Other categories that CO_2 storage could possibly fall under are deep drillings,[126] or installations for the disposal

---

121   Directive on the assessment of the effects of certain public and private projects on the environment, Annex I (9, 10).

122   Ibid., Annex I (21).

123   Ibid., Annex I (16).

124   Ibid., Annex II (6c).

125   Ibid., Annex II (10i).

126   Ibid., Annex II (2d).

of waste.[127] It seems that what is most likely to catch CO_2 storage is the provision in Annex II that covers 'any change or extension of projects listed in Annex I or Annex II, already authorized, executed or in the process of being executed, which may have significant adverse effects on the environment; Projects in Annex I, undertaken exclusively or mainly for the development and testing of new methods or products and not used for more than two years'.[128] It seems that an environmental assessment is currently required for the erection of floating installations such as oil and gas rigs.[129] If these rigs are modified to inject CO_2 into the seabed then it appears an EIA will be required.

If a project falls within Annex II of the Directive this does not necessarily mean that an EIA is required. It depends whether the project or activity is likely to have significant effects on the environment by virtue of factors such as its nature, size or location. The Directive contains explicit guidance on when an EIA is required. Schedule 3 now contains selection, or screening criteria to which the decision maker must have regard. These are grouped together under general headings of:

- the characteristics of the development;
- the location of the development;
- the characteristics of the potential impact.

If an EIA is required the statement should include information on the direct and indirect effects of a project on a variety of factors, including human beings, fauna, flora, the environment and material assets and the cultural heritage. The developer must submit certain specified information to the authority dealing with the application. This crucially includes information on alternatives studied – which could mean the developer having to argue why other CO_2 reduction or storage alternatives were not chosen. The developer must also consult and make information available to statutory consultees, other authorities likely to be concerned with the project, and members of the public. The Directive also implements the ESPOO Convention in respect of the EC and Member States affected are given rights to participate in the decision making process. Members of the European Economic Union must also be supplied with information about the project but they have no rights to participate in the decision making process. In the case of CO_2 storage projects in the North Sea the UK therefore has obligations under European law (the Directive) to engage Member States and under international law for countries such as Norway who are signatories to the ESPOO Convention but who are not members of the EC. The results of all of this consultation are taken into account in the authorisation procedure of the project. In the event of a dispute over whether an EIA is required, it rests with national courts and possibly the European Court of Justice to interpret the Directive.

---

127   Ibid., Annex II (11b).

128   Ibid., Annex II (13).

129   E.g. see The UK Offshore Petroleum Production and Pipe-lines (Assessment of Environmental Effects) Regulations 1999.

Strategic Environmental Assessment (SEA)

The EC SEA Directive became law in July 2004. SEA is a process for predicting and evaluating the environmental implications of a policy, plan or programme and the SEA is a key input to decision making. Authorities which prepare and/or adopt a plan or programme that is subject to the Directive will have to prepare an environmental report on the plan or programme's likely significant effects on the environment.[130] The SEA Directive is similar to the EIA Directive but the actual assessment is done earlier at a broader, more strategic level. This contrasts with EIA which is carried out for a specific development or activity. Both an EIA and SEA can be carried out and they are not mutually exclusive.

The SEA Directive has quite complicated criteria and the EC has issued guidance on when SEA is applicable (EC, 2003). It seems that whether the SEA Directive applies depends firstly on whether there is a plan or programme, and secondly whether it falls under one of the categories listed in the Directive. The Directive applies to plans and programmes which public sector bodies and a limited number of private sector bodies (principally privatized utility companies) are 'required' to produce and/or adopt.[131] No detailed explanations of the terms 'plan' or 'programme' are given, but it is likely that these will be interpreted widely. For a plan or programme to be 'required', essentially an authority must have no discretion as to whether or not it prepares the plan or programme.

An SEA is mandatory if it falls under one of the categories listed in the Directive. Four project categories could be relevant to CO_2 capture and storage:

- plans and programmes which are prepared for energy;
- waste management;
- water management;
- plans or programmes requiring appropriate assessment under the Habitats Directive.[132]

Other plans or programmes which set the framework for development consent of projects (here not limited to project types listed in the EIA legislation) require SEA if they are determined by screening to be likely to have significant environmental effects. Minor modifications to plans and programmes in the categories which generally require mandatory SEA and those for small areas at local level only require SEA where screening determines that they are likely to have significant environmental effects.[133] Screening can be carried out on a case-by-case basis and/or by specifying categories of plans or programmes.[134] The Directive sets out criteria to

---

130 Ibid., Article 3(1).
131 Ibid., Article 2.
132 Ibid., Article 3(2).
133 Ibid., Article 3(3).
134 Ibid., Article 3(5).

be used for screening and these include such things as effects on internationally and nationally designated sites.

The assessment takes place during the preparation of the plan and programmes and before their adoption. An environmental report should be prepared which identifies, describes and evaluates the likely significant environmental effects on the environment of implementing the plans or programmes and reasonable alternatives. Measures to avoid, mitigate or compensate for serious adverse impacts must be included, as must a description of proposed monitoring measures. The public and environmental authorities must be consulted and can give their opinion. All the results from the consultation process are integrated and taken into account in the course of the planning procedure. After the adoption of the plan or programme the public is informed about the decision and the way it was made – particularly what environmental considerations were integrated into the plan or programme. In the case of significant transboundary effects, the affected Member State and its public are informed and have the possibility to make comments which are integrated into the national decision making process.

There has already been a significant amount of activity by the UK Government into how SEA could come into play in offshore waters (House of Commons, 2004). The Department of Trade and Industry (DTI) have already begun undertaking a series of SEAs for the offshore oil and gas sector based on the requirements of the Directive. An SEA has already been carried out to inform offshore oil and gas licensing with the purpose to make sure these are developed in an environmentally sensitive manner. The DTI are examining the geological structures of the area by conducting seismic surveys and exploration drilling. This is to determine where hydrocarbons could accumulate and be retained by examining the size of the reserves and extent of the reservoirs. The SEA will also examine areas of the continental shelf to identify areas that may require special protection or consideration after examining the environmental effects on seabed fauna, whales, dolphins and other marine mammals. The DTI is also carrying out an SEA looking at offshore renewables such as wind farms and the risks and uncertainties with developing these.

It is not completely clear whether an SEA for geological CO_2 storage will be a legal requirement under the SEA Directive. This will in part be dependent on whether the Government prepares any plans or programmes that come within the scope of the Directive. The government is currently conducting SEAs for offshore oil and gas installations and renewables and is hoping to tie the results of these SEAs together. There is an argument that the remit of these SEAs could be extended to include carbon dioxide storage as there are obvious overlaps. It seems likely that there could be a Government commitment to do an SEA for CO_2 storage in the near future.

At a meeting in Kiev in May 2003, the parties to the ESPOO Convention adopted a new international convention called the Strategic Environmental Assessment (SEA) Protocol. Thirty five countries signed the Protocol, including the UK, together with the European Union. The SEA Protocol is not yet in force, but when it is ratified will require its parties to evaluate the environmental consequences of their official draft plans and programmes. It is similar to the European SEA Directive but will have greater application – in the sense that it will also apply to countries not belonging to the EC. This is important in relation to geological CO_2 storage projects because the UK may have to consult North Sea neighbouring countries such as Norway (which is not an EC Member State). The Protocol could take many years before receiving the required number of ratifications allowing it to enter into force.

4.6 CO_2 Storage and Habitat Protection Legislation

Habitat Protection under International Law

The UK Government has ratified a number of international conventions which concern habitat protection. The most relevant of these is the Biodiversity Convention of 1992, which came into force in 1993. This Convention requires parties to adopt national strategies, plans and programmes for the conservation and sustainable use of biological diversity. It requires Contracting Parties to integrate the conservation and sustainable use of biological diversity into relevant sectoral or cross-sectoral plans, programmes and policies.[135] The aim of this is to establish protected areas to conserve and protect ecosystems, habitats, and threatened species.

The Convention recognizes the traditional sovereign rights of states to exploit their own resources and their responsibility to ensure that activities within their jurisdiction do not cause damage beyond the limits of national jurisdiction.[136] The application of the Convention extends to marine waters, and at the second conference of the parties to the Convention in 1995 it was agreed that marine biodiversity should be a priority area for action. The Jakarta Mandate on Marine and Coastal Biological Diversity was subsequently adopted and this sets out a strategy for marine biodiversity with special emphasis on integrated marine and coastal area management and the precautionary approach. The Convention and Mandate provide support that some assessment of the impact of geological carbon storage on marine biodiversity could be required. Whether or not such an activity will be in conflict with this Convention depends on what such an assessment of biodiversity reveals – i.e. what impact CO_2 geological storage will have on the biodiversity around the storage sites.

Even if there is an impact on biodiversity this does not necessarily mean that the Convention will prohibit CO_2 storage from taking place. The Convention states that 'Contracting Parties, as far as possible and as appropriate, shall take into account the environmental consequences of its programmes and shall initiate action to prevent or minimize conditions that present an imminent or grave danger or damage to biological diversity'.[137] The use of the words 'as far as possible' and as 'appropriate' weakens the legal status of the Convention. It seems likely that CO_2 storage could still take place as long as there is not a significant impact on habitats and the Government

---

135   Biodiversity Convention, Article 6.
136   Ibid., Article 3.
137   Ibid, Article 14(1).

has taken into account these environmental consequences and proposed some form of mitigation measures.

The OSPAR Convention was discussed in detail in the section 4.3.3. While the OSPAR Convention has marine pollution as its main focus, it also contains important provisions in Annex V aimed at the protection and conservation of the ecosystems and biological diversity of the maritime area. Contracting Parties to the OSPAR Convention must 'take the necessary measures to protect and conserve the ecosystems and the biological diversity of the maritime area, and to restore, where practicable, marine areas which have been adversely affected; and cooperate in adopting programmes and measures for those purposes for the control of the human activities identified by the application of the criteria in Appendix 3.'[138] This appears to place a legal duty on Contracting Parties, although it uses similar weak language as the Biodiversity Convention – e.g. 'where practicable'. The Biodiversity Committee of the OSPAR Commission has to draw up plans and programmes designed to achieve programmes and measures for the control of human activities.[139] They can then impose measures for instituting protective, conservation, restorative or precautionary measures related to specific sites or a particular species.[140] It is possible that the OSPAR Commission could seek to protect habitats or species in an area where CO_2 disposal is planned, and thus block potential projects in certain locations, although very little seems to be have been done under these provisions so far.

In March 2002, the UK signed up to the Bergen Ministerial Declaration.[141] This agreed to 'the strengthening of cooperation in the spatial planning processes of the North Sea states related to the marine environment' to prevent and resolve the potential problems created by conflicts 'between the requirements for conservation and restoration of the marine environment and the different human activities in the North Sea'.[142] The Declaration also invites the OSPAR Biodiversity Committee, 'to investigate the possibilities for further international cooperation in planning and managing marine activities through spatial planning of the North Sea states taking into account cumulative and transboundary effects'.[143] The Declaration agrees that close cooperation of regional governments, local authorities and other stakeholders, is important for future development of a marine planning system in the North Sea.[144] Similarly, OSPAR's own work is examining how the role of spatial planning will help to improve co-operation and management of the range of different activities that take place in coastal waters.

---

138   OSPAR Convention, Annex V, Article 2.

139   Ibid, Article 3(1)(a).

140   Ibis, Article 3(1)(b).

141   Ministerial Declaration of the Fifth International Conference on the Protection of the North Sea. Bergen, Norway 20–21 March 2002.

142   Ibid., Section XI, paragraph 76.

143   Ibid., Section XI, paragraph 77(ii).

144   Ibid., Section XI, paragraph 78.

Habitat and Species Protection under European Law

The EC adopted the Habitats Directive, in 1992.[145] The Directive seeks to preserve/ restore natural habitats and wild fauna and flora, by obliging Member States of the EC to provide a comprehensive network of special areas of conservation for endangered and vulnerable species and habitats.[146] This nature network established under the Habitats Directive in conjunction with the Birds Directive, consists of sites of international importance. The Habitats Directive provides measures to protect conservation areas[147] and measures to protect species.[148] The Annexes to the Directive list the broad categories of natural habitat types and the specific animal and plant species of interest. The network will consist of sites containing the one-hundred and sixty nine natural habitats types listed in Annex I of the Directive and sites containing the six-hundred and twenty three habitats of the species listed in Annex II.

The Habitats Directive calls for the creation of a network of protected areas known as Natura 2000, to consist of Special Areas of Conservation. EC Member States are required through a statutory, administrative and/or contractual act to propose sites for designation as special areas of conservation, drawn up by reference to the criteria laid down in Annex III, and send these to the EC. There is also provision for EC designation in exceptional circumstances where a site hosts a priority natural habitat type or priority species. A mandatory duty of care is placed on the Member State in which that habitat is found.

The Habitats Directive covers marine biodiversity in its scope. The UK Government was once of the opinion that the Directive did not extend beyond its territorial waters.[149] This cumulated in a court case in 1999 with Greenpeace taking the UK Government to court, where the judge ruled in favour of Greenpeace. The current UK policy is that the Habitats and Birds Directives should apply to waters up to two-hundred nautical miles from the coast and to adjacent designated areas of continental shelf. One implication of the Greenpeace case is that the UK regulations implementing the Directive[150] will need to be revised so that they cover the continental shelf. The UK Government is currently taking steps to implement the Habitats Directive in offshore waters and draft Regulations were consulted upon in 2003 (DEFRA, 2003). The draft Offshore Marine Regulations (Natural Habitats) Regulations do not appear to have been approved by Parliament as yet. Several Member States, have also struggled to designate special areas of conservation in sea areas outside their territorial waters. At the current time, a number of coastal areas

---

145   Council Directive 92/43 on the Conservation of natural habitats and of wild fauna and flora.

146   Ibid., Article 2.

147   Ibid., Articles 3–11.

148   Ibid., Articles 12–16.

149   R v Secretary of State ex parte Greenpeace, Queens Bench Division, 5 November 1999 CO/1336/99.

150   Conservation (Natural Habitats) Regulations 1994.

around the UK are designated as special areas of conservation, but no sites on the continental shelf are designated. The UK will probably begin to designate sites in this area after the regulations come into force.

Although marine biodiversity falls within the scope of the Directive, marine habitats and species are poorly represented in the Directive itself, which has an almost exclusive focus on territorial and coastal habitats. To enable identification of offshore habitats and species, and comply with the requirements of the Directive, the Joint Nature Conservation Committee (JNCCC) have been asked by the UK government to provide information on special areas of conservation. The 'Offshore Natura 2000 Project' is being conducted by JNCC under a steering group consisting of representatives from sponsoring government departments. Under this review special areas of conservation may be put forward for habitats of conservation importance (listed in Annex I to the Habitats Directive) or for species of conservation importance (listed in Annex II).

Some areas of possible Annex I habitat in offshore waters have been mapped using existing British Geological Survey geological seabed map interpretations. This does not cover all of the UK continental shelf and for a number of areas of potential Annex I habitat there are no, or limited, biological data. The JNCC have identified a number of habitats that occur in UK offshore waters and these are produced in Table 4.3.

Table 4.3   Habitats from the EC Habitats Directive occurring in UK offshore waters

Annex I habitats (from Directive 92/43/EC amended by 97/62/EC)
Sandbanks which are slightly covered by seawater all the time
Reefs
Submarine structures made by leaking gases
Submerged or partially submerged sea caves

Source: Taken from the Joint Nature Conservation Committee website

Shallow sandbanks are concentrated off north and north-east Norfolk, in the outer Thames Estuary, off the south-east coast of Kent and off the north-east coast of the Isle of Man. Reef habitat occurs in the English Channel, Celtic Sea, Irish Sea and west and north of Scotland extending far out into the North Atlantic. Reef is scarce in the North Sea. Shallow sandbanks are found in UK offshore waters. In the northern North Sea, 'pockmarks' containing carbonate structures deposited by methane-oxidising bacteria occur, these structures may fit within the definition of the Annex I habitat of 'submarine structures made by leaking gases'. No sea caves have yet been identified in UK offshore waters. There are only a limited number of species listed in Annex II of the Habitats Directive which are known to occur in UK offshore waters. The JNCC have identified a number of species that occur in UK offshore waters and these are produced in Table 4.4.

Table 4.4   Species from the EC Habitats Directive occurring in UK offshore waters

Annex II species(from Directive 92/43/EC amended by 97/62/EC)
Harbour porpoise *Phocoena phocoena*
Bottlenose dolphin *Tursiops truncatus*
Common (or harbour) seal *Phoca vitulina*
Grey seal *Halichoerus grypus*
Loggerhead Turtle *Caretta caretta*
Lamprey *Petromyzon marinus*
Sturgeon *Acipenser sturio*

Source: Taken from the Joint Nature Conservation Committee website

It is possible that CO_2 storage may impact on the marine eco-system as a result of injections into the seabed, or the laying of pipelines and other seabed infrastructure used during operations. The possibility of CO_2 escaping and its effects on biodiversity will also require closer scientific consideration. The most foreseeable impact on habitats and species will be dependent on the location of the platforms and storage sites, as well as the routes followed by pipelines (if they are the method of transportation). The question therefore is whether these habitats or species that are present in these areas are likely to be affected by licensing CO_2 storage. Clearly if they are not, then they are unlikely to be protected as special areas of conservation. Even if an area has been designated as a special area of conservation, and none have as yet in UK's offshore waters, this still does not provide absolute protection against interference – only restrictions. The Directive states that in the absence of alternative solutions, Member States may permit interference for 'imperative reasons of overriding public interest' – which expressly include social or economic interests. If Member States rely on this clause then appropriate compensatory measures must be provided to preserve the coherence of Community habitats. 'Priority' habitat types, however, generally may only be interfered with for environmental, human health or public safety reasons. Other reasons of overriding public importance may only be invoked 'further to an opinion from the Commission'. It is possible that one might argue that CO_2 storage is in the 'overriding public interest', and/or they are interfering with the site for 'environmental, human health or public safety reasons'.

A further example of the limitations of the Habitats Directive, can be seen in the case brought by Greenpeace in 1999.[151] Greenpeace argued that reefs fell within

---

151   R v Secretary of State ex parte Greenpeace, Queens Bench Division, 5 November 1999 CO/1336/99.

Annex I of the Directive and the oil and gas licences issued by the UK Government would have an adverse effect on these. The Directive requires Member States to establish 'a system of strict protection of animals listed in the Directive'.[152] The Directive also prohibits the following activities, (i) all forms of deliberate capture or killing of specimens of the species in the wild, (ii) deliberate disturbance of the species especially during breeding seasons, (iii) deterioration or destruction of breeding sites. Greenpeace argued that these prohibited activities were merely illustrative of the overarching requirement to establish a strict system of protection. But the court agreed with the Government and the oil companies that they are not an exhaustive list of the means of protection. The court then considered the meaning of the word 'deliberate' which qualified the prohibition on capture, killing and disturbance of species. The word is not defined in the Directive, and, as the court noted, it is not a concept normally used in UK law. Greenpeace argued that when an oil company conducted operations which they knew were likely or possible to result in killing or disturbance, that was a deliberate act which fell within the Directives prohibition. Again the court agreed with the Government and the oil companies and did not think it was deliberate disturbance. But the prohibited act covering the deterioration or destruction of breeding sites is not qualified by the word 'deliberate'. The question in the Greenpeace case then was whether this was an absolute prohibition, which was therefore incompatible with the general defence under the UK regulations which related to incidental actions. The judge found that whilst the Directive requires a prohibition, it did not follow that it obliged Member States to create criminal offences.

4.7   Conclusions

Increasingly, geological carbon dioxide storage has been gaining attention as a potential technological solution to climate change, both in the UK and internationally. At national and international level there has also been growing discussion as to whether the storage of CO_2 in this manner is consistent with existing international laws, but there has been no consensus as yet. This is because the current legal framework is often ambiguous because laws were not drafted with this mitigation option and its technologies in mind.

Most of the focus on the storage of CO_2 under existing international laws to date has been on marine laws. A review of these laws does not offer definitive answers as to the correct legal position, but certain observations may be made. Firstly, it is not clear whether geological reservoirs and formations are caught under the definitions of sea, seabed or subsoil under the conventions. Secondly it is also unclear as to whether CO_2 should be treated as a waste under the conventions. On balance, the author considers that geological reservoirs and formations are probably within the remit of the conventions' scope and that CO_2 is also probably a waste. The method

---

152   Council Directive 92/43 on the Conservation of natural habitats and of wild fauna and flora, Article 12.

by which the CO_2 reaches the storage site also strongly influences the legality of such projects. Under each of the marine conventions it seems likely that if CO_2 is transported by ship then disposed of from an offshore installation (such as an oil rig), this will be prohibited. In the case of a pipeline carrying CO_2 to an installation, this will be prohibited under the London Convention and the 1996 Protocol. This is not always the case under the OSPAR Convention, where the prohibition against dumping only applies to installations carrying out activities concerning hydrocarbons. Since it is not a hydrocarbon it is permissible under the OSPAR Convention to pipe CO_2 to offshore installations provided they have not already been used for activities involving hydrocarbons. Under OSPAR, however, states have general environmental obligations with respect to land-based pipelines.

Recent studies into the legality of CO_2 storage have tended to call for immediate reviews into the current drafting of the marine conventions. This was a recurring theme in the recent report of the IEA; an example comment being – 'the contracting parties to these agreements need to interpret, clarify or, as the case may be amend these treaties with a view to account for some form of controlled carbon storage. There is significant room for such interpretation and clarification under these treaties' (IEA, 2005). The IEA thought that the contracting parties should also proactively 'take into consideration not only their marine environment protection objectives, but also their objectives regarding climate change mitigation', especially 'if they want a coherent international framework for carbon storage to be developed' (IEA, 2005). The IPCC was less forceful in its recent report, but still concluded that 'it will be essential to resolve these [marine laws] issues if CCS [carbon capture and storage] is to become part of the portfolio of mitigation options' (IPCC, 2005).

Although it is correct to point out that there are uncertainties under the current legal regime for marine protection, this is only because they were not drafted with the relatively new concept of CO_2 storage in mind. What, in the authors' opinion, is perhaps equally important in relation to taking geological dioxide storage as a mitigation option forward, is the status of CO_2 storage under the UNFCCC and Kyoto Protocol. At the current time it seems clear that both the UNFCCC and Kyoto Protocol allow for projects that reduce greenhouse gases at their sources and CO_2 storage can in theory be counted as an emission reduction. However, methodologies and rules for accounting of greenhouse gas reductions still need to be developed and approved. Without agreement on these, no CO_2 storage projects will be undertaken, because it is unlikely that costly projects will be financed without any incentive e.g. that they are counted as an emission reduction.

Focusing on changes to marine legislation alone is not the way forward, when greater clarification and certainty is also needed in relation to climate change legislation and rules. However, because of the slow pace of international law it is probably correct to raise all of the legal issues concerning CO_2 storage projects and consider them all in advance. What is needed before the holy grail of *greater legal certainty* (authors' emphasis) is international consensus as to whether a large enough group of the international community want offshore geological CO_2 storage as a significant mitigation option. Only when looking at difficult environmental choices such as climate change in the context of other international environmental laws, and considering all of the options including geological carbon dioxide storage, can politics reach a solution that can then be backed up by introducing adequate international laws.

4.8 References

Brubaker, R.D., and Christiansen, A.C. (2001), *Legal Aspects of Underground CO_2 Storage: Summary of Developments under the London Convention and North Sea Conference*, Pre-Project Report, The Fridtjof Nansen Institute, Oslo.

Cabinet Office (2002), *The Energy Review*, A Performance and Innovation Unit Report, UK Government, London.

Churchill, R. (1996), 'International Legal Issues Relating to Ocean Storage of CO_2: A Focus on the UN Convention on the Law of the Sea', in *Ocean Storage of CO_2, Workshop 3, International Links and Concerns*, IEA Greenhouse Gas R&D Programme, Cheltenham, pp. 117–126.

DEFRA (2002), *CO_2 Sequestration and Storage – Legal Issues*, DEFRA Legal Services, internal document, DEFRA, London.

DEFRA (2003), News Release, *DEFRA Launches Consultation on Extending Habitat and Birds Directive*, 6 August, DEFRA, London.

DTI (2001), *Review of the Case for Government Support for Cleaner Coal Technology Demonstration Plant*, Final Report, DTI, London.

European Commission (2003), *Implementation of Directive 2001/42 on the Assessment of the Effects of Certain Plans and Programmes on the Environment*, Brussels.

Garner, B.A. (ed.) (1999), *Black's Law Dictionary*, 7th edition, West Group Publishers.

GESAMP (1997), Report of the Twenty-Seventh Session of GESAMP, Nairobi, Kenya, April 1997, *GESAMP Rep. and Studies* 63.

House of Commons – Environment, Food and Rural Affairs Committee (2004) *Marine Environment*, Session 2003–04 Sixth Report, UK Parliament, London.

IEA (2005), *Legal Aspects of Storing CO_2*, International Energy Agency, Paris.

IMO (1999), *Report of the Twenty-Second Meeting of the Scientific Group to the London Convention*, International Maritime Organisation, London.

IMO (2000), *Twenty Second Consultative Meeting of Contracting Parties to the Convention on the Prevention of Marine Pollution by Dumping of Wastes and Other Matter 1972*, 18–22 International Maritime Organisation, September 2000; published 25 October 2000; LC 22/14, London.

IMO (2004), *Twenty-Sixth Consultative Meeting of Contracting Parties to the Convention on the Prevention of Marine Pollution by Dumping of Wastes and Other Matter*, 1–5 November 2004; published 17 December 2004, International Maritime Organisation, LC 26/15, London.

IMO (2005), *Invitation to Consider the Legal Questions Associated with CO_2 Sequestration in Geological Formations under the London Convention and Protocol*, International Maritime Organisation, 31 March, LC.2/Circ.439, London.

IPCC (2005), *IPCC Special Report on Carbon Dioxide Capture and Storage*, prepared by Working Group III of the Intergovernmental Panel on Climate Change, B. Metz, O. Davidson, H.C. de Coninck, M. Loos and L.A. Meyer (eds), Cambridge University Press, Cambridge.

Johnston, P., Santillo, D. and Stringer, R. (1999), *Ocean Disposal/Sequestration of Carbon Dioxide from Fossil Fuel Production and Use: An Overview of Rationale, Techniques and Implications*, Greenpeace Research Laboratories, Technical Note 01/99, Exeter.

Kiss, A.C. and Shelton, D. (1999), *International Environmental Law*, 2nd edn, Ardseley, New York.

McCullagh, J. (1996), 'International Legal Control over Accelerating Ocean Storage of Carbon Dioxide', in IEA Greenhouse Gas R&D Programme, *Ocean Storage of CO_2 Workshop 3, International Links and Concerns*, pp. 85–115.

OSPAR Commission (2002), *Disposal of CO_2 at Sea*, OSPAR Summary Record 02/21/1-E.

OSPAR Commission (2003), *Summary Record*, 03/17/1-(A-B)-E, Bremen, 23–27 June.

OSPAR Commission (2004), *Summary Record*, 04/23/1-E, Reykjavik 28 June – 1 July.

OSPAR Group of Jurists and Linguists (2003), 'Compatibility with the OSPAR Convention of Possible Placements of Carbon Dioxide in the Sea or the Sea-Bed', Agenda Item 3, JL 03/3/1-E, Hamburg, 12 May.

Purdy, R. and Macrory, R. (2004), *Geological Carbon Sequestration: Critical Legal Issues*, Tyndall Centre Working Paper Number 45, Manchester.

RCEP (2000), *Energy – The Changing Climate*, Royal Commission on Environment and Pollution, Twenty-second Report, Cm 4749, TSO, London.

Snelders, C.A.M. (2002), *CO_2 Storage in the Seabe : A Solution to the Greenhouse Effect seen from the Perspective of International Law*, Thesis for Masters degree in International Law, Open Universiteit, Heerlen, Netherlands.

UK Government (2003), *Energy – The Changing Climate*, The UK Government Response to the Royal Commission on Environmental Pollution's Twenty-Second Report, February 2003, Cm 5766, TSO, London.

Addendum to Chapter 4

Since writing this chapter there have been a number of significant developments concerning the legality of CCS. The first significant development is that the 1996 Protocol to the London Convention entered into force in March 2006. This is important because it goes beyond the provisions of its predecessor and aims to provide greater protection for the marine environment.

The second significant development is that after a number of legal and technical reviews, a number of Contracting Parties to the London Convention and Protocol are already seeking to amend the 1996 Protocol. Australia, co-sponsored by France, Norway and the UK, put forward a proposal in 2006 recommending an amendment to Annex I of the Protocol, thus bringing the regulation of CO_2 into line with the regulation of other substances eligible for dumping or storage. The proposal would allow for carbon dioxide streams from CCS consisting 'overwhelminginly of CO_2' to be stored in geological formations (IMO, 2006).

The next meeting of the Contracting will be held in November 2006 and to amend the Protocol will require a two-thirds agreement of those present at the meeting. The amendment will then enter into force immediately for any party agreeing to it, and for all other Parties (whether they agreed to it or not) after a period of one hundred days following the relevant meeting, unless a declaration against acceptance is made by a Party within that period. It is debateable whether a two-thirds majority will support amending the Protocol, but it seems that if the vote goes against those pushing for CCS then it is very likely it will be back on the agenda for amendment in 2007. If CO_2 storage does receive the two-thirds majority go-ahead in November 2006, then it is also feasible that the London Convention and other regional agreements such as the OSPAR Convention might also be considered to be amended.

The final significant development since the chapter was written has been the implied acceptance that CCS projects will take place at European level. The European Commission has recognised that if CCS projects are undertaken there will need to be regulatory frameworks at international and national level covering capture, transport and storage sites, as well as agreements in place on monitoring and liability. The European Commission therefore announced plans in May 2006 for a draft legislative proposal for an enabling CCS framework. (European Commission, 2006).

References

European Commission (2006) *Contract: Technical support for an enabling policy framework for carbon capture and geological storage*; Official Journal Ref: 2006/ s102-108792.
IMO (2006) *CO₂ Sequestration in sub-seabed formations: Consideration of proposals to amend Annex I to the London Protocol*; LP 1/6, 28 April.

Chapter 5

The Public Perception of Carbon Dioxide Capture and Storage in the UK

Simon Shackley, Clair Gough and Carly McLachlan

5.1   Introduction

The study of the public perceptions of carbon dioxide capture and storage (CCS) and its perceived acceptability is at an early stage and comprises only a handful of studies (Curry *et al.*, 2005 and 2005a; Gough *et al.*, 2002; Huijts, 2003; Itaoka *et al.*, 2005; Palmgren *et al.*, 2004; Shackley *et al.*, 2004). One of the few experiences of a related mitigation approach is the Ocean Field Experiment in Hawaii, which was a proposed experimental release of CO_2 into the ocean for the purposes of studying its fate and distribution. A case-study of the proposal has illustrated how bureaucratic obstacles, a few dedicated activists and slow recognition of the need for public outreach derailed the project (de Figueiredo, 2002).

We know, however, that public perceptions can have a very significant, and frequently unanticipated, effect upon major planned projects involving new technologies and infrastructure. Examples include: the effects of public opinion on the planned disposal of the Brent Spa oil platform in the mid-1990s, and the on-going debates over genetically modified organisms (GMOs), and nuclear waste disposal (Smith, 2000; Irwin and Michael, 2003; Hunt and Wynne, 2000). Closer to the issue of CO_2 storage, proposals for underground natural gas storage schemes have generated public opposition in some localities in the UK, despite similar facilities operating very close by with out apparent concern (Gough *et al.*, 2002). In two such cases, concerns about uncertain and difficult to assess risks to health and safety emerged. The effect that such perceived risks could have on local property prices also caused concern. The local media played an active role in disseminating the concerns about these natural gas storage proposals to a wider public audience (Overwyrefocus, Nogasplant websites, no date).

In response to such reactions, there has been a general shift by decision-makers to gathering better prior intelligence of possible public reactions to major new technologies, and also towards more consultative and 'deliberative' decision-making styles (RCEP, 1998; House of Lords, 2000). Clearly, possible public reactions to CCS, and how developments in the area might proceed so as to take such reactions into account, are important areas of research.

Research on the public perceptions of CCS is, however, challenging because of:

a) the relatively technical and 'remote' nature of the issue, meaning that there are few immediate points of connection in the lay public's frame of reference to many of the key concepts;
b) the early stage of the technology, with very few examples and experiences in the public domain to draw upon as illustrations.

An in-depth research approach is frequently useful for understanding public perceptions of an unknown technology, whereby technical information is provided in an incremental fashion to the target public sample. Methodologically, focus groups and in-depth discussion groups can therefore make a valuable contribution. The disadvantage is that only small samples can be surveyed using in-depth methods, as opposed to surveys which can, ideally, offer a representative sample and their findings subject to more robust statistical testing. In the study we report on here, we combine the strengths of both methods, by designing a questionnaire administered through face-to-face interviews on the basis of prior discussions in extended in-depth groups which we call 'Citizen Panels' (see Figure 5.1).

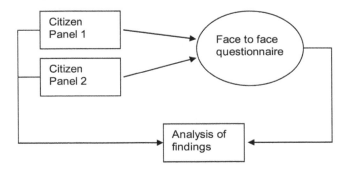

Figure 5.1   Methodology

5.2   Research Objectives and Methodology

The main objectives of the research were:

- to explore and understand what the public perceptions of off-shore CO_2 capture and storage (CCS) are, both when first presented with the idea and when more background information is provided;
- to explore and understand perceptions of the key risks and concerns surrounding CCS and what information, policies and processes would make CCS more and less acceptable to the public.

We only considered off-shore CO_2 storage because our earlier research suggested that, where there is a viable off-shore storage option it is much preferred to on-shore storage. Given the availability of potentially highly suitable off-shore storage sites in the UK we decided that involving the respondents in a discussion of on-shore storage would be distracting and create unnecessary confusion.

We ran two Citizen Panels over two five week periods in the last quarter of 2002 and the first quarter of 2003. Each Panel met for ten hours in total. A Citizen's Panel is a moderated group of between eight and ten individuals who meet over an extended period to discuss a set of related issues and who provide an informed opinion on those issues at the end of the Panel. Similar to a focus group, the key differences are that there is a gradual build-up of information on the particular topic of discussion, usually with expert witnesses, whom the Panel has the opportunity to question. The methods are described in detail in Kasemir *et al.* (2003). We recruited two demographically distinct Citizen Panels. The York group was all-male with six participants from socio-economic group B and three participants from socio-economic group C. The Manchester group was all-female and composed of six participants from group C and three participants from group B.

Our intention in recruiting distinct groups was to provide a clear and interesting contrast between the two Citizen Panels. Expert witnesses representing a range of viewpoints gave presentations and then took part in a Question and Answer (Q&A) session. The expert speakers were from: the British Geological Survey (BGS), BP, IEA Greenhouse Gas R&D Programme, Tyndall Centre for Climate Change Research and Friends of the Earth.

We then designed a questionnaire by drawing upon the citizen panel findings. The citizen panel work led us to exclude ocean and onshore storage options. It also highlighted the need to: measure the extent of belief in, and concern regarding, climate change; make an explicit comparison between CCS and other low-carbon energy options; and to analyse the effect of introducing information about the purpose of CCS upon the sample's responses. We also drew upon other climate change questionnaires (e.g. Lorenzoni, 2003; Shackley *et al.*, 2001). We tested the design of the questionnaire with a small sample (5 people) and modified it in the light of their comments and suggestions. A team of six interviewers conducted 212 face-to-face interviews over two days in August 2003 at the Liverpool John Lennon International Airport. The acceptance rate was between 40 and 50 per cent. The questionnaire took between 10 and 20 minutes to complete, depending on the level of interest of the respondent. Table 5.1 outlines the information which was provided to the respondents and the chronological order in which it was presented. Researchers filled in response forms as the respondents verbally answered the questions. On these response forms there were various instructions for the researchers and specified sections to read to the respondents.

In order to ascertain the extent to which the socio-economic profile of the travellers at Liverpool John Lennon International Airport reflects the population of the UK as a whole, we compared the information collected on annual income of the sample to the information on the income bands of the UK as a whole. Although

Table 5.1 Information provided to the survey respondents and the order in which it was provided

Question	Information to respondent	Guide/actions for researcher
		Half day training session
	In this section we are interested in your first impression of an idea. It may seem odd that there is little information given but your immediate response is what interests us here. The Government is currently looking at putting Carbon Dioxide in to underground storage sites under the North Sea. This process is called Carbon Storage. Carbon Dioxide is the gas which is produced by burning coal, gas and petrol.	
1. What is your initial reaction to this idea?		
	The words Climate Change are generally used by scientists to describe the way that the climate of the world may be changing as a consequence of human activities. If human beings are influencing the climate, there will be changes in temperature, rainfall and weather patterns, which will affect the natural world and human beings.	
2. Do you believe that human activities are affecting the climate? 3. Are you concerned about climate change? 4. In general, would you say that climate change receives too much or too little attention by politicians at the present time? 5. Do you think that the public should be actively involved in deciding what should be done about climate change ? 6. Do you think that policies to combat climate change should be decided mostly by government experts and scientists?		
	Most experts believe that in order to have a significant impact upon climate change we must achieve a 60% reduction in emissions such as carbon dioxide. By using Carbon Storage the United Kingdom could significantly reduce its carbon dioxide emissions while continuing to use fossil fuels. This could allow society to continue to use existing levels of fossil fuels for many decades to come. It could also act as an 'in-between' strategy while longer-term solutions are further developed, such as renewable energy technologies. A simple diagram of CCS was then shown to the respondents (showing capture at a power station, pipelines to the coast, pipelines along the sea bed and injection to a depleted oil or gas well or aquifer).	
7. Do you think that there may be any negative effects of doing this? 8. Do you think there may be any positive effects of doing this?		Do not prompt the respondent, classify their responses based on training and pre-agreed set definitions.
Remainder of questions. No new information presented.		

our sample covered a wide range of household income levels, compared to the distribution of UK household incomes (Department of Work and Pensions, 2003) it was skewed towards higher incomes (i.e. above £30,000). The highest incomes (i.e. above £40,000) were the most over represented group. This distribution is to be expected given that research by the Civil Aviation Authority has shown that those in the top three income brackets are more than four times as likely to fly as those in the lowest three brackets (Environment Audit Committee, 2003).

Whilst clearly not statistically representative of the UK population, it is possible to calculate the accuracy of the responses, relative to the target population of the airport on the days that the survey was conducted, given that the sample size of 212 represents approximately 2 per cent of the people (10,500) travelling to and form the airport over the period in which the survey was conducted (CAA, 2004). The Normal approximation to the Binomial distribution was then used to calculate the confidence intervals (Upton and Cook, 1997): $ps \pm 1.96\sqrt{(ps(1-ps)/n)}$, where ps is the proportion of the sample who respond in a particular way to a given question, and n is the sample size. The value obtained is the error at 95 per cent probability.

5.3   Main Findings of the Survey

None of the respondents were familiar with CO_2 Capture and Storage prior to the interview. We found that, on first contact with the idea of CO_2 capture and storage, and without any information on its purpose, most people (48±7 per cent) are neither for nor against CCS or say that they do not know, with a significant number (38±6.5 per cent) expressing slight or strong reservations (see Figure 5.2). Only 13 per cent (±4.5 per cent) volunteered support for the idea.

When more detailed information was provided on the reasons *why* CCS is being proposed, i.e. as a way of reducing CO_2 emissions into the atmosphere, support increased substantially. Half of the survey respondents (±7 per cent) developed a more positive attitude towards CCS, though a sizeable minority (35±6.5 per cent) did not change their opinion and 16±5 per cent became more negative. Respondents shifted primarily from the 'don't know' or 'neither support nor not support' categories to the 'slightly support' category (see Figure 5.3).

We also asked respondents about their support for CCS relative to other low- or zero-carbon energy options (wind, solar, wave and tidal, nuclear, energy efficiency and higher energy bills). We found that support for CCS was somewhat greater when compared to the main other options than when considered in isolation. A larger number of respondents also said that they did not know, or were neither in favour nor against, when asked specifically about their support for CCS than for other decarbonisation options (at 35±6.5 per cent compared to 24±6 per cent). This suggests that CCS is more favourably regarded when it is compared alongside the other principal decarbonisation options, than when it is considered in isolation. This might reflect the tendency to perceive options more negatively when they are considered in isolation, rather than compared to the other main options for achieving a given objective (in this case low carbon emissions).

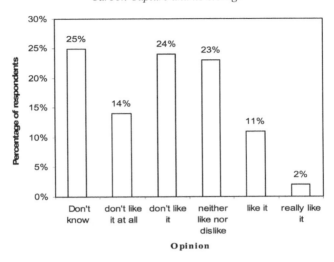

Figure 5.2   Initial reaction to CCS without any information on its purpose

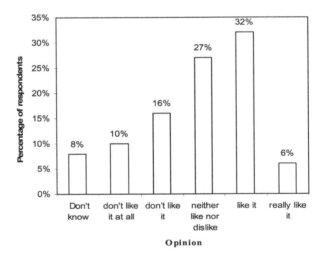

Figure 5.3 Opinion of CCS at the end of the survey

Overall, support for CCS on the basis of this survey can best be described as moderate or lukewarm compared to strong support in general for wind, solar and energy efficiency (see Figure 5.4). Whilst over 80±5.5 per cent of respondents 'strongly supported' wind, solar and energy efficiency (and with a further 10±4 per cent or more slightly supporting these options) 12±4.5 per cent of the sample strongly supported CCS with a further 43±7 per cent offering slight support, and 23±6 per cent slightly or strongly against it. CCS is much preferred, however, to nuclear power and to higher energy bills (see Figure 5.4).

The large majority of our sample believed that human activities are causing climate change (either strongly or moderately). Only 7.5±3.5 per cent of the sample disagreed or did not express a view either way. There was also a generally moderate to high level of concern about climate change, though there was also a substantial number (22±5.5 per cent) who did not express a view either way, and a further 15±5 per cent who were not concerned about climate change. This suggests that whilst many people now accept that human activities are a major cause of climate change, there is less consensus on whether climate change is a problem, though still over 60±7 per cent say there are 'very concerned' or 'concerned'. The high level of belief and concern regarding anthropogenic climate change meant that these factors could not be used to explain preferences with regards to CCS. There were no significant differences in the perceptions of CCS in our sample according to the other demographic variables measured (age, income, gender). Note, however, that we did not request the respondents to rate the importance of climate change as an issue compared to other contemporary socio-economic, political and environmental 'problems'. This may have led to greater concern being expressed for climate change than would have been the case if an explicit comparison with such other issues had been made.

Curry *et al.* (2005) conducted a survey of public attitudes towards energy and the environment in the UK through YouGov, an online polling company. The sample size was 1,056. A number of questions concerned perceptions of a range of carbon mitigation options including CCS. The findings are broadly in agreement with our survey results. Over 70 per cent of respondents believed that action needs to be taken to address global warming and the majority of that group thought that immediate action is necessary. Very few respondents had previously heard of CCS. As in our survey there was a strong preference for renewable energy technologies as a way of tackling global warming, with 80 to 90 per cent of the sample in support, similar to the percentage in our own survey. We identified some what greater support for CCS in our survey (at c. 50 per cent of the sample compared to c. 30 per cent in Curry *et al.* (2005)), though in both surveys about 20 per cent of the sample was opposed to use of CCS. Curry *et al.* (2005) identified some what greater support for nuclear power than in our survey (at c. 35 per cent in favour compared to c. 25 per cent in our sample) and less negative opinion towards nuclear (at c. 30 per cent compared to c. 55 per cent against in our sample). Curry *et al.* also explored the difference that provision of information on potential costs of different options made to the respondents' perceptions of the alternative mitigation options. There was, in response to this further information, a relatively small shift towards support for nuclear and CCS and away from renewables. What is striking is the similarity of Curry *et al.*'s conclusion that: 'Carbon capture and storage … received a slightly net favourable response, whereas nuclear energy and iron fertilisation [not explored in our survey] were viewed more negatively' (2005:14), and the conclusions of our survey work.

Carbon Capture and its Storage

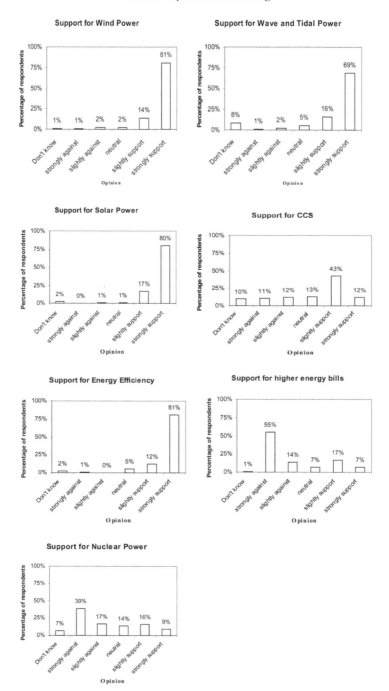

Figure 5.4 CCS compared to other low- or zero-carbon options

Perception of Negative and Positive Attributes of CCS

When asked, unprompted, if they could think of any negative effects of CCS respondents' most frequent answer was leakage (49 per cent) (see Figure 5.5). (Note that these answers are not expressed as a percentage out of 100 per cent since respondents were not restricted in the number of issues that they identified.) The next most frequently mentioned items were ecosystems (31 per cent), the new and untested nature of the technology (23 per cent) and human health impacts (18 per cent). Whilst these practical, physical and environmental risks were the most frequently mentioned, there were also a number of negative attributes mentioned in relation to CCS as a part of climate change abatement policy.

Avoiding the real problem (13 per cent), short termism (12 per cent) and the policy demonstrating reluctance to change from government (11 per cent) were all mentioned regularly. Grouping these last three responses into a general concern that CCS is treating the 'symptoms' not the cause of excessive CO_2 emissions, this would constitute, at 36 per cent, the second most frequently mentioned negative aspect of CCS after leakage. (Whilst these responses are not entirely independent the fact that the same individual might have identified two or three of the three areas is itself indicative of the extent of their concerns). 46 people (22 per cent) did not offer any response. When asked if they could think of any positive effects of CCS, by far the most frequent response was its role in abating climate change (58 per cent of all responses). The notion that using CCS could 'buy time' to develop other solutions was the next most frequently mentioned at 7 per cent.

More certainty about the environmental and safety risks of CCS in the long-term would help people to come to a clearer decision about the desirability of CCS. Many respondents indicated that they would like more information and more certainty in the assessments of CCS with regards to the above issues. When we asked about who should regulate the implementation of CCS, the Government was the most common answer (46 per cent), followed closely by the Environment Agency (43 per cent), environmental groups (34 per cent) and the oil industry (32 per cent). It is interesting that environmental NGOs were regarded by many respondents as an important part of the regulatory system, in many cases along with government, the Environment Agency and the oil industry.

A number of other surveys have now been undertaken of public perceptions of CCS in different countries (Curry *et al.*, 2005a; Palmgren *et al.*, 2004; Huijts, 2003; Itaoka *et al.*, 2005). It is difficult to make detailed comparisons between the various surveys because of differences in research design and implementation, phraseology, sampling methods and the underlying rationale of the research. Similar to the findings of our work, Curry *et al.* (2005a) found a low level of knowledge of CCS amongst a representative sample of the US public, whilst Huijts (for the Netherlands) and Itaoka *et al.* (for Japan) found higher levels of knowledge. Our survey showed a reasonably strong move towards more support for CCS upon provision of more information. This was not corroborated by Itaoka *et al.* or by Palmgren *et al.* (the latter of which showed a decrease in acceptance of CCS upon provision of more information). The general perception of CCS in our survey was somewhat more positive than in the other surveys but there are methodological and sampling issues that might explain this difference (apart from, or in addition to, national political or cultural difference). Clearly, more systematic and directly comparable research will need to be conducted to shore-up comparative analyses of public perceptions.

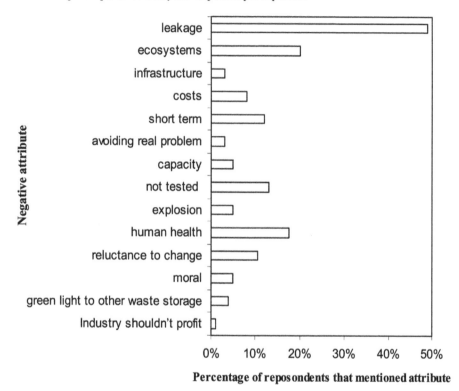

Figure 5.5   Negative attributes of CCS mentioned unprompted by
              respondents

5.4   Main Findings of the Citizen Panels

To recap, the Manchester nine strong panel was all-female and composed of two-thirds socio-economic group C and one-third group B, with an average age of 32 and an age range of 21 to 52. The York nine strong panel was all-male and composed of two-thirds group B and one-third group C, with an average age of 42 and an age range of 21 to 58. Socio-economic group B makes up approximately 14 per cent of the UK population and is made up of middle management, more senior officers in the civil service and local government and owners of small to medium sized businesses. Socio-economic group C makes up approximately 26 per cent of the UK population and is composed of junior management, owners of small establishments and non-manual office and retail workers. The York group was predominantly professionals and managers, whereas the Manchester group was predominantly administrators though with three professionals, so there was some overlap in the socio-economic composition.

The schedule for each group is shown in Table 5.2 indicating which topics were covered when and how, including the involvement of external experts. The selection of experts was intended to provide a range of perspectives on CCS from industry, environmental NGOs, geology and 'systems' views of energy. We should note that our selection of the geological experts was intended to reflect the present consensus amongst many geologists who have looked at the issue in detail, e.g. on the scale of the likely risks (e.g. as reflected in IEA GHG, 2004).

The following discussion will dwell at somewhat greater length on the Manchester panels, in part because some of the same findings emerged from the York panel. Hence, the discussion of the York panel will focus in particular upon the differences which emerged compared to the Manchester panel.

The discussion will include selected quotes from panel members to illustrate the argument. The Manchester panel session 1 is referred to as MS1, session two as MS2, and so on, whilst the York panel session 1 is YS1, YS2, etc. The number following the MS and YS label refers to the page number of the transcription from the recording made of each panel.

Table 5.2   Programme for each panel

Manchester Citizen Panel	York Citizen Panel
Session 1: Warm-up discussion on quality of life	*Session 1:* Expert presentation on climate change and round-table discussion
Session 2: Expert presentation on climate change and round-table discussion	*Session 2:* Expert presentation on CCS and initial discussion
Session 3: Expert presentation on CCS and initial discussion	*Session 3:* Two contrasting expert perspectives on CCS and discussion
Session 4: Two contrasting expert perspectives on CCS and discussion	*Session 4:* Criteria Weighting for selection of storage sites
Session 5: Participants summing-up	*Session 5:* Weighting of different decarbonisation options

The Manchester Group

As with other focus group work in the UK, we found that local and visible issues such as waste are those which are most evident in everyday experiences of the 'environment' (Darier *et al.*, 1999, 1999a). The Manchester group did respond, however, with surprise and concern at the account of climate change, and its possible impacts from the global to the local scales, which was provided in the second session.

The panel was generally quite positive about the idea of CCS after Session 3, when it was presented by a British Geological Survey (BGS) scientist. There appeared to be three underlying reasons for their positive perception:

1. Confidence in the ability of government to undertake appropriate assessment and regulation.

    > They [government] never go in blind do they. They never just do things just for the sake of doing things.   *(Sarah, MS3:18)*

2. Participants could not initially identify any real negatives associated with CCS, in part because the location of the CO_2 storage off-shore meant that local opposition was unlikely.

    > ... It [CCS] just seems a nice little neat way of doing it. Cause you know it's 'not in my back yard' is it? That would get round all the critical NIMBYs like me wouldn't it?   *(Elizabeth, MS3:19)*

    > Its not in anyone's backyard.   *(Jo, MS3:19)*

3. Avoidance of damage from climate change through CCS. The potential risks of climate change were regarded as greater than those arising from CCS.

    > You just look at the costs and benefits and Greenpeace will weigh them up for themselves but my personal opinion from what I've seen is that it [CCS] is more beneficial than ...   *(Sara, MS3:19)*

    > We're doing more damage by not doing anything.   *(Sue, MS3:19)*

Note, however, that the other options for reducing CO_2 had not been presented to the group at this point. Some participants struggled at this stage to see why CCS would be opposed:

> why would Greenpeace oppose it? I just don't see why they would ...
> *(Samantha, MS3:18)*

Despite this generally positive initial reaction to CCS various concerns were raised during the panels, in response to presentations by the invited experts.

Integrity of reservoir stores   Various technical questions emerged during the geologist's presentation regarding the nature of the geological stores. For example, it was asked where the water in an aquifer would be displaced to. The possible effect of past human utilization of fossil fuel reservoirs was also raised as a problem for their future integrity (MS3:5). After the geologist had left the room, Samantha commented that:

... Yeah it [oil & gas field] does store it for all these years but then drilling into it when they get the oil out ..... you just don't know where ... The holes have been filled, but she [the geologist] said the cement wasn't that great ... So they need to research into that or ...

... they were saying that with some of the sites ... they can't be 100 per cent sure if there is a fault or not, so what if they start pumping all the CO_2 down there?

(Samantha, MS3:13–14)

One participant picked up upon the geologist's use of the term 'bubble' to describe the CO_2 storage in the underground aquifer. This participant, Sue, was concerned that a large bubble of CO_2 might burst with catastrophic consequences.

Sue: ... you mentioned the world 'bubble' ... its looking for an escape ... If it's that large, that much of an area that its [the CO_2] going to be in, its going to be more than just catastrophic isn't it really? [if it bursts]

Geologist: Well it's not explosive ... Underground it's not a bubble and I probably shouldn't have used that word ... We term it the CO_2 bubble but it's not really a bubble because it's not existing on its own ... it's actually in the rock pore spaces ... there will be dissolving in some of the water, it will be reacting with some of the minerals in the rock.

(MS3:6,7,8)

Sue later returned to the issue when the geologist had left the room:

... me saying bubble! She scared the living daylights out of me, [saying] there's this bubble! *(MS3:24)*

This exchange indicates the importance of the terminology which is used by experts in communicating with the public. Associated with this dialogue were questions about the fate of the CO_2 in the Sleipner field aquifer, seismic images of which had been shown to the group by the geologist. One participant was concerned that the CO_2 in the aquifer appeared to have 'risen quite a lot' (MS3:7), raising a concern for her about long-term integrity. Such movement in just 5 years led her to wonder what would happen to the CO_2 in 200 years. Samantha[1] picked up on the time scale issue to return to her concern about the integrity of large geological structures over hundreds of years:

... they can't possibly foresee whether it's going to ... whether something else might happen ... I mean they won't have done a seismographic on the whole of the section because it's too expensive ... they'll have only done it on a few parts.

(Samantha, MS3:15)

---

1 Samantha was one of the more vocal of the participants. She was a 21 year old administrator with no particular prior socio-economic or educational reason for her high level of engagement in the issue being discussed that we could ascertain.

Questions were also asked about the subsurface potentially changing and its subsequent impact on storage integrity. These questions were, in general, not answered in detail due to a lack of time, a change of focus in the conversation or simply that the expert speaker could not address all the points raised. Samantha, in particular, expressed concern over the level of uncertainty and the difficulty of extrapolating findings from small test sites over small time periods to potentially huge sites over hundreds of years.

Technical fix Several participants were concerned that CCS would result in society becoming complacent in addressing other ways of reducing CO_2 emissions because the problem would be perceived as having been 'fixed'. The group pointed to the, in general, low level of recycling in the UK as evidence of 'laziness' in responding to environmental problems. Any 'solution' which meant that individuals or other sections of society did not have to make wider changes would allow such 'laziness' to continue. It was widely felt in the panel that CCS might well constitute such a 'technical fix' that would stop or delay other desirable actions and steps.

> Sue: ... if we decide that we were going to do this [CCS] it would be easier ... because we're lazy to an extent aren't we? ... We don't recycle enough ... It's been taken out of our hands and somebody else is doing it for us then ...
>
> Facilitator: Is that good or bad?
>
> Sue: It's a bad thing because that's us being lazy.
>
> Heather: Because people would just think, like a quick fix think, 'oh, they've fixed it, so I don't have to do anything'.   *(MS3:20)*

A further development of this argument was that CCS might even create a 'false sense of security' in our ability to cope with climate change and carbon abatement and that this could, perversely, result in an increase in CO_2 emissions.

> Elizabeth: ... people will just think, 'oh, well that's alright then, that's kind of been fixed ... There's this invisible body out there that's taking care of it ... I'll just shove my carbon emissions ...'. We'd become more blasé wouldn't we just like the car drivers? [driving more dangerously in response to safer design and seat belts]. We'd think, well, this has all been sorted.
>
> Heather: Maybe we just shouldn't tell anyone about it then. Maybe it should be a secret?
>
> Samantha: ... can you imagine if there was a leak or something and everyone would go, 'oh my god, the government are hiding this'.   *(MS:21–22)*

This above exchange shows a 'risk compensation' type argument being used (Adams 1995): as peoples' perceptions of the risks of carbon emissions are potentially reduced due to application of CCS, they may compensate by higher CO_2 emitting behaviours and lifestyles. The idea of withholding information from the public to avoid such CO_2 compensation was seen by many other participants as highly risky for the government's image. It was generally agreed within the group that individuals were partly to blame for CO_2 emissions and therefore had to share some of the responsibility for reducing CO_2 emissions into the atmosphere. It was also felt that individuals *could* make a change in their lifestyles, but that some sort of 'crisis mentality' might be required before they would be prepared to do so.[2]

A shift to more uncertain and ambiguous perceptions A critical presentation by an academic energy expert followed that by an oil industry representative in session 4, and challenged the panel members to re-think their earlier generally positive endorsement of CCS. For example:

> Last week to me it [CCS] sounded really good, it sounded like the only option (…) and then this week it just sounds as though there's more options, (…) and then are we prepared, as a day to day person doing your day to day job … are we prepared to make sacrifices we would have to make, like you say if we want to reduce that 60 per cent, we're not going to be able to do it are we?   *(Sue, MS4:10)*

This comment was elicited in response to the academic's analysis of the benefits of greater energy efficiency compared to CCS (including the energy penalty associated with the latter). Thinking about CCS, and its energy penalty, in a systems way relative to energy efficiency was not a perspective that had emerged within the group prior to then. The second part of Sue's quote refers to the extent to which individuals would be prepared to change their lifestyles (possibly dramatically) in order to reduce CO_2 emissions and hence avoid the need for extensive reliance on CCS. This need for lifestyle change had been an implicit part of the academic's preferred approach. The oil industry official had stressed that achieving a 60 per cent reduction in CO_2 emissions was a huge challenge and had made the argument that people had to decide between options such as nuclear and CCS. Without major new low- or zero- carbon dioxide emitting supply side options, he argued that it would be necessary for a large scale reduction in energy consumption, which would require a major shift in the use of energy and in lifestyles.

He asked the panel whether they could envisage reducing CO_2 emissions arising from their own lifestyles by 50 per cent. Several participants, e.g. Sue in the above quote, acknowledged in response that such lifestyle changes could be difficult to envisage. Such doubts about the feasibility of lifestyle changes made the comparison

---

2   The group (and the facilitator) did not make any explicit distinction in the above discussion between 'individual' behaviours and behavioural change, and more 'collective' behaviours and their potential for change. On the other hand, the group were implicitly distinguishing between actions taken voluntarily by individuals (e.g. recycling, or reduction of energy consumption, use of public transport, etc.) and the actions of government which did not involve voluntary change by individuals. Some of the tension in the discussion reflected the need for collective governmental actions given individual apathy on the one hand, but the suspicion of policy by dictate from the centre on the other.

of options more complex and uncertain for the group than for the academic presenter, who was more confident in the prospects of energy demand reduction through regulatory, legislative, fiscal and lifestyle changes.

In effect, the industrialist raised for the group an uncertainty about how effective household or individual actions to reduce CO_2 emissions by a large proportion of the population would be *in reality*. The panel had itself raised similar concerns during session three. Whilst there appeared to be admiration for the academic's belief in energy demand reduction, the industrialist perhaps validated the view on the panel that energy demand reduction was very difficult in current socio-economic conditions.

As a consequence of having been presented with an assessment of the alternatives to CCS, together with information on their some what unknown effectiveness at the present time, the panel was left more uncertain and undecided than they had been after session 3, at which point there was a reasonably positive position on CCS. Furthermore, the fact that both experts were perceived by the panel as open-minded and not overtly pushing a single option also highlighted the uncertainty associated with choosing an appropriate option.

> I think we all expected these guys to come in and go 'right its green, green, green, green', ... and you to come and go 'you've got to do it this way ... there's no other way'. I think that's what we expected, I know I did, I didn't expect anything else but its more of an option. *(Sue, MS4:10)*

Towards the end of Session 4, a 'rapprochement' between CCS and lifestyle change as options for reducing CO_2 emissions appeared to emerge in the thinking of several participants.

> ... if we were going to do the storage way [i.e. CCS], could we not do both? Could we, as the general public, do more [in reducing energy use] so that you wouldn't have to dig up as much ground [for laying CO_2 pipes] ... *(Sue, MS4:9)*

> ... that would be a bit of a sight wouldn't it? Everyone working together ... we'll have 50 per cent of it [CCS], we'll have 50 per cent of that [demand reduction]. *(Sue, MS4:11)*

The industrialist certainly encouraged such a consensual approach, but Sue herself, and other panel members, recognized the practical difficulties associated with such 'joined-up' thinking and working.

> ... they're usually working against each other aren't they ... Greenpeace against whoever, but then it would be nice for all to work together and say well if we do it together we'll get more out of it. *(Sue: MS4:11)*

Emergence of three positions For most panel members, the extent of the challenge had impressed upon them the difficulty of achieving a 60 per cent reduction by lifestyle change alone.

I certainly thought about it a lot this week, really ... I've had a lot of discussions with (...) but it seemed interesting to me, people who're not involved in it – I've had the same reaction, I changed my view (...), once I'd listened to [the industrialist], and then it came across that we're not all going to be able to do it on our own, ... we've gone that far.

I don't think you're going to get as dramatic change in everyone's lifestyle as we want to ... [get] each household [to reduce CO_2] by 60 per cent.   *(Sue, MS5:2)*

Whilst most participants shared this somewhat negative assessment of lifestyle change on the part of the public, there were perhaps three categories of response to it.

1. The view that because of the reasonably long timescales involved, it would still be possible for the 60 per cent reduction to come from lifestyle change and introduction of energy efficiency and renewable energy technologies. This group was not in favour of CCS, because it was placing CO_2 out of effective control and taking an unnecessary risk if the 60 per cent reduction could be done through safer means.

    I would prefer to do it ourselves rather than store anything underground that I can't see and can't control, don't know what's happening with it.   *(Sue, MS5:6)*

    Those holding this view were in favour of measures imposed upon companies, councils and, ultimately, households to make sure that emissions of CO_2 were steadily brought under control, i.e. incentives, fines, taxation instruments, etc.

2. Against this argument was the position that control is in any case 'in your head' (meaning that it is a psychological belief that the carbon dioxide has been safely stored and controlled and that you do not necessarily need to see something literally to have an understanding of it, or confidence in it). Holders of this view saw the benefits of CCS as outweighing the risks. A worst case-scenario, for this group, would involve CO_2 escaping from storage reservoirs in large volumes, but it was argued that this CO_2 would have been in the atmosphere in any case.

3. A third view was one which expressed more ambivalence about the role of CCS. Like perspective (1) there was strong support for government to take more action on energy efficiency, demand reduction and renewable energy. Any support for CCS was tempered by the perception that fossil fuels are, in any case, running out, hence other sources of energy are going to be required. In addition, it was felt that the risks of CCS are somewhat too high and uncertain at the current time and that there need to be more definitive answers to questions about the various risks encountered. However, a possible role for CCS as part of a wider package of decarbonisation measures was identified, provided that some of the key questions about the risks could be better addressed.

The York Group

Our discussion of the York Panel highlights some of the key areas of difference from the Manchester Panel, hence is some what briefer. The concept of using CCS as a 'bridging strategy' to more renewable energy was presented by one of the expert witnesses and well understood and fairly well supported by the panel. However the need for longer term solutions was also stressed.

> It [CCS] just bides time to figure out, if it doesn't work or if we get a bit more time ... hopefully get some more ideas about how to sort it out ... it just does buy a bit more time doesn't it even if we don't put much [CO_2] down there?   *(Andrew YS2:18)*

The majority of the group were reasonably supportive of CCS due to the significant impact it could have over a short period of time, and to the fact that no other technology could do this.

> It's not an ideal solution but it looks like a solution that can achieve drastic reductions in the short term.   *(Russell YS5:6)*

Safety issues were ranked consistently highly by the entire group, and it was agreed that minimum standards would need to be met before any project was undertaken.

> If you can lock it [the CO_2], fine, I can understand why they are putting so much time in to this technology. If you can't, if there's still potential of it leaking, that strikes me as a bomb, because if it goes pop ...   *(Mark YS2:18)*

As in the Manchester group, images of explosions such as this were common throughout the panel discussions despite expert witnesses informing them of the inert nature of CO_2. CCS was also perceived as an 'end of pipe' technology and this was discussed frequently as a negative attribute by the panel.

> We are treating the symptoms not the causes [of] doing this.   *(Mark YS2:21)*

The uncertainty associated with this technology (given that it has not been used on a long term basis or on a wide scale before) generated a number of concerns throughout the meetings.

> In one hundred, two hundred years time we may actually be paying the penalty for putting a pollutant back in to the earth ... It could do anything couldn't it? ... but that is the risk that we're taking today isn't it, doing all of this?   *(John YS2:15)*

This uncertainty and the long term implications conjured-up analogies with previous events.

> ... that's what they did with asbestos ... suddenly they did a test for it and it has become catastrophic really ... we have used lots of chemical products that we've found out after the time – we shouldn't have done it.   *(John YS2:15)*

It was stressed by one participant that the idea of CCS needs to be very carefully and thoroughly explained if it is to be supported. If it is not properly assessed, with input from local participation, then he suggested that there could be serious consequences:

> I can imagine if it was discussed quite widely in the community there would be a lot of alarm. These things do provoke that sort of reaction. *(Andrew YS2:15)*

CCS (in particular) and also the Government's approach to climate change (in general), were frequently criticised for not being radical enough, though the panel was not well informed about what government is actually doing.

> Have the government tried to sell any green issues apart from a few silly adverts about filling your kettle? They don't seem to do much. *(Graeme YS2:4)*

> They don't seem to be thinking outside the box do they? They are actually trying to find somewhere to store it, to keep it, when surely can we not try to use it [CO_2] as another form of energy that dissipates naturally creating heat or whatever. *(John YS2:22)*

The above quote is a prime example of how several members of this, rather technically minded group, were very keen to suggest solutions to problems but were not able to explain the feasibility of such solutions or how they might be implemented in practice. These few individuals tended to assume that their suggested technologies or approaches were feasible but had simply not been thought of by others, or had not been the subject of sufficient research activity. The idea of using CO_2 in the process of making something else was often discussed by the panel. This technological optimism demonstrates a 'utilitarian' type approach, preferring to re-use and recycle CO_2, rather than just storing it. For this reason the York panel was strongly in favour of using CO_2 for Enhanced Oil Recovery (EOR). We suspect that the emphasis and interest of these panellists in novel technologies not having being explored sufficiently by 'the experts' reflects a distrust in those experts and in their capacity to be truly innovative. A number of moral and emotional arguments against CCS were also expressed, though by only two participants.

> Our deep irrational fears ... , I think you've come up with deep irrational fears about injecting mother earth. I think that is an irrational fear but I feel it as well.
> *(Graeme YS2:25)*

These concerns are often linked for these few participants with a sense of responsibility to not cause problems for future generations.

> I just don't like the idea about pumping another pollutant back in to the earth, we're doing it all the time aren't we? We are storing up another problem for the future. *(John YS2:23)*

There was widespread support for what one participant referred to as an 'encyclopaedia of facts' that would be presented in a digestible format for the public and the media. Some members of the group suggested that the encyclopaedia could include different opinions on the same issue to allow for inevitable disagreement and uncertainty. The groups were asked to consider how trust in the information presented in the encyclopaedia could be developed. One suggestion was that it would be a 'living document' where the encyclopaedia could be 'challenged' by lay members of the public or experts. Information would have to be given to defend an entry in the encyclopaedia or that entry would have to be altered. It was stressed that this needed to be a very visible process if the public was to be convinced that the encyclopaedia was a reliable and unbiased source of information. How this 'challenging process' could actually be carried out was also discussed. Some support for using the internet was expressed, as was some severe scepticism of the internet as it is not felt to include everybody. The use of moderated web-based discussion fora was also strongly supported by one member of the group.

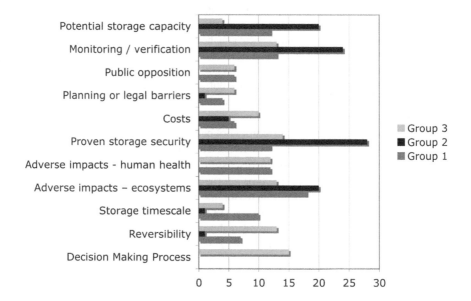

Figure 5.6   Weighting of criteria for assessment of geological storage sites by sub-groups (1–3) in York Citizen's Panel

In contrast to the Manchester Panel, the York Panel included a formal weighting process by which the participants rated a number of criteria for assessing geological CO_2 storage sites. These same criteria had been developed and weighted by a number of expert geologists (Gough and Shackley, 2006). The panel split into three sub-groups to conduct this weighting exercise (see Figure 5.6). We attempted to create a broadly 'pro-CCS' sub-group, an 'agnostics' sub-group and a 'doubters' sub-group, drawing upon our knowledge of the opinions of Panel members. There was considerable variation between the three sub-groups, though the pattern of the scoring showed many similarities across the sub-groups. The same criteria tended to score most highly across all three groups, namely proven storage capacity, monitoring and verification, proven storage security, adverse impacts – human health, and adverse impacts – ecosystems. Generally less important criteria were public opposition, planning and legal barriers, costs and storage time scale. The explanation for the low weighting of public opposition was not that the sub-groups necessarily regarded public opposition as unimportant, but rather that opposition was seen by all three sub-groups as politically manageable, or as some thing which could be ignored. As representatives of two sub-groups put it:

> Public opposition was something that we thought would be there but could be managed.
> *(George YS4:2)*

> Being cynical we felt that public opinion could be swept aside, as recent examples show [referring to the large demonstrations against the war in Iraq in 2003].   *(David YS5:5)*

For at least one sub-group costs, reversibility and storage time scale were of moderate importance. One sub-group also added the criterion of decision-making processes and gave it a moderately high weighting. It is interesting to note that all three sub-groups rated the security, safety and risk criteria most highly (and despite the different overall perceptions of CCS within the three sub-groups). This reinforces the finding from the survey work that public perceptions of CCS will be strongly influenced by the extent to which potential risks and hazards have been seen to be properly identified, assessed and appropriate responses and safety measures implemented.

The criteria weights can be compared to those from a group of six expert stakeholders (three geologists, one academic, one NGO representative and one local policy official) engaged in a similar exercise reported in Gough and Shackley (2006). There was strong convergence between the citizens' weightings and those of the stakeholders with respect to four highly weighted criteria: monitoring/verification, proven storage security, potential capacity and ecosystem impacts. There was also convergence with respect to one lowly weighted criterion – costs, and for two moderately weighted criteria: human health impacts and storage timescale. The stakeholders tended to weight planning or legal barriers and public opposition more highly than the citizens'. There appears, therefore, to be agreement across the experts and citizens that safety, monitoring, capacity and ecosystem impacts are all very important issues in selecting appropriate CCS storage locations; and furthermore that cost issues are very much secondary in site selection. Disagreement between the experts and stakeholders relates in particular to the importance of public opposition and planning issues. The citizens' tended to regard opposition and planning as potential barriers that are relatively easily over-come by governments and companies, whereas the stakeholders considered such barriers as a more serious issue. This may reflect the extent to which the lay public lacks confidence in its own capacity (through opposition) to effect change in the routine practices of decision-making by government and companies. Alternatively, it may reflect the political and policy sensibilities of the stakeholders, who are keen to acknowledge the role of public opinion in the context of debates over scientific and technological risks such as GMOs and BSE.

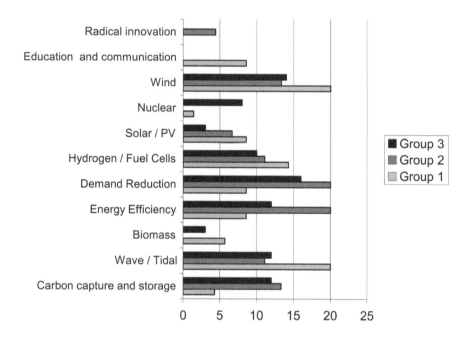

Figure 5.7   Suggested allocation of RD&D resources for different low-carbon energy options by sub-groups (1–3) in York Citizen's Panel

The same three sub-groups were also asked to allocate points to represent the level of funding support for a range of carbon mitigation options (see Figure 5.7). This was used as a surrogate indicator for the mitigation options which they would most like to see developed in order to reduce CO_2 emissions, though there are potential problems with this (e.g. some options could be perceived as already having received large amounts of funding in the past, hence less deserving now, even though perhaps still important for carbon abatement). There had been a reasonably detailed discussion of these different options during Session 3 with the two external experts. In many respects the exercise reinforces the findings from the survey, namely that there is a strong preference for energy efficiency, demand reduction and renewable energy. The sub-groups allowed a more detail discussion than in the survey vis-à-vis different renewable energy options. Solar (photovoltaics) does rather poorly because it was perceived to have less utility in the UK. The effectiveness of biomass was regarded as not yet proven (the strength of this opinion may have been stronger in the York area because of its close proximity to the failed Arbre biomass energy project). Nuclear does reasonably well for one sub-group, which considered that nuclear technology might be necessary in the long-term 'solution' of climate change, but very poorly for the other two (which only supported funding to close down the nuclear industry!). CCS received fairly high ratings from two of the sub-groups, whilst the 'doubters' sub-group gave a moderately low score (though still more than nuclear).

There appeared to be greater consensus between the three sub-groups on the role of wind, energy efficiency, demand reduction, wave and tidal energy than on the possible role of CCS. This may reflect the greater focus in the Panel on CCS rather than on a detailed examination of the impacts of the other carbon mitigation options. However, many of the other options are already implemented and hence their impacts more widely documented and understood. It is noticeable that two of the sub-groups scored CCS at approximately the same level as hydrogen, wave and tidal, two other technologies which are less well known and with which there is minimal experience. Hence, the higher scores for wind, demand reduction and energy efficiency may reflect to some extent familiarity through implementation.

5.5   Summary and Discussion

The citizen panel participants had the advantage over the survey respondents of lengthy discussions between themselves and with expert witnesses. Their ability to cross-examine experts does appear to have influenced their perceptions and to have provided some greater reassurance on the potential risks than was available to the questionnaire respondents. This might, however, be a function of the particular experts chosen and the panel might have responded differently if a 'sceptical geologist', for instance, had spoken to the group, i.e. one who might have posed more sceptical questions about the integrity of geological reservoirs for storing CO_2. Leakage of carbon dioxide from reservoirs was a commonly expressed concern amongst both panels, as was the perception of CO_2 as a potentially explosive substance. In their interactions with the panels, the experts tended to present the risks of leakage (and adverse consequences if it should occur) as low to very low.

The questionnaire highlighted leakage as *the* major issue of concern, whereas leakage was only one amongst a range of concerns discussed within the citizen panels and, on balance, not the overriding one. We speculate that the reason for this is that the citizen panels were first presented with information about CCS from an academic geologist who shaped the context for discussion. Not only did this expert raise many other issues that were less 'intuitive' than leakage (e.g. providing information on the time scales over which natural gas can remain in geological formations) she also allayed fears of leakage by characterising them as low to very low (provided appropriate controls are in place). Many of the survey respondents, bereft of the inputs from the expert geologist, may have employed a cognitive model in which the principal risk arising from storage of a gas underground is the escape of that gas back to the atmosphere. Support for the existence of such a cognitive model is provided by panel participant Sue, who picked-up on the geologist's use of the term 'bubble', possibly using it to construct a cognitive model of CO_2 underground in a gaseous form ready to escape. The expert geologist was quick to revise this respondent's impression of a CO_2 bubble waiting to burst.

The two Citizen Panels, whilst composed of very different demographic samples, came to rather similar views about CCS, albeit it through sometimes different reasoning processes and arguments. In both Panels there was a reasonable level of consensus surrounding the potential need for CCS given the scale of the decarbonisation challenge and the uncertainty and difficulty of achieving a 60 per cent reduction in emissions through behavioural and lifestyle change and other routes. Support for CCS was conditional upon the implementation of a range of other decarbonisation options – in particular renewable energy and energy efficiency. An integrated approach towards decarbonisation was generally preferred by both Panels in which all options were considered, including social change as well as the 'harder' technological options. There were rather similar concerns regarding the possible risks, though the York group relied more upon 'analogous' cases of environmental and health and safety risks than the Manchester group. The results of the Citizen Panels are also broadly consistent with the findings from the survey and we can describe support for CCS as 'moderate' or 'lukewarm' compared to strong support in general for wind, solar and energy efficiency. From the citizen panels and questionnaire we suspect that gender, socio-economic status and education all play a role in influencing perceptions of CCS, though our analysis suggests that their role is not large.

On the basis of our analysis of, and reflection upon, the findings of the citizen panels and survey, we would suggest that a basic concern about climate change and recognition of the need for massive CO_2 emission reductions might well be a prior requirement for the consideration of CCS as a legitimate option for evaluation. We suggest that there are (at a minimum) three conditions which provide the context for regarding CO_2 capture and storage as a potential option:

- acceptance of the basic underlying science of human-induced climate change;
- acceptance of the seriousness of the potential threat of climate change impacts to society and the environment in the UK and more generally; and
- acceptance of the need to make very large reductions in carbon emissions (e.g. 60 per cent cuts) over the next 50 years.

The survey found that not only was there a high level of belief that human activities are causing climate change (78 per cent) but that the majority of respondents (62 per cent) were 'concerned' or 'very concerned' about climate change. This confirms the findings of many existing surveys from the UK, and Europe more generally, which have indicated fairly widespread concern over the problem of global climate change, and a prevailing feeling that the negative impacts outweigh any positive effects (Poortinga and Pidgeon, 2003; Eurobarometer, 2003; Hargreaves *et al.*, 2003; Shackley *et al.*, 2001). On the other hand, some survey and focus group research in the UK suggests a less homogeneous, and more sporadic, perception of the occurrence and seriousness of global climate change (and in particular the need for large reductions in CO_2 emissions) (Darier *et al.*, 1999a; Lorenzoni 2003; Hargreaves *et al.*, 2003). In partial support of these latter findings, we found that even amongst the most climate change aware of our citizen panel participants, no one comprehended the enormous scale of the challenge of a 60 per cent reduction in carbon dioxide emissions, and there was in general a lack of awareness and knowledge of what different carbon mitigation options had to offer (cf. Curry *et al.*, 2005a; Palmgren *et al.*, 2004).

Three Broad Positions vis-à-vis CCS 'Pro-', 'Anti-' and 'Ambivalent' were identified in the Citizen Panels. The Citizen Panels elucidated broadly different perspectives on CCS which did appear to relate, at least to some extent, to underlying beliefs and different sets of values.[3] A small minority was in favour of CCS, mainly for utilitarian reasons that it is an effective use of geological reservoirs and removes CO_2 so reducing the risks of global climate change, which are regarded as larger than the risks of CCS itself. Another small minority was opposed to CCS, mainly for moral reasons that it is basically wrong to 'inject mother earth' with an industrial waste by-product. Humans have responsibility, according to this perspective, for changing their ways – through new technologies and lifestyle changes – such that CO_2 emissions are not produced in the first place.

The third, and most common perspective, was essentially ambivalent – at times in favour, at other times against, CCS. The citizen panels were opposed to regarding CCS as a single 'fix it' solution and expressed concerns that such use of CCS would be to treat the symptoms rather than the causes of climate change. There was a sense that CCS could 'let us off the hook' of making more fundamental, deep-rooted changes and this avoidance of change was perceived generally negatively. There was also concern expressed that CCS would divert R&D resources and attention away from renewable energy technologies, demand reduction and energy efficiency. This concern was largely allayed when the level of new resources being directed to renewable energy R&D, demonstration and support schemes was indicated, alongside the moderate amount going into CCS R&D at present.

Whilst many in this third group were initially sceptical of CCS for the above reasons, they became more favourably inclined as the scale of the decarbonisation challenge was appreciated, as the risks of CCS were more thoroughly discussed, and as the risks and opportunities associated with the other major decarbonisation options were also discussed. The majority view tended to find more support for CCS when the latter was combined with other options which had a (seemingly) more favourable cost-benefit profile than CCS itself, in particular renewable energy,

---

3 The exact relationship between these perspectives and beliefs and values would require precise definitions of 'beliefs' and 'values' and empirical research to identify and measure beliefs and values.

energy efficiency, energy demand reduction, and (more speculatively) the hydrogen economy (based at least initially on fossil fuels with decarbonisation). This finding strongly supports the need to embed CCS within a portfolio of decarbonisation options. It may also support promoting CCS as a 'bridging strategy' to other low- or zero-carbon energy sources.

5.6   Implications for Policy and Research Needs

Zaller (1992) argues that the lay public does not have well formed opinions on most issues which are not of immediate salience or relevance to their everyday life and livelihood. Converse (1964) similarly criticises the 'expectation of opinionness' which is an underpinning assumption of much survey research. Opinions and perceptions are, instead, shaped by (*inter alia*) the media and other marketing efforts of stakeholders. There are several very good examples of such shaping having taken place, e.g. in the case of disposal of the Brent Spa platform, Greenpeace was successful in convincing the media, and consequently the general public, that disposal at sea would incur unacceptable environmental risks (Smith 2000). A further example is the role of the media and campaign groups in shaping perceptions of GMOs in Europe in the late 1990s. Feedbacks between the media and public opinion are also documented, and have been formalised in the theory of risk amplification (Jaeger *et al.*, 2001), which maintains that risk perceptions can become amplified through media presentations, and subsequent stakeholder responses.

The implication of such theory and real-cases is that because there is not (for most) a strong *a priori* belief in favour or against CCS, public opinion on CO_2 storage could, at some future stage, be strongly shaped by stakeholder groups, including the media or NGOs, who come themselves to formulate a strong opinion. As Wynne (1995, 1996) notes, bereft of sufficient technical knowledge, the public may come to rely upon their sense of trust in the organisations involved, and in their past institutional performance, when assessing a new technology such as CCS. Research is not able to anticipate how public perceptions might change, possibly dramatically and rapidly, in response to pro-active stakeholder and media interventions and real-world events, though it can provide lessons from the past and guidance on 'good practice' in the communication of risks and uncertainty (Powell and Leiss 1997).

With the above proviso clearly in mind, the results suggest that public reactions to CCS could be reasonably supportive of the technology, provided that its purpose is well understood and that the key risks are acknowledged. This research suggests that proponents of CCS need to put their case clearly in the context of reducing the risks of global climate change, and the concomitant need for large long-term reductions in CO_2 emissions to the atmosphere. The use of CCS as part of a portfolio of decarbonisation options which range from new technologies, to lifestyle change, should be stressed, rather than presenting CCS as a 'stand alone' option. A partnership approach to control and regulation of CCS would be generally welcomed, in which government, industry and environmental NGOs each have a role to play.

With respect to public decision-making, the citizen panels could be reconvened to explore more specific CCS proposals or projects once they are on the table or in the pipeline. Alternatively, a new 'bespoke' panel could be established to discuss a specific CCS proposal or project, e.g. drawn from the local occupants and stakeholders near to the proposed project.

5.7 References

Adams, J. (1995), *Risk*, UCL Press, London.

CAA (2004), www.caa.co.uk/erg/erg_stats/default.asp.

Converse, P. (1964) 'The Nature of Belief Systems in Mass Publics', in D. Apter (ed.), *Ideology and Discontent*, Free Press, USA.

Curry, T., Reiner, D., de Figueiredo, M. and Herzog, H. (2005), *A Survey of Public Attitudes towards Energy and Environment in Great Britain*, Unpublished MS, University of Cambridge, UK.

Curry, T., Reiner, D., Ansolabehere, S. and Herzog, H. (2005a), 'How Aware is the Public of Carbon Capture and Storage?', in E. Rubin, D. Keith and C. Gilboy (eds), *Proceedings of 7th International Conference on Greenhouse Gas Control Technologies, Volume 1: Peer-Reviewed Papers and Overviews*, Elsevier Science, Oxford, pp. 1001–1009.

Darier, E., Gough, C., de Marchi, B., Funtowicz, S., Grove-White, R., Kitchener, D., Pereira, A., Shackley, S. and Wynne, B. (1999), 'Between Democracy and Expertise? Citizen's Participation and Environmental Integrated Assessment in Italy (Venice) and St.Helens (UK)', *Journal of Environmental Policy and Planning*, **1**(2), pp. 103–21.

Darier, E., Shackley, S. and Wynne, B. (1999a), 'Towards a "Folk Integrated Assessment" of Climate Change?', *International Journal of Environment and Pollution*, **11**(3), pp. 351–72.

Department of Work and Pensions (2003), *Family Resources Survey 2002-2003*, available at www.dwp.gov.uk/asd/frs.

Energy White Paper (2003), *Our Energy Future: Creating a Low Carbon Economy*, Cm5761, HMSO, London.

Environment Audit Committee (2003), *Budget 2003 and Aviation, Ninth Report of Session 2002–3*, House of Commons, London.

Eurobarometer (2003), *Energy Issues, Options and Technologies: A Survey of Public Opinion in Europe*, Energy DG, European Commission, Brussels.

de Figueiredo, M.A. (2002), *The Hawaii Carbon Dioxide Ocean Sequestration Field Experiment: A Case Study in Public Perceptions and Institutional Effectiveness*, Masters Thesis, Massachusetts Institute of Technology, Boston.

Gough, C. and Shackley, S. (2006), 'Towards a Multi-Criteria Methodology for Assessment of Geological Carbon Storage Options', *Climatic Change*, **74**(1–3), pp. 141–174.

Gough, C., Taylor, I. and Shackley, S. (2002), 'Burying Carbon Under the Sea: An Initial Exploration of Public Opinion', *Energy & Environment*, **13**(6), 883–900.

Hargreaves, I., Lewis, J. and Speers, T. (2003), *Towards a Better Map: Science, the Public and the Media*, Economic and Social Research Council, Swindon, UK.

House of Lords (2000), *Science and Society*, Third Report, House of Lords Science and Technology Select Committee, Houses of Parliament, London.

Huijts, N. (2003), *Public Perception of Carbon Dioxide Storage*, Masters Thesis, Eindhoven University of Technology, Netherlands.

Hunt, J. and Wynne, B. (2000), *Forums for Dialogue: Developing Legitimate Authority through Communication and Consultation*, A Contract Report for Nirex, University of Lancaster, Lancaster.

IEA GHG (2004), *Report of Risk Assessment Workshop held in London, UK, February 2004*, IEA Greenhouse Gas R&D Programme Report No. PH4/31 July, Cheltenham.

Irwin, A. and Michaels, M. (2003), *Science, Social Theory and Public Knowledge*, Open University Press, Milton Keynes.

Itaoka, K., Saito, A. and Akai, M. (2005), 'Public Acceptance of CO_2 Capture and Storage Technology: A Survey of Public Opinion to Explore Influential Factors', in E. Rubin, D. Keith and C. Gilboy (eds), *Proceedings of 7th International Conference on Greenhouse Gas Control Technologies. Volume 1: Peer-Reviewed Papers and Overviews*, Elsevier Science, Oxford, pp. 1011–1019.

Jaeger, C., Renn, O., Rosa, E. and Webler, T. (2001), *Risk, Uncertainty and Rational Action*, Earthscan, London.

Kasemir, B., Jager, J., Jaeger, C. and Gardner, M. (eds) (2003), *Public Participation in Sustainability Science: A Handbook*, Cambridge University Press, Cambridge.

Lorenzoni, I. (2003), *Present Futures, Future Climates: A Cross-Cultural Study of Perceptions in Italy and in the UK*, PhD Dissertation, University of East Anglia, Norwich, UK.

Nogasplant website: http://www.nogasplant.co.uk.

Overwyrefocus website: http://www.overwyrefocus.co.uk/gas_storage/gas_articles.htm.

Palmgren, C., Granger Morgan, M., Bruine de Bruin, W. and Keith, D. (2004), 'Initial Public Perceptions of Deep Geological and Oceanic Disposal of CO_2', *Environmental Science and Technology* **38** (24), pp. 6441–50.

Poortinga, W. and Pidgeon, N. (2003), *Public Perceptions of Risk, Science and Governance*, Centre for Environmental Risk, University of East Anglia, Norwich, UK.

Powell, D. and Leiss, W. (1997), *Mad Cows and Mother's Milk: The Perils of Poor Risk Communication*, McGill-Queen's University Press, Montreal.

RCEP (1998), *Setting Environmental Standards*, 21st Report of the Royal Commission on Environmental Pollution, London.

Shackley, S., Kersey, J., Wilby, R. and Fleming, P., (2001), *Changing by Degrees: The Potential Impacts of Climate Change in the East Midlands*, Ashgate, Aldershot.

Shackley, S., McLachlan, C. and Gough, C. (2004), *The Public Perception of Carbon Dioxide Capture and Storage*, Tyndall Working Paper 44, Tyndall Centre, Manchester.

Smith, J. (2000), 'After the Brent Spa', in J. Smith (ed.), *The Daily Globe: Environmental Change, the Public and the Media*, Earthscan, London.

Upton, G. and Cook, I. (1997), *Understanding Statistics*, Oxford University Press, Oxford.

Wynne, B. (1995), 'Public Understanding of Science', in S. Jasanoff, G. Markle, J. Petersen and T. Pinch (eds), *Handbook of Science and Technology Studies*, Sage, Thousand Oaks, CA, pp. 361–88.

Wynne, B. (1996), 'May the Sheep Safely Graze? A Reflexive View of the Expert-Lay Knowledge Divide', in S. Lash, B. Szerszynski and B. Wynne (eds), *Risk, Environment and Modernity*, Sage Publications, London, pp. 44–84.

Zaller, J. (1992), *The Nature and Origins of Mass Opinion*, Cambridge University Press, Cambridge.

Chapter 6

A Regional Integrated Assessment of Carbon Dioxide Capture and Storage: North West of England Case Study

Simon Shackley, Karen Kirk, Carly McLachlan, Clair Gough
and Sam Holloway

6.1   Introduction

In this chapter we undertake an integrated assessment of CCS in the North West region of England. This integrated assessment consists in addressing the following questions.

1. What is the potential for CO_2 storage in oil and gas field and saline aquifers beneath the East Irish Sea? (The adjacent offshore area with CO_2 storage potential).
2. How do different energy scenarios of the North West region for 2050 compare in terms of their perceived benefits and disadvantages against a set of pre-defined criteria?
3. How do different types of stakeholders (private sector, public sector, non-governmental organisations) evaluate the scenarios and what does this tell us about those stakeholders and their thought processes regarding different energy futures?

In addition to the geological assessment we have used a multi-criteria assessment (MCA) methodology (Stirling and Mayer, 2001; Stewart and Scott, 1995; Brown *et al.*, 2001) to examine the trade-offs between different scenarios of the future of the energy system regionally. Each of the energy scenarios has a different role for CCS, ranging from no contribution, to a major contribution to 2050. We decided to focus the study upon a sub-national region because it is a scale for governance which can, potentially, overcome the problems that have arisen between the centre and the local levels in the UK in the last few decades (Stoker, 2004; cf. Wilbanks and Kates, 1999). For example, the region has a specific characterisation in terms of its portfolio of power stations, its opportunities for renewable energy development and in terms of the availability and closeness of suitable off-shore geological storage sites for CO_2. A regional focus also reduces the complexity of considering energy scenarios at the national scale, for example the respondent can focus upon a handful of power stations rather than having to grapple with hundreds of power stations at the national scale.

We created the framework of the scenarios, and the criteria for their assessment, through an earlier project, which is described elsewhere (Gough and Shackley, 2006). It was necessary to use scenarios of the energy system because: a) we were looking at the long-term (to 2050) and over these periods of time the energy system will change, possibly dramatically; b) given that CCS is just one element in a complex energy system, it is necessary to create alternative visions of the relative extent to which CCS will be employed in a new energy system. We begin with the geological assessment of the East Irish Sea for CO_2 storage.

6.2   Characterisation of the Rocks Beneath the East Irish Sea

Reservoir Unit

The main reservoir rocks in the East Irish Sea Basin form the Sherwood Sandstone Group. The Sherwood Sandstone Group extends westwards over most of the East Irish Sea Basin from onshore UK (Figure 6.1), and is the equivalent of the Bunter Sandstone Formation in the southern North Sea. It is more than 2000 m thick in the centre of the East Irish Sea Basin and has an average thickness of 1450m (Jackson *et al.*, 1987).

Most of the hydrocarbon discoveries are entirely within the uppermost unit of the Sherwood Sandstone Group; the Ormskirk Sandstone Formation. Discoveries also extend downwards into the upper parts of the St Bees Sandstone formation. The top of the Ormskirk Sandstone Formation lies at depths of 250–3000m. It has an average thickness of 250m. The Ormskirk Sandstone Formation demonstrates all of the required characteristics for CO_2 storage including closed structures (traps for buoyant fluids), high porosity and permeability and it is overlain by an effective seal, the Mercia Mudstone Group.

Porosity and Permeability of the Ormskirk Sandstone

There are huge porosity and permeability variations in the Ormskirk Sandstone. These are the result of diagenesis – the post-burial alteration of the original sandstone (Levison, 1988; Meadows and Beach, 1993). Porosities range from 8–30 per cent, and permeabilities from 0.05–10000 mD. The most important diagenetic effect is the precipitation of illite within the pore spaces. Thin illite crystals (small clay plates) can grow perpendicular to the sandstone grain faces, making it more difficult for fluids to pass through the reservoir (see Figure 6.2). They clog up the pore throats (pathways between pore spaces) affecting the permeability rather than the porosity of the reservoir (Ebbern, 1981).

The illite-affected layer in the South Morecambe field is >304 metres thick in the north of the field, but only about 137 metres thick in the south of the field, where it passes downwards into sandstones characterised by poorly developed fibrous illite.

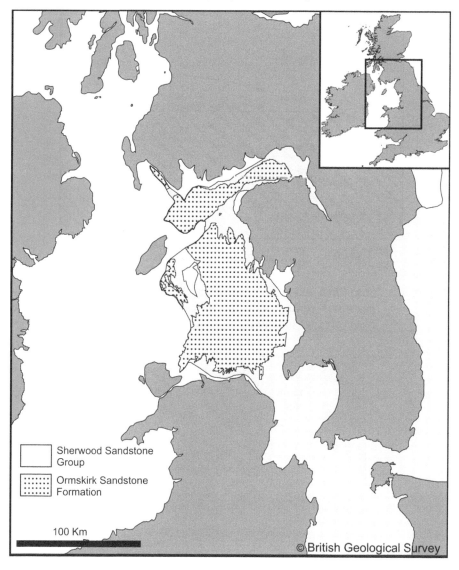

Sherwood Sandstone
Group

Ormskirk Sandstone
Formation

100 Km

© British Geological Survey

Figure 6.1   Map showing the extent of the Sherwood Sandstone Group and the Ormskirk Sandstone Formation

It is not, therefore, completely ubiquitous outside the oil and gas fields, giving hope that there might be reasonable permeability in some of the non-hydrocarbon-bearing closures. However, the presence of platy illite in the Sherwood Sandstone may greatly limit the amount of CO_2 which can be stored within the non-hydrocarbon-bearing closed structures that are described below.

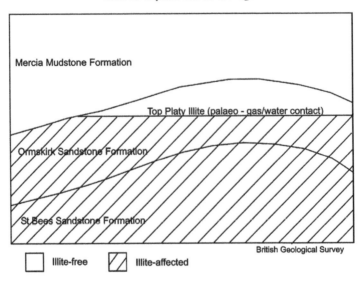

Figure 6.2   Simplified diagram illustrating the relationship of the palaeo-gas/water contact with the illite-free/illite-affected areas of the reservoir (not to scale)

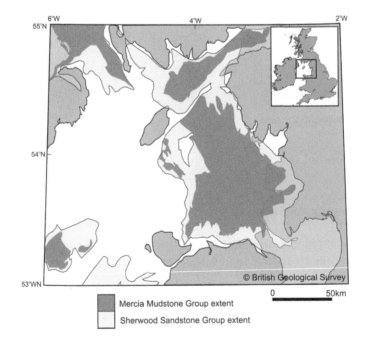

Figure 6.3   Extent of the Mercia Mudstone Group in and around the Irish Sea (adapted from Jackson *et al.*, 1995)

Caprock/seal

The Mercia Mudstone Group forms an effective seal over the top of the Ormskirk Sandstone (Figure 6.3); this is proven by the hydrocarbon discoveries in this area. Up to 3200m thick in the East Irish Sea Basin (Jackson *et al.*, 1987), it comprises silty mudstones interbedded with commonly thick units of halite (rock salt). Rock salt comprises some 35 to 55 per cent of the Basinal Mercia Mudstone succession, and occurs at 5 levels (Jackson *et al.*, 1995). It is almost impermeable unless fractured.

6.3 CO_2 Storage Potential

The major CO_2 storage potential in the East Irish Sea Basin is within:

- hydrocarbon fields;
- structures within which gas could potentially be trapped but which do not contain hydrocarbons (reservoir rocks that contain saline water in their pore spaces; the so-called saline aquifers).

CO_2 Storage Capacity of the Oil and Gas Fields in the East Irish Sea Basin

Both oil and gas fields are found in the East Irish Sea Basin (Table 6.1 and Figure 6.4), indicating that reservoir rocks present are capable of storing buoyant fluids. Oil and gas generation probably started in Jurassic times, but the Jurassic accumulations of oil and gas that are believed to have developed in most of the fields are thought to have escaped during the first period of basin uplift and erosion – at the end of the Jurassic times (Bastin *et al.*, 2003; Stuart and Cowan, 1991; Cowan and Boycott-Brown, 2003; Yaliz and Chapman, 2003; Yaliz and Taylor, 2003; Yaliz and McKim, 2003). The oil and gas presently found, for example, in the South Morecambe gas field was probably trapped in late Cretaceous to early Tertiary times (Bastin *et al.*, 2003) and thus may have been stored for some 50 million years or so. Table 6.1 shows the estimated CO_2 storage capacity of the gas and oil fields in the East Irish Sea Basin. The storage capacity of the gas fields was estimated according to the Equation 6.1 (Wildenborg *et al.*, 2004):

Equation 6.1

$$M_{CO2} = (V_{GAS} \,(stp) \,/\, Bg) \times \rho CO_2$$

Where:
M_{CO2} = CO_2 storage capacity (10^6 tonnes)
stp = standard temperature and pressure
V_{GAS} (stp) = volume of ultimately recoverable gas at stp (10^9 m^3)
Bg = gas expansion factor (from reservoir conditions to stp)
ρCO_2 = density of CO_2 at reservoir conditions (kg m^{-3})

The volume of ultimately recoverable gas at standard temperature and pressure and the gas expansion factor were obtained from the DTI oil and gas website, Meadows *et al.* (1997) and Gluyas and Hichens (2003). The density of CO_2 was calculated from the reservoir temperature and pressure of the individual fields. The main problem with Equation 1 is that it assumes that all the pore space originally occupied by the ultimately recoverable reserves of natural gas could be filled with CO_2. There is uncertainty about this because the reservoir may compact slightly as the pore fluid pressure within it decreases. Moreover water invasion may reduce the pore space available for CO_2 storage. Ideally, the percentage of the pore space originally occupied by natural gas that could subsequently be filled with CO_2 would be determined using numerical reservoir simulations. However, no reservoir simulations were available and in their absence the following factors have been used to adapt Equation 1 (from studies by Bachu and Shaw (2003) on oil and gas fields in Alberta).

1. In gas fields with depletion drive, i.e. those where the wells are opened up and the pressure in the gas field simply depletes as it would if the gas were being produced from a sealed tank, it is assumed that 90 per cent of the pore space could be occupied by CO_2.
2. In gas fields with water drive, i.e. those where water encroaches into the pore space formerly occupied by the produced natural gas reserves, it is assumed that 65 per cent of the pore space could be occupied by CO_2.

The Douglas oil field has no gas cap and is being water-flooded to maintain reservoir pressure during production. This means that little of the pore space formerly occupied by the produced oil reserves will be available for CO_2 storage when water flooding ceases. The absence of a gas cap does not necessarily imply that the field is not gas tight. Gases originally present are believed to have been removed by a process called water washing, in which gas is dissolved by ground waters causing a low gas/oil ratio (Yaliz and McKim, 2003). The storage capacity for Douglas was calculated by assuming that at some stage in its development the field would undergo enhanced oil recovery using CO_2 as an injectant. A 7 per cent incremental oil recovery could be achieved as a result of enhanced oil recovery using CO_2. The Lennox oil field consists of a thin oil column (44 m) overlain by a thick gas cap (232 m). In the early stages of production, the oil and its dissolved gas were produced. The dissolved gas was separated from oil, and injected into the gas cap, along with gas from the Douglas field, to maintain pressure during production of the oil. Gas production was scheduled to begin in 2004 (Yaliz and Chapman, 2003). It is likely that water drive will be observed when gas production starts. Because the oil was to all intents and purposes replaced by gas during its production, the CO_2 storage capacity of the Lennox field was calculated by treating the field as a gas field with water drive.

Table 6.1 CO_2 Storage capacities of the oil and gas fields in the East Irish Sea Basin

Field name	Ultimately Recoverable Reserves (billion cubic metres)	CO_2 storage capacity (Mt)
South Morecambe	146.8	734.4
North Morecambe	27.9	139.1
Hamilton	14.33	65.9
Lennox (gas cap)	10.31	42.6
Millom	6.07	24.3
Hamilton North	5.34	22.7
Dalton	2.87	11.5
Bains	1.36	5.4
Calder		
Darwen		
Hamilton East		
Crossans		0.0
Subtotal		1045.9
Douglas (oil)	0.00224	1.7
Total		1047.6

To put these figures in context, the nearby Connah's Quay Combined Cycle Gas Turbine (CCGT) power plant on the Dee Estuary (Figure 6.4) emitted 4.3 Million tonnes of CO_2 in 2002.[1] The estimated total CO_2 storage capacity available in the East Irish Sea Basin oil and gas fields amounts to some 243 years of emissions from this plant. However, given a plant lifetime of 25 years, and the fact that CO_2 capture would itself create significant extra emissions, it is clear that only the North Morecambe and South Morecambe fields have the potential to store the lifetime emissions from such a plant. The location of oil and gas fields accessible from the Point of Ayr and Barrow Shore Terminals is shown in Figure 6.4. Additional significant potential currently accessible from the Barrow terminal includes Millom and Dalton. A combination of the Hamilton, Lennox and Hamilton North fields, all currently accessible from the Point of Ayr terminal, could also be used. It is unlikely that any of the fields smaller than Dalton would be used for CO_2 storage from a large power plant such as Connah's Quay as they would have a very short operational

---

1   Connah's Quay is actually located in North Wales, although the gas it uses comes from the (offshore) North West region. Fiddlers Ferry is a coal power plant in the North West which emits between 5.8 and 6.8 million tonnes of CO_2 per annum. Assuming a figure of 6.5 Mt CO_2 per annum, the geological sites in the East Irish Sea could store the entire emissions of the power station for 161 years.

lifetime. Moreover, under the assumptions made above, little CO_2 would be stored as a result of EOR in the Douglas field.

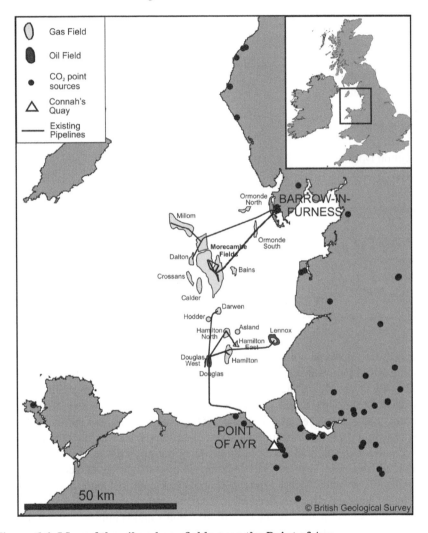

Figure 6.4 Map of the oil and gas fields near the Point of Ayr

Further Information on the Oil and Gas Fields of the East Irish Sea Basin

The Morecambe field (Figure 6.4) is volumetrically the second largest gas field on the UK continental shelf. It contains 12.1 per cent of the proven UK gas reserves and is divided into north and south fields by a deep narrow graben filled by the Mercia Mudstone Group. Morecambe's main reservoir lies within the Ormskirk Sandstone Formation, which on average is 250 m thick, though in crestal parts of the South

Morecambe field, the top 200 m of the St Bees Sandstone Formation is above the gas-water contact (1143m). The north and south Morecambe fields together comprise approximately 83 per cent of the available storage capacity in the oil and gas fields of the East Irish Sea.

The development plan for south Morecambe is based on using the facility for high rate seasonal gas supplies during the winter months; the first gas was produced in January 1985. South Morecambe has 34 producing wells, the top of the gas reservoir is 900 m below sea level and the original recoverable reserves were 5.1 trillion cubic feet (tcf). It is believed that this reservoir shows tank-like behaviour in that there appears to be no evidence of water influx or of further significant gases being introduced from the surrounding rocks (Bastin *et al.*, 2003). The design life for the facilities here is 40 years, and they were installed in 1990. This site is therefore not likely to be available for storing CO_2 until at least the year 2030.

Production at the Millom and Dalton fields began in 1999, with estimated combined recoverable reserves of 300 billion cubic feet (bcf). The expected lifespan of these two fields is 20 years. The earliest therefore that they would be expected to be available for CO_2 storage is approximately 2019. The Rivers fields (Calder, Crossans, Darwen, Asland and Hodder) are only now being developed and coming on line. Production started at Calder in 2004, whilst Crossans and Darwen are to be implemented by 2007. The five fields are estimated to contain a combined total of 250 bcf of gas.

The Liverpool Bay fields (Douglas, Douglas West, Lennox, Hamilton, Hamilton East and Hamilton North) have estimated initial recoverable gas reserves of 1.2 trillion cubic feet of gas, and more than 160 million barrels of oil. Production began in 1996. The Bains field has estimated gas reserves of 50 bcf and only started production in 2002. Bains has no production platform and is remotely operated from south Morecambe. The gas produced from Bains and south Morecambe are mixed and transported onshore. It is unlikely that the Rivers, Liverpool Bay and Bains fields will be available for storage for many years yet.

Oil and Gas Composition

Table 6.2 summarises the composition of the oil and gas fields in the east Irish Sea. A gas containing high levels of hydrogen sulphide (H_2S) is referred to as a 'sour gas' and a gas containing low levels is referred to as a 'sweet gas'. Both the H_2S and CO_2 in gas are highly corrosive and therefore specialist chromium steels are required for transportation pipelines and for the casings in the production wells. The gas from the Rivers fields contains high levels of H_2S. The Douglas Field, which produced the first offshore oil from the East Irish Sea basin, contains high levels of H_2S and other sulphur compounds that are removed during processing (Yaliz and McKim, 2003). North Morecambe contains high levels of CO_2 (approx 6 per cent), and due to the corrosive effects a new pipeline had to be installed. The CO_2 is removed during processing on the north Morecambe terminal (Cowan and Boycott-Brown, 2003). Therefore, there are fields in the east Irish Sea where the infrastructure is already sufficient to cope with the corrosive effects expected whilst injecting CO_2.

Table 6.2   Composition of the oil and gas in fields in the East Irish Sea Basin

Field name	Field type	Gas/oil composition
Millom	Gas	Sweet
North Morecambe	Gas	High CO_2 and N_2
South Morecambe	Gas	Sweet
Dalton	Gas	Sweet
Hamlton North	Gas	Sweet
Darwen	Gas	Sour
Hamilton East	Gas	Sweet
Hamilton	Gas	Sour
Calder	Gas	Sour
Bains	Gas	Sweet
Crossans	Gas	Sour
Lennox	Gas	Sour
Asland	Gas	Sour
Douglas	Oil	High H_2S
Lennox	Oil	High H_2S

Storage Capacity of Saline Aquifers in the East Irish Sea

The combined total of CO_2 that it is estimated can be stored in the closed structures of the saline aquifer is 630 million tonnes (Mt). This is the equivalent of 146 years of emissions from Connah's Quay power plant. Closed structures were identified from a map of the top Ormskirk Sandstone Formation derived from seismic data (British Geological Survey, 1994) and are shown in Figure 6.5. Because water has to be displaced from the pore space in aquifers, and the reservoir is heterogeneous, much of the pore space can be bypassed by migrating CO_2 when it is injected into such structures. This results in a less than perfect sweep of CO_2 through the pore space and relatively low CO_2 saturation of the reservoir rock. Based on reservoir simulation of closed structures in the Bunter sandstone (Obdam *et al.*, 2003), it is expected that the maximum CO_2 saturation of the pore space that could be achieved is approximately 40 per cent. Other parameters used in calculating CO_2 storage in the saline aquifers are given below:

Average surface temperature	10°C
Geothermal gradient	25°C km^{-1}
Porosity	15%
Pressure gradient	1.1 bar m^{-1}

Although the reservoir unit demonstrates all of the necessary properties required for geological storage, several of the structures do not lie at depths greater than 800m.

The CO_2 therefore will not be in its dense supercritical phase where it occupies less space. This does not mean however that CO_2 cannot be stored; it just means that less will be stored (Brook *et al.*, 2003). The fact that the aquifers do not contain gas (or oil) suggests that either they are not gas-tight or they do not lie on the migration path of any oil and gas generated in the basin. Further work is required to establish which of these reasons accounts for the absence of oil and gas in the non-hydrocarbon-bearing structures.

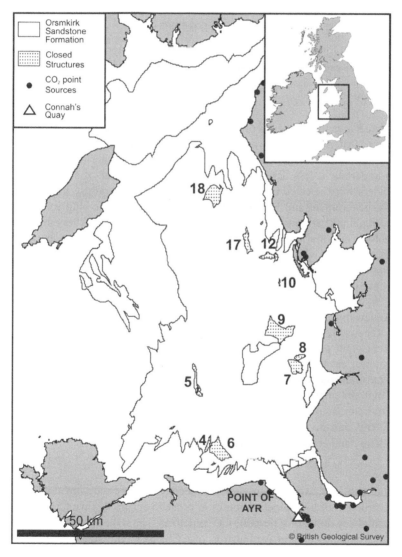

Figure 6.5 **Map of the closed structures in the saline aquifer near the Point of Ayr**

6.4   Summary of Geological Assessment of the East Irish Sea

The East Irish Sea Basin has considerable CO_2 storage potential, particularly in its gas fields. On depletion these have a CO_2 storage capacity estimated to be in the order of 1047 million tonnes. More than half of this lies in the South Morecambe field, and more than 1040 million tonnes lies within fields with an estimated storage capacity of more than 10 million tonnes. There is considerable further potential for CO_2 storage in the Ormskirk Sandstone aquifer. Here 630 million tonnes of CO_2 might be stored in mapped closed structures. However, further work is required to prove whether or not these structures are gas-tight.

6.5   Multi-criteria Assessment of Different Energy Futures

Methodology

We selected a range of key regional stakeholders to perform the Multi Criteria Assessment (MCA). The aim was to conduct detailed in-depth interviews (typically lasting 1.5 to 2 hours) with a range of key stakeholders from across the region rather than to undertake a less detailed survey of a larger number of stakeholders. The stakeholders were selected to represent key interests and expertise from the energy business, government and NGOs within the region. The following list details the organisational affiliation of the interviewees. We have used the letters in attributing comments or information in the following account to these interviewees. A and A* are work colleagues who conducted the MCA together and came to a consensus score between them. We have included them as a single individual in the data analysis (as A) but have distinguished between them where quotations have been employed.

    A: Renewable energy business manager
    A*: Renewable energy business manager
    B: Renewable energy business development manager
    C: Nuclear energy business manager
    D: Environmental regulator
    E Environmental Non-Governmental Organisation Manager
    F: Environmental Non-Governmental Organisation Manager
    G: Energy and environment consultant
    H: Energy and environment official in regional government

In addition to this list, we also interviewed a specialist on the environmental assessment and construction of long pipeline routes in order to ensure that we had included all the potential key impacts of bespoke CO_2 pipelines. The stakeholders were presented in advance with: a) a short explanation of CCS; b) a one page summary of each of the five scenarios, including a summary in words and numbers of the energy mix and a map showing the location of any new power plants and pipelines. At the start of the interview, respondents were asked whether they wished to clarify any of the information provided or specific facts concerning CCS. Respondents were then asked to score each of the scenarios against the nine criteria that the project team previously developed (Gough and Shackley, 2006). These criteria and their meaning are described in Table 6.3.

Stakeholders were also given the opportunity to add their own criteria, though only one respondent opted to do this (see Table 6.3). Scoring took place by asking the respondents to assign 100 points across the five scenarios for each of the criteria, allowing the relative performance of each scenario to be indicated. Respondents were then asked to weight the importance of the nine criteria, again having a total of 100 points to assign. The criteria scores for each scenario were then multiplied by the relevant weighting and summed to give an overall score. The scorings and weightings were entered directly during the interview into an Excel spreadsheet. Respondents were invited to revisit, and if necessary, change their scorings once they had seen the final ranking of the scenarios. On completion of the MCA process respondents were asked for their opinion on two further issues: whether enhanced oil recovery (EOR) using CO_2 would act to increase their acceptance of CCS; and the desirability or otherwise of CO_2 being exported from the region for storage elsewhere. Throughout the MCA process, respondents were invited to enter into discussion to explain their approach to the scoring and weighting. We recorded, and transcribed, the interviews. All of the respondents performed the MCA exercise as we had intended, with the exception of E, who found the scoring method too prescriptive and opted not to score the two Spreading the Load scenarios at all, preferring instead to use the remaining three scenarios (which E found to be more intuitively understandable).

Description of the Scenarios

The five energy scenarios are: Nuclear Renaissance, Fossilwise, Renewable Generation, Spreading the Load (High CCS) and Spreading the Load (Low CCS). All of these scenarios have been designed to achieve an approximately 60 per cent reduction in CO_2 emissions by 2050, though we have not at this stage undertaken the necessary quantification to guarantee that this level of emissions reduction would ensue. Such a quantification is far from straightforward in the UK because: a) the energy system is highly integrated (Anderson *et al.*, 2005); b) key energy data, such as electricity consumption, is not available at the regional scale. Hence, at this stage, the scenarios are *indicative* to allow comparison of alternative pathways and this was explained to, and accepted by, the stakeholder interviewees.

Table 6.3   The criteria used in the assessment of the scenarios

Criterion	Explanation of the Criterion
Cost	The financial outlay required to implement the scenario using respondent's own knowledge of current, and estimate of future, costs (including capital, operation and maintenance costs)
Infrastructure	The disruption and level of change in infrastructure required now and in the future (as distinct from costs)
Adverse lifestyle impacts	Perceived effects (if any) arising from the scenario upon peoples' sense of well-being (excluding overt public opposition to individual power plant developments)
Security of energy resource	The security of the fuel inputs to the energy system (i.e. coal, gas, oil, nuclear fuels and renewables)
Environmental impacts	The environmental impacts from the scenario excluding CO_2 (which is controlled equivalently in each scenario)
Public opposition	The public opposition to the scenario (from local to national)
Reliability of supply	The extent to which the scenario implies a challenge in delivering a constant and reliable electricity supply (e.g. 'keeping the lights on' when faced with problems of intermittency)
Risk of major disaster	The risk of large-scale failure with adverse consequences for the environment and/or human health and safety
Lock-in	The extent to which decisions taken in the shorter term may come to limit the opportunities for changing those decisions in the longer term
ADDITIONAL RESPONDENT CRITERIA	
Deliverability	The extent to which the scenario could be delivered in practice, given the future (short term) perceived direction of policy, economic, social and technological drivers

We have assumed that the overall demand for electricity remains constant across all five scenarios. This assumption was made in order to make the key supply-side features of the scenarios more readily comparable. If there were differences in both supply *and* demand then it would have been difficult to have clearly differentiated respondents' assessment of the change in demand from the changes to the supply-side. A more elaborate assessment exercise would have been required in order to consider both the supply- and demand-side changes (as in Anderson *et al.*, 2005). The scenarios reflect a reasonable range of the key options on the future generation side for the UK, though are by no means comprehensive. A larger number of scenarios would have been unwieldy in the stakeholder process and hence a decision had to be taken on the basis of prior discussions with stakeholders on presenting a suitable range of alternatives (Gough and Shackley, 2006).

At this stage, the scenarios also refer only to the electricity generating sector. Whilst it would have been possible to extrapolate the storylines to the transport sector and to the direct use of fuels in the domestic sector, this would have created a further level of complexity involving consideration of new fuels such as hydrogen which have multiple potential generation routes. It was decided to limit the complexity of our scenarios, in part because we did not have the resources to extend the technical modelling, and also to ensure that the respondents would understand them reasonably quickly in the time prior to the interview.

The scenarios were formulated to be applicable to the North West region of England, home to 6.7 million people and including the major metropolitan areas of Greater Manchester and Merseyside (including Liverpool). It also includes large rural areas in Cumbria, Cheshire and Lancashire. In terms of energy, the North West has some features which make it distinctive in the UK context. These include: a strong presence of the nuclear industry, with most of the UK's nuclear fuel preparation and re-processing facilities within the region; a number of gas and oil fields in the eastern Irish Sea (Morecambe Bay and Liverpool Bay) and associated Combined Cycle Gas Turbine (CCGT) plants; potential for large amounts of renewable energy, especially on- and off-shore wind; and relatively little coal-powered generation (with one large coal power station in the region) (NWRA, 2005). Overall, the North West region is approximately in balance vis-à-vis its energy imports and exports, though this is largely achieved through exporting gas supplies and importing coal-generated electricity from outside of the region (Carney, 2005).

The North West regional version of the scenarios was developed with these distinctive features in mind. A summary of each of the scenarios is provided in Table 6.4. The Nuclear Renaissance scenario (no CCS) therefore reflects the strong presence of the nuclear industry with a large increase in capacity through refurbishment of existing nuclear power plants and construction of two new plants on existing nuclear installation sites. Nuclear power provides over 80 per cent of electricity in this scenario, with the balance being made up by equal amounts of gas CHP and renewables.[2] The Renewable Generation scenario reflects a massive

---

2   In these scenarios we have referred to changes in capacity as a surrogate for actual electricity generation. Capacity is, of course, not the same as actual generation because of load

increase in the amount of off-shore wind and significant development of on-shore wind. Renewable electricity contributes 45 per cent of capacity, with the remainder being made up in equal measure from CHP (partly bio-CHP) and gas CCGT. There is a small amount of CCS occurring from gas plants in the period up to 2050, as renewable energy replaces fossil-fuel based power.

The Fossilwise scenario sees a doubling of coal-fired power stations, this fuel coming to supply 53 per cent of overall electricity generating capacity. Gas CCGT constitutes a further 24 per cent and gas-fired CHP 18 per cent. There is very little renewable electricity and no nuclear power in this scenario. CO_2 is collected from the re-furbished and new build power stations and piped for storage in the depleted gas fields of the eastern Irish Sea several miles out to sea. We selected the locations for new build power stations based upon submitted planning proposals available in the public domain, but focusing upon the major areas of demand in the southern part of the region (see Figure 6.6) (Electricity Association, 2003).

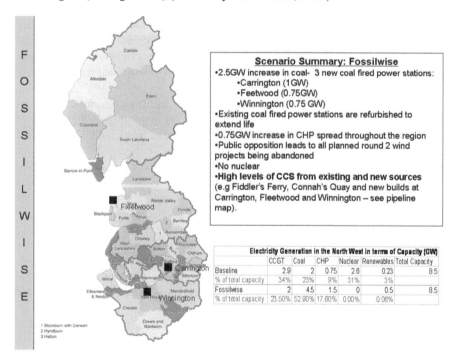

Scenario Summary: Fossilwise
- 2.5GW increase in coal- 3 new coal fired power stations:
  - Carrington (1GW)
  - Feetwood (0.75GW)
  - Winnington (0.75 GW)
- Existing coal fired power stations are refurbished to extend life
- 0.75GW increase in CHP spread throughout the region
- Public opposition leads to all planned round 2 wind projects being abandoned
- No nuclear
- **High levels of CCS from existing and new sources** (e.g Fiddler's Ferry, Connah's Quay and new builds at Carrington, Fleetwood and Winnington – see pipeline map).

Electricity Generation in the North West in terms of Capacity (GW)					
	CCGT	Coal	CHP	Nuclear	Renewables	Total Capacity
Baseline	2.9	2	0.75	2.6	0.23	8.5
% of total capacity	34%	23%	9%	31%	3%	
Fossilwise	2	4.5	1.5	0	0.5	8.5
% of total capacity	23.50%	52.90%	17.60%	0.00%	0.06%	

Figure 6.6   Fossilwise scenario summary

---

factors, whether the latter are influenced by intermittency or commercial factors. We have made an allowance for the intermittency of renewables by increasing the overall capacity of the regional electricity supply system in proportion to the penetration of renewables within the scenario. Hence, the capacity figures reflect the need for greater installed capacity due to intermittency.

The routes for the CO_2 pipelines were determined using the Geographical Information Systems (GIS)-based model described in Chapter 3. The GIS model works out the optimal pipeline route given the source, and the reservoir for the storage of, the CO_2, the costs of the pipeline and the need to avoid sites with nature designations such as National Parks, Areas of Outstanding Natural Beauty and EU designated sites such as Special Protection Areas and Special Areas of Conservation (SPAs, SACs). The route also remains at a minimum of 1km from villages and larger habitations in line with guidelines from the British Standards Institute (BSI, 2002). For simplicity we also assumed that there would be a single pipeline from each power station to the depleted gas field reservoir and, furthermore, that the pipeline would enter the Irish Sea at the existing landing sites for the natural gas pipelines (north and south) (see Figures 6.4 and 6.7). In the Fossilwise scenario we also allowed for an additional pipeline to be constructed from one of the new coastally-located power stations (see Figure 6.7).

Table 6.4   Quantitative summary of the five scenarios (in 2050)

	Electricity Generation Capacity in the North West (GW)					
	CCGT	Coal	CHP	Nuclear	Renewable Generation	Total Capacity
Baseline	2.9	2	0.75	2.6	0.23	8.5
% of total capacity	34%	23%	9%	31%	3%	
Fossilwise	2	4.5	1.5	0	0.5	8.5
% of total capacity	23.5%	52.9%	17.6%	0.0%	6.0%	
Nuclear Renaissance	0	0	0.75	7.5	0.75	9
% of total capacity	0.0%	0.0%	8.3%	83.3%	8.3%	
Renewable Generation	3	0	3	0	5	11
% of total capacity	27.3%	0.0%	27.3%	0.0%	45.4%	
Spreading Load I (high CCS)	2	2.5	1	2	2	9.5
% of total capacity	21.0%	26.3%	10.5%	21.0%	21.0%	
Spreading Load II (low CCS)	1.5	1	2	2.5	3.5	10.5
% of total capacity	14.3%	9.5%	19.0%	23.8%	33.3%	

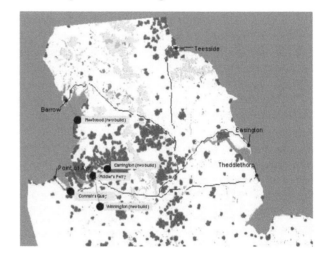

North West will fill its storage capacity around 2075

Figure 6.7   Fossilwise CCS pipeline routes

Not surprisingly the Fossilwise scenario has the largest amount of CCS, though there is sufficient capacity in the (depleting) oil and gas reservoirs of the East Irish Sea to store all of the CO_2 generated from the electricity sector within the region under this scenario until 2050. The capacity of these reservoirs would be exceeded approximately one decade beyond 2050, however. This implies that under the Fossilwise scenario CO_2 would have to be exported for storage in reservoirs elsewhere, probably in the North Sea, in the second half of this century. Alternatively, or in addition, the CO_2 could be stored in the aquifers found in the East Irish Sea for a further 50 to 60 years, though these reservoirs are much less well characterised than the oil and gas fields, and the estimate of their storage capacity is highly uncertain.

Clustering of the Weightings

In discussing the results we begin by examining the criteria weightings of respondents A to H. It appears that there is a reasonably clear division amongst the stakeholders with respect to the pattern of criteria weightings. However, we did not identify a consistent pattern of scoring across the nine criteria. In some cases, similar patterns of scoring did emerge for a few respondents with respect to a few criteria but not across all criteria. Furthermore, the similarity in scoring was not consistent between respondents across all criteria, but only across a sub-set of them. In other words, clusters of consensus emerged for some criteria, for some respondents, but not consistently across all criteria for these same respondents.

Table 6.5   Aggregation of the criteria weightings

Respondent	Cluster	Sum of Business-Focused Criteria Weightings	Sum of Environment and Socially Focused Criteria Weightings
B	Energy as	78	27
C	Business	65	50
A (A*)		64	55
G		56	58
E	Environment and	5	100
F	Society First	45	70
D		50	67
H		49	65

On the basis of our own interpretation of the interviews we categorized the criteria into two groups as shown in Table 6.5. The *business-focused criteria* are those which are important for the commercial activity of supplying energy to customers. They are: costs, infrastructural change, security, reliability and (for respondent B) deliverability. The *environment and socially-focused criteria* are those which relate to the wider environmental, social and political impacts arising from change in the energy system. They are: adverse lifestyle impacts, security, environmental impacts, public opposition, disaster and lock-in. It will be noticed that the issue of security overlaps both categories which reflects the fact that energy insecurity has huge adverse implications for business and for society at large. Some consideration of energy security issues is increasingly part of the commercial decision-making context for energy supply firms due to factors such as conflict in the Middle East, the high oil price and the depletion of the UK's gas and oil reserves.[3] In Table 6.5 we have added up the weights for both categories, including the security criterion for both categories. Since there are six environment and socially-focused criteria and four business-focused criteria, when they are added up there is something of a bias towards the former. However, the selection of the criteria reflects the priorities of an earlier cohort of stakeholders (Gough and Shackley, 2006) in addition to which

---

3   It might be argued that reliability is a criterion which spans both business and environment / societal concerns, since 'keeping the lights on' is clearly a vital social service. However, the way that we defined the scenarios and the criterion of reliability in Table 6.3 implies that the energy system will have to find a way of being reliable given the changes that it has undergone. E.g. it is assumed that the problem of intermittency has been solved in some way in the renewables scenario. Hence, the respondents were encouraged to score the reliability criterion in terms of how challenging it would be within that scenario to provide a reliable energy system. For this reason, we have categorised this as a business criterion.

the current group of stakeholders were invited to add additional criteria. It can be seen that the respondents fall into two clusters, with B, C and A clearly weighting business criteria more highly, and E, F D and H weighting environment and social issues more highly. Respondent G is less easy to categorise and effectively spans both clusters. Given the fact that there are more environment and social focused criteria than business ones, we decided to locate G in the business-focused criteria category.

Environment and Socially Focused Criteria Cluster

Turning in more detail to the environment and socially-focused criteria cluster, E stands out as having an exceptionally high weighting on the environment criterion (80 out of 100 points) (see Figure 6.8), the scoring against which therefore comes to dominate the performance of the scenarios for E. E was uncomfortable at being asked to score elements that were not directly related to nature conservation (the respondent's particular area of expertise). This response reflects E's role in an ENGO where he feeds specialist knowledge on the impacts of development upon biodiversity into public policy making decisions.

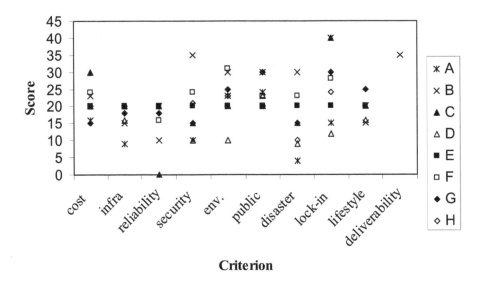

Figure 6.8   Criteria weighting

F, D and H illustrate a more even balance in their weightings between criteria (see Figure 6.8). This includes recognition of the business criteria such as cost and reliability, though the meaning given to the cost criterion was not always the investment, operation and maintenance cost (which is what we intended). One respondent was thinking rather of cost implications of the scenarios for the fuel poor. The respondents in this cluster were also in general not confident in making judgements on the costs of the technologies in the various scenarios, with one stating that 'it is up to others to make the economic arguments' (H).

Energy as a Business Cluster

Whilst, overall, respondents A (A*), B and C shared a focus upon the business-related criteria, there were some important differences in their criteria weighting. Opinion on the importance of infrastructural change varied, with some regarding this as effectively the same as the cost criterion, whilst others considered the logistics of infrastructural change to be the key factor. Environmental impact is weighted highly by A, at a medium level by G and at a low level by B and C. The difference in weighting reflects the importance accorded to the non-climate change environmental impacts such as fuel extraction, preparation, transportation and waste disposal. Public opposition was weighted highly by C and G, at a medium level by B and a low level by A. The key difference here seemed to be the extent to which the respondents believed that public opinion was not so important or at least 'malleable' and hence could be overcome in one way or another through appropriate government policies and given appropriate political will (the view of A and B). As A put it:

> What people think about it [the future energy supply mix] is neither here nor there for me because in the end the population of this country is going to have to be told what is good for them. You can't be democracy about everything. Government should govern. (A)

B, on the other hand, argued that public opinion was likely to be highly fragmented, hence there would be no clear consensus on what the public was opposed to. This is contrasted to criteria over which there would be a much stronger consensus, e.g. reliability. C and G, meanwhile, felt that public opposition could well have a major influence on the viability of future energy technologies such as nuclear (C works in the nuclear industry).

Clustering of the Scoring

We assessed each respondent's scoring for each criteria across the five scenarios. We identified high and low scores and then compared the pattern of scoring across the nine criteria. We then sought clusters from consistent patterns of high and/or low scoring by respondents. If there was no consistency in the scoring across all criteria, we looked for clusters related to a smaller number of criteria.

Clustering of the Nuclear Renaissance scenario scorings Scoring varied most between respondents for Nuclear Renaissance on the three criteria of: costs, security and environment (see Figure 6.9). Three clusters emerged: Nuclear Sceptics (B, A and F) who thought that the nuclear scenario would be very expensive, with high environmental risks and would not perform well in energy security terms; Nuclear Advocates (C) and Nuclear Ambivalents (D, E, G, H). The nuclear ambivalents scored nuclear poorly with respect to environment, and relatively highly with respect to security. The ambivalents were divided with respect to costs.

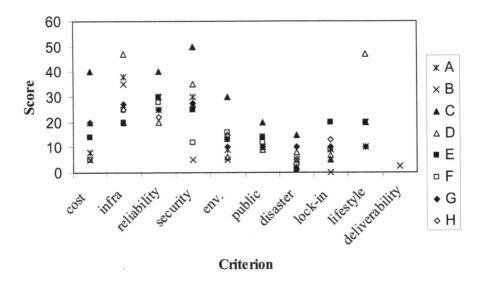

Figure 6.9   Scoring of the Nuclear Renaissance scenario

Clustering of the Fossilwise Scenario scorings The scorings for Fossilwise were more complicated than for Nuclear Renaissance and are summarised in Table 6.6 and Figure 6.10. The blanks in Table 6.6 indicate criteria against which there is little variation in scoring between the respondents. Since we are mainly interested in the comparison between respondents, we have not included those respondents in Table 6.6 whose scores are in the mid-range. The bold letters indicate those respondents around whom we can begin to identify prototype clusters.

There is a high level of consensus between participants on two criteria: that the costs of Fossilwise are lower than those of other scenarios, but also that the scenario does not perform so well in terms of energy security due to the need for imported gas and (to a lesser extent) coal. C disagrees on the cost consensus, regarding the construction of new fossil powered generation with CCS to be much more expensive than the other respondents. G disagrees with the consensus view that fossil fuels perform poorly with respect to security, arguing that supplies are more readily available to the extent required in the scenario than is assumed by the other respondents.

Table 6.6   Summary of the scorings for the Fossilwise and Renewable Generation scenarios

Criterion	Fossilwise		Renewable Generation	
	Higher performance	Lower performance	Higher performance	Lower performance
Costs	A,B,D,E,F,G,H	C	B,F,A,H	C,G
Infrastructure			H	A,B,D,G
Lifestyle			B,G,H	C,D
Security	G	A,B,C,D,E,F,H	A,B,D,E,F,H	C,G
Environment	D	A,C,E	A,B,C,D,E,G,H	
Public opposition	A,C,E			C,G
Reliability				A,B,C,E,F,G,H
Disaster	C,G,H,D	A,E	A,B,D,E,F,G,H	C
Lock-in	C,H,G,F	A,B,D,E	A,B,D,F,G,H	C,E

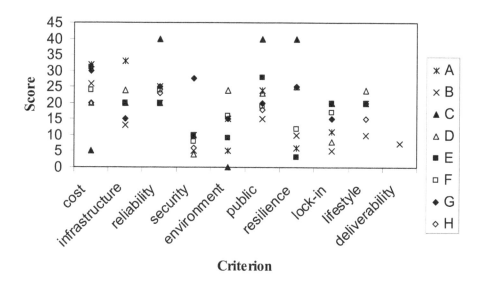

Figure 6.10   Scoring of the Fossilwise scenario

Respondents A and E provide the nucleus of a *Fossilwise Sceptics* cluster, rating the scenario poorly in terms of the (non-CO_2) environmental impacts, disaster and lock-in, although they see little potential for public opposition. It is some what more difficult to identify a *Fossilwise Advocates* cluster, though D and G do (between them) appear to have a positive view of Fossilwise with respect to security, environment, disaster and lock-in. It is perhaps surprising to note that only D thought that Fossilwise performed well with respect to the environment given that the problem of CO_2 emissions had (by definition) been removed. What is more, other emissions associated with coal burning are subject to control by the EU's Large Plant Combustion Directive (as well as by the requirements of the CO_2 capture process), though we did not provide this information to respondents. For a few respondents, the poor environmental performance of Fossilwise arose from the impacts of the extraction of coal in other parts of the world, though most respondents did not include impacts arising outside of the UK.

Clustering of the Renewable Generation scenario scoring There is a high degree of consensus with respect to the scorings for Renewable Generation on five criteria, namely a high score for security, environment, disaster and lock-in; and a low score for reliability (see Table 6.6 and Figure 6.11). There appears, perhaps surprisingly, to be more consensus over the performance of Renewable Generation than there is disagreement. Nevertheless, a *Renewables Advocates* cluster emerges (A, B, F and H) on the basis of the above criteria but also with respect to costs and (to some extent) adverse lifestyle impacts. C is a consistent *Renewables Sceptic* who is joined in his scepticism at times by G.

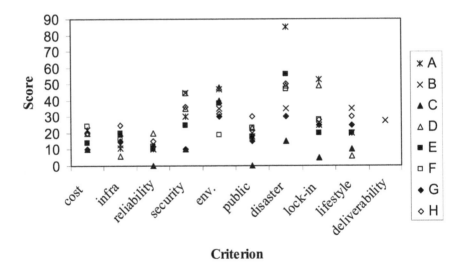

Figure 6.11   Scoring of the Renewable Generation scenario

Respondent H was alone in believing that the infrastructural performance of Renewable Generation was high. H accepted that the large-scale implementation of Renewable Generation would necessitate extensive infrastructural change.

However, his perception was that a large-scale change of the infrastructure was, in any case, necessary under all future energy scenarios. Hence through stimulating the development of such a modified infrastructure, the Renewables Generation scenario could be regarded positively with respect to this criterion.

Clustering of the Spreading the Load (High CCS) scenario scorings The scorings for Spreading the Load (High CCS) are summarised in Table 6.7 and Figure 6.12. There are no clear clusters for or against this scenario, in part because it does not evoke strong pro- or anti- assessments on the part of stakeholders. Both B and C rate the scenario quite favourably in terms of its having few adverse impacts on lifestyles (it is perhaps the closest of all the scenarios to current baseline conditions). The only really poor scoring of the Spreading the Load (High CCS) scenario is by A in relation to disaster, a result of A's perception that CCS entails high risks and given that this scenario entails a reasonably high level of CO_2 storage.

Clustering of the Spreading the Load (Low CCS) scenario scorings Opinion on the Spreading the Load (Low CCS) scenario is more varied than that for the high CO_2 storage version of Spreading the Load (see Table 6.7 and Figure 6.13). Respondents A and D demonstrate elements of a *Sceptical Spreading the Load (Low CCS)* perspective, in particular with respect to infrastructure and disaster. There is stronger evidence of support for this scenario on the part of B, F and G. The scenario scores particularly well with respect to environmental impacts, public opposition, security and lock-in.

Table 6.7   Summary of the scorings for the Spreading the Load High CCS and Low CCS scenarios

Criterion	Spreading the Load (High CCS)		Spreading the Load (Low CCS)	
	Higher performance	Lower performance	Higher performance	Lower performance
Costs			C	
Infrastructure		A		**A,D**
Lifestyle	B,C		G,H	D
Security		B,D,H	**B,F**	
Environment		C	**G,B,F**,H	
Public opposition			**G,B,F**,D	
Reliability				C
Disaster		D,A,H	**B,F**	**A,D**,H
Lock-in	B,D,C		G,B,C	

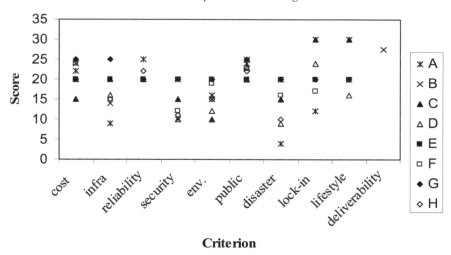

Figure 6.12   Scoring of the Spreading the Load (High CCS) scenario

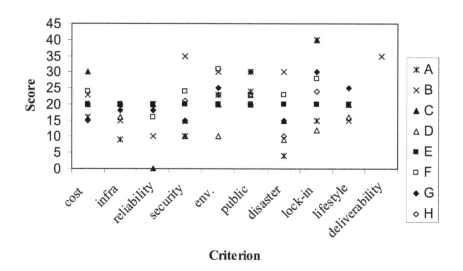

Figure 6.13   Scoring of Spreading the Load (Low CCS) scenario

Aggregating the Results for Scoring of Individual Respondents

We now summarise each individual's scoring for each scenario (i.e. with no criteria weightings, so that each criteria is assumed to be of equal importance).

Respondent A: Not only is the Renewable Generation scenario favoured, but A is quite sceptical of the other scenarios with Fossilwise scoring slightly above the others in second place.

Respondent B: Favours equally the Renewable Generation and the Spreading the Load (Low CCS) scenarios. Spreading the Load (High CCS) comes in third, whilst Nuclear and Fossilwise are both much less preferred.

Respondent C: Strong advocate of Nuclear Renaissance, but also demonstrating reasonably high levels of support for Fossilwise, Spreading the Load (Low and High CCS), but sceptical of Renewable Generation.

Respondent D: Favours Renewable Generation but also quite favourably inclined to Nuclear and Fossilwise.

Respondent E: Favours Renewable Generation with other scenarios more or less evenly scored.

Respondent F: Supportive of Renewable Generation and of Spreading the Load (Low CCS), but sceptical of Nuclear Renaissance.

Respondent G: Supports a range of scenarios, Fossilwise, Spreading the Load (Low and High CCS), and Renewable Generation but less enthusiastic about Nuclear Renaissance.

Respondent H: Renewable Generation is a long way in front, with Nuclear and Fossilwise the least favoured.

Combining the Effects of the Scoring and Weightings

Including the criteria weights affects the overall ranking of the scenarios by the respondents (see Figure 6.14). We now describe the influence of the weightings for each respondent.

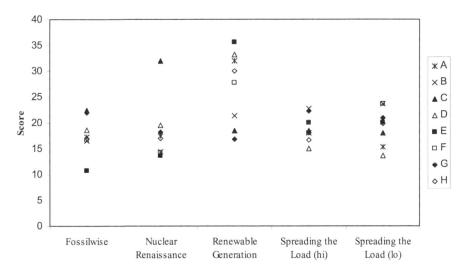

Figure 6.14 Final ranking of scenarios

Respondent A: The inclusion of weightings does not make a large difference to the scenario rankings.

Respondent B: The attractiveness of the Renewable Generation scenario is reduced through including the effect of the weightings, whilst the Spreading the Load (Low CCS) and Fossilwise scenarios both become more attractive.

Respondent C: The weightings render the Nuclear Renaissance scenario even more attractive. The Spreading the Load (Low CCS) comes out more favourably due to its nuclear component, whilst Renewables Generation comes out less favourably.

Respondent D: Renewable Generation is scored more highly with the weighting, whilst nuclear is slightly worse off.

Respondent E: The Renewable Generation scenario does much better with inclusion of weightings, whilst Fossilwise and Nuclear Renaissance do worse.

Respondent F: The weightings make the Renewable Generation scenario slightly more favourable.

Respondent G: The weightings do not make a large difference, though Nuclear does some what better and Renewable Generation does some what worse.

Respondent H: The inclusion of weightings does not make a large difference to the scenario rankings.

The weightings make a major difference in the ranking of the scenarios for respondents B, C and E. Both B and C exemplify business-focused criteria weightings.

Strategies of Scoring and Weighting

From the analysis of the respondents' explanations of the scoring and weightings during the interview stage, it would appear that there are two different ways in which the scoring and weighting are undertaken by respondents. Some of the respondents seem to have a clear idea of the pattern of scenario rankings that they wish to obtain at the end of the exercise as they are proceeding with the MCA process. We term this approach *strategic scoring and weighting*. Those who employ a strategic approach appear to have worked out, to a greater or lesser extent, how the MCA tool works and how to use the scorings and weightings to achieve their own preferred ranking of the scenario options. Other respondents seem to use the MCA in a much more explorative, experimental fashion, and generate scenario rankings which appear on occasions to surprise them. We term this approach *explorative scoring and weighting*. For these respondents there is less of a sense of approaching the MCA with a view to confirming a pre-defined favourite, even though the respondents may indeed have their own subjective preference. Below we provide evidence of strategic and explorative approaches to both weighting and scoring.

Strategic weighting and scoring We identified two different ways in which the weighting and scoring of criteria was performed strategically. *Delegating Responsibility on Specific Issues:* Respondent E massively weighted environment relative to the other 8 criteria. This ensured that the weighting of the environmental criteria dominated the ranking of the scenarios. E justified this approach by effectively *delegating responsibility* to others for the assessment of the economic and other more socially-focused criteria. This perhaps reflects E's role in representing the environmental interests in planning applications and policy developments more generally. Much of E's work is taken up in commenting on the environmental aspects of development applications on behalf of an environmental Non-Governmental Organisation and communicating this to planning authorities and other policy makers and stakeholders. Hence E appears to have extended such a delegatory approach to the conduct of this MCA exercise, despite the researchers' explanation that the purpose of the MCA is to grapple with the trade-offs that individual stakeholders might not normally have to confront. To a much lesser extent, D also shows some of the same approach, with the environmental impact criterion again given a high weighting, which D explained in terms of his responsibilities as an environmental regulator. However, in the case of D the initial environmental criterion weighting (60 out of 100 points) was revised downwards (to 33 points) as the purpose and rationale of the MCA exercise began to become more apparent. Hence, D's response appears to be somewhere between the strategic and explorative approaches.

Early Identification of a Clear Favourite ... and of Less Favoured Options Respondent C clearly preferred Nuclear Renaissance to the other scenarios and used the scorings and weightings to favour the high nuclear scenario. Other options, in particular Renewable Generation, were clearly less favoured by C and down-scored, even on some of the more socially-oriented criteria such as security, public opposition, risk of major disaster and lock-in. Most respondents gave higher scores to the Renewable Generation scenario on such criteria. Fossilwise tends to score reasonably highly for C (outperforming nuclear with respect to public opposition, risk of major disaster and lock-in, but not doing so well as nuclear with respect to costs, security and environmental impacts). The other scenarios are evaluated by C largely with respect to the amount of nuclear relative to renewables (more nuclear and less renewables tending to improve the performance of the scenario). C's stronger weighting of the business-focused criteria acts to exacerbate what is already a strong preference for the Nuclear, followed by the Fossilwise, scenarios.

Clearly, it is not just the advocate of nuclear power who has clear likes and dislikes. D, who works in environmental regulation, is an example of a respondent who expressed a strong belief that the energy system would have to change quite radically in order for it to become sustainable in the future. Even with respect to infrastructural change, for example, the scoring of the Renewable Generation scenario was not lower than for the other scenarios; in fact the scenario was scored jointly first (the most favourable scoring with respect to this criterion for the Renewable Generation scenario of all respondents). The explanation given by D is that the infrastructure will have to change in any case because of the necessary transition to sustainability, implying that modification for distributed and intermittent generation is a secondary issue. It is interesting to note that those respondents most directly involved in renewable energy development (A and B) scored Renewable Generation either lowest or very nearly lowest with respect to infrastructural change.

D is also very optimistic about low levels of public opposition to renewables, considerably more so than A, who is a renewables energy developer. D considered that there would be less public opposition to the Renewable Generation scenario than to any of the other scenarios, basing his assessment on general opinion surveys of the acceptability of renewables versus other energy technologies. By contrast, renewable energy developer A scored renewables as second worst vis-à-vis public opposition, with only Nuclear Renaissance scoring more poorly. B, who is also employed in the renewable energy business, is somewhere in between D and A in the scoring of this criterion, with Nuclear Renaisssance and Fossilwise scoring worse with respect to public perceptions, but the two Spreading the Load scenarios doing better.

Respondent A presents a third interesting example of strategic scoring and weighting in favour of one option (Renewable Generation), and sceptical of some other options. In this case it is interesting to compare the response of A with that of B, since both work in the renewable energy business. Respondent A falls within the business-focused weightings cluster, yet whilst he shares the high weighting of reliability and costs with B, A considers infrastructure to be one of the least important criterion. To some extent this might be a function of the way that the criterion has been interpreted, with A regarding it as effectively equivalent to the cost criterion, hence downplaying it here as a separate criterion.

The weighting of the security of supply criterion is a further example where A and B differ quite markedly, with A including it as one of the top three criteria, whilst B weights it very low. It is unclear from the accompanying interviews why such a difference occurred amongst two individuals both involved in renewable energy development. The effect of a high weighting for the criterion is to enhance the performance of scenarios with more renewables and nuclear and to downplay that of fossil-fuel dependent scenarios.

With respect to public opposition, B gave a mid-level weighting, whilst A gave it the lowest weighting of all the criteria. The rationale behind the weightings by A and B was the same however: namely that the public does not have sufficient knowledge and information to make an informed decision. Hence, it would be necessary to find a way of persuading the public to accept renewable energy. It was assumed by both A and B that the public could, in general, be so persuaded.

A's more highly weighted criteria are costs, reliability, security and environment. In A's scoring Nuclear Renaissance does well with respect to security and reliability, but falls down on the cost and environmental impact criteria, whilst Fossilwise does well on cost and reliability, but falls down on security and environmental impact criteria. Fossilwise receives the lowest score of all the scenarios with respect to environmental impacts, with mention made of the problem of NO_x and SO_x and ash. Respondent A was not referring to the environmental impacts of coal extraction in other parts of the world, an issue that was explicitly raised by a few other respondents and which accounted for the poor scoring of the Fossilwise scenario by D.

On the other hand, Renewable Generation does exceptionally well for A against the environmental criterion. Respondent A stated that he considered wind turbines to be an attractive addition to the rural landscape, so his scoring might reflect a sense that renewables will actually have a net positive environmental impact. Renewable Generation, whilst not performing as well with respect to reliability, also manages to score well against the security and cost criteria and accordingly comes out as the preferred option for respondent A.

Whilst difficult to 'prove', the pattern of scoring by A appears to demonstrate aspects of strategic scoring in favour of renewables. Where renewables score well, they tend to do so by a very large margin compared to the other scenarios, e.g. with respect to environment, lock-in and risk of major disaster. Yet where renewables do not score as highly, then their relative disadvantage compared to other scenarios is much less evident, i.e. there is more 'bunching-up' of the scores across the five scenarios, e.g. for public opposition and for reliability (where the Renewable Generation scenario is indeed the lowest scoring, but it is not the lowest by very much). From analysis of the results and of the interview transcript it appears to us that respondent A has a clear pre-defined preference for the Renewable Generation scenario and then uses the scorings to ensure that this scenario comes out as the clear favourite. The criteria weightings of A do not make such a large difference to the rankings, though as noted above the pattern of the weightings will tend to work in favour of the Renewable Generation scenario.

Explorative Strategies of Scoring and Weighting Respondents B and G both demonstrate a more explorative approach to the weightings and scorings. For B the older and more conventional energy generation types tended to score most highly on the more highly weighted business-focused criteria. Conversely, the scenarios with higher levels of renewable energy tended to score poorly with respect to those criteria: i.e. costs, infrastructural change and reliability. The high renewables scenarios perform very well under the criteria which are least weighted: adverse lifestyles, security of supply and environmental impacts. The spreading the load scenarios do better on one criterion – that of deliverability. The cumulative effect of this is that whilst B has actually scored the Renewable Generation scenario highly across many criteria, his overall ranking is quite strongly influenced by the way in which the criteria have been weighted. Hence, using equal criteria weights, the Renewable Generation scenario comes out as slightly preferred over the other scenarios; including weighting, the two Spreading the Load scenarios overtake the Renewable Generation scenario. This type of response illuminates that for a respondent such as B many of the benefits and advantages of renewables (beyond their zero-carbon status) arise from the 'social' and 'public good' quality of such forms of energy generation rather than because of their 'commercial' benefits. Conversely, putting their carbon status to one side, many of the benefits of fossil fuels and nuclear power reside in their good performance in commercial terms, including reliability, costs and infrastructure.

Respondent G is the only respondent who did not fall easily into *either* a business-focus or environment/society focus response to criteria weighting. G's scoring is also characterised by relatively little differentiation between the scenarios. Not surprisingly, therefore, there was little difference in the scenario rankings, although the weightings did, to some extent, bring out the business-advantages of nuclear power (reliability, security and infrastructure) and the disadvantages of renewables with respect to reliability and costs.

6.6   Discussion and Conclusions

This study has involved the participation of a small, though varied, number of stakeholder respondents. The sample was never intended to be representative of energy stakeholders in the North West region of England. Clearly, we cannot conclude too much about stakeholder opinion more widely on the basis of the evidence from such a small sample. Rather, the current study was designed to explore the combined use of future energy scenarios and multi-criteria assessment to investigate the perceptions of stakeholders at the regional scale. As such, the value of the work does not lie in the number or representativeness of the respondents but rather in the insights which emerged from even a limited number of stakeholders.

The Renewable Generation scenario was preferred by five of the eight respondents, whilst the remaining three respondents preferred Nuclear Renaissance, Spreading the Load (Low CCS) and Spreading the Load (High CCS). It is also interesting to note that the Renewable Generation scenario was a very clear favourite for four of the five respondents who preferred this particular scenario. The respondents for whom the Renewable Generation scenario was most popular were a renewable energy developer, a sustainable development planner, an environmental regulator and two environmental campaigners. The nuclear professional rated the Nuclear Renaissance scenario most favourably whilst the Spreading the Load (Low CCS) was preferred by a renewable energy professional and the Spreading the Load (High CCS) by an energy consultant. To a large extent, the preferred scenario of the respondent reflects what would be anticipated intuitively on the basis of their occupation and professional background. I.e. we tend to expect that environmental professionals will favour a scenario containing the most renewable energy, and that nuclear energy professionals would prefer nuclear energy. However, we also note that one of the respondents who works in renewable energy development actually preferred the Spreading the Load (Low CCS) scenario, so illustrating the danger of assuming that occupation or profession will overly determine preferences.

Opinion (expressed through criteria scoring and overall scenario ranking) across all respondents varied most markedly with respect to the Nuclear Renaissance and Renewable Generation scenarios, with much less variation expressed with respect to the other three scenarios. In other words, both these scenarios elicited strong reactions from the respondents, either positive or negative. Such reactions reflect commonly observed opinion amongst stakeholders regarding both renewable and nuclear energy. Opinion on Spreading the Load (High CCS) was most uniform across respondents, followed by Spreading the Load (Low CCS) and Fossilwise. It is interesting to note that where there is greater convergence in criteria scoring of a scenario between individuals, there tends to be less difference in the individual criteria scoring of each respondent. Conversely, where there is greater divergence between respondents, there tends also to be greater divergence in the criteria scoring within an individual's response. This suggest that there are specific aspects of nuclear and renewable energy technologies which elicit strong opinions, rather than strong opinions emerging across all the criteria with respect to those technologies, e.g. environmental impacts and risk of major disasters in the case of nuclear energy.

The Spreading the Load (High CCS) deliberately adopts a mixed-approach to the supply-side, and it is reasonable to assume that this accounts for the somewhat subdued reaction to the scenario. The same reasoning applies to the response to the Spreading the Load (Low CCS) scenario, though the higher amount of renewable energy it includes compared to Spreading the Load (High CCS) scenario probably accounts for the some what stronger reactions. Fossilwise elicits moderately strong positive and negative reactions, depending on the respondents and the specific criterion of interest.

Having identified two strategies for using the MCA tool, how can we explain these different responses? And within the strategic approach, how can we explain the different stances taken towards particular favoured options? Since the purpose of an MCA approach is to explore trade-offs between options, it might be argued that the strategic approach to scoring and/or weighting is, at best, somewhat missing the point and, at worse, a misapplication of the MCA tool to get the 'desired answer'. We would not accept such an argument, however, since an inclusive approach to engaging stakeholders cannot be overly selective or prescriptive. Since we are interested in the subjective assessments of stakeholders we cannot censor opinions on methodological grounds that they have not been reached in what the researcher considers to be the 'correct' way. MCA as a tool is not per se concerned with the validity of the reasons why respondents have the views that they do, though of course asking the respondents why they give the answers they do can provide material for such interpretation subsequently and is desirable in terms of transparency. Therefore, we do not feel able to say whether either the strategic or explorative approaches to MCA are 'better' or 'worse', though clearly it is useful to be aware of such stakeholder differences in approaching MCA.

It might be argued that the strategic approach would be found most strongly amongst respondents who were clearly energy experts, and hence have already come to a clear opinion on the relative strengths and weaknesses of different generation types. The two respondents who illustrated the explorative approach, however, were undeniably experts in energy assessment, more so than four of the other respondents. On the other hand, one of the strategic-behaving respondents was not an energy expert and only involved in energy related issues as part of a much wider portfolio of duties.

It is probably not a sufficient explanation to argue that a strategic approach illustrates the pursuit of self or organisational interest, the reason being that all of the respondents have some institutional or commercial agenda or mission to pursue, as was clear in all of the interviews. Nevertheless, businesses usually have a more clearly defined and articulated organisational mission than public bodies and regulatory agencies, which typically have to attempt to balance competing interests. Hence, it may have been easier for business interests to employ a strategic approach. It is interesting to observe that the two explorative-mode respondents were an energy consultant and a public-private sector facilitator of renewable energy developments. Both occupations require considerable diplomacy and mediation skills between competing interests, and consideration of energy from a wide-range of perspectives. Development of such skills is likely to make such respondents more comfortable with utilising an explorative approach. There may also be differences in perceived identity when performing the MCA. Some respondents in the public sector took on the mantle of a 'public servant', trying to look for the answer that was in the best public interest. Such respondents will inevitably use the MCA differently and more 'neutrally', though we have to be careful to remember that this is an issue of assumed identity as the public servant of which we have no empirical evidence.

Amongst the strategic weighters/scorers, the influence of particular commercial agendas, and/or personally-held values, can nonetheless be quite evidently witnessed; hence, the pro-nuclear and pro-renewables stances adopted by some respondents. What, then, can be concluded regarding the application of the MCA tool to evaluate stakeholder perceptions of CCS in the context of different energy scenarios more generally?

One important finding is that different respondents employed a different knowledge and information base in justifying their opinions. Sometimes respondents with particular specialist knowledge, e.g. of reliability, had a different perspective from other respondents who were not themselves experts in that topic. At other times, respondents used secondary sources of information which have been challenged in the academic literature (e.g. the finding of public opinion surveys that 'the public' is generally in support of renewable energy (Devine-Wright, 2005; Upham and Shackley, 2006)). We wonder, therefore, whether greater consensus on the information and knowledge 'baseline' against which assessments are being conducted could be attempted. This would mean that all assessments are being made in relation to the same agreed information and knowledge on the technical, organisational, commercial, social or political conditions and context. Even if a consensus was not possible between respondents, a description and explanation of the different opinions regarding the knowledge baseline would be informative for the respondents. For example, where there is disagreement on a technical issue, it is better for the respondents to be aware of this when scoring the scenarios, rather than proceeding on the basis of information which they assume (incorrectly) to be correct and widely accepted.

In order to operationalise this more collective approach to assessment it would be necessary to pay particular attention to the selection of appropriate experts who can between them reasonably reflect the range of technical opinion. A more collective approach to the assessment of the information and knowledge baseline would also be necessary, involving workshops and possibly use of methods such as Delphi.

Clearly, further work is also required to extend the number and type of stakeholders included. The results here also provide some potential lines of enquiry to follow-up, e.g. regarding the role of pre-defined 'interests' and the use of different strategies of scoring and weighting.

6.7   References

Anderson, K., Shackley, S., Mander, S. and Bows, A. (2005), *Decarbonising the UK: Energy for a Climate Conscious Future*, Tyndall Centre Technical Report 33, Tyndall Centre, Manchester.

Bachu, S. and Shaw, J. (2003), 'Evaluation of the CO_2 Sequestration Capacity in Alberta's Oil and Gas Reservoirs at Depletion and the Effect of Underlying Aquifers', *Canadian Journal of Petroleum Technology*, (**42**) 9, pp. 51–61.

Bastin, J.C., Boycott-Brown, T., Sims, A. and Woodhouse, R. (2003), 'The South Morecambe Gas Field, Blocks 110/2a, 110/3a, 110/7a and 110/8a, East Irish Sea', in J.G. Gluyas and H.M. Hichens (eds), *United Kingdom Oil and Gas Fields, Commemorative Millenium Volume*, Geological Society, London, Memoir **20**, pp. 107–18.

British Geological Survey (1994), *East Irish Sea (Special Sheet Edition)*, 1:250000, British Geological Survey Edinburgh.

Brook, M.S., Shaw, K.L., Vincent, C.J. and Holloway, S. (2003), *Storage Potential of the Bunter Sandstone in the UK Sector of the Southern North Sea and the Adjacent Onshore Area of Eastern England*, British Geological Survey Commissioned Report, CR/03/154, Nottingham.

Brown, K., Adger, N., Tompkins, E., Bacon, P., Shin, D. and Young, K. (2001), 'Trade-off Analysis for Marine Protected Area Management', *Ecological Economics*, **37**, pp. 417–34.

BSI (2002), *Code of Practice for Pipelines, Part 2: Pipelines on Land: Design, Construction and Installation*, BS 8010-2.8:1992, British Standards Institute.

Carney, S. (2005), *Greenhouse Gas Regional Inventory Project (GRIP)*, Tyndall Centre, Manchester.

Cowan, G. and Boycott-Brown, T. (2003), 'The North Morecambe Field, Block 110/2a, East Irish Sea', in J.G. Gluyas and H.M. Hichens (eds), *United Kingdom Oil and Gas Fields, Commemorative Millenium Volume*, Geological Society, London, Memoir **20**, pp. 97–105.

Devine-Wright, P. (2005), 'Local Aspects of UK Renewable Energy Development: Exploring Public Beliefs and Policy Implications', *Local Economy*, **10**(1), pp. 57–69.

DTI oil and gas website: http://www.og.dti.gov.uk/information/index.htm.

Ebbern, J. (1981), 'The Geology of the Morecambe Gas Field', in L.V. Illing and G.D. Hobson, *Petroleum Geology of the Continental Shelf of North-West Europe*, Proceedings of the second conference, pp. 485–93.

Electricity Association (2003), *Electricity Handbook*, Electricity Association, London.

Gough, C. and Shackley, S. (2006), 'Towards a Multi-Criteria Methodology for Assessment of Geological Carbon Storage Options', *Climatic Change*, **74**(1–3), pp. 141–174..

Gluyas, J.G. and Hichens, H.M. (eds) (2003), *United Kingdom Oil and Gas Fields, Commemorative Millenium Volume*, Geological Society, London, Memoir **20**, pp. 87–96.

Jackson, D.I., Mullholland, P., Jones, S.M. and Warrington, G. (1987), 'The Geological Framework of the East Irish Sea Basin', *Petroleum Geology of North West Europe*, pp. 191–203.

Jackson, D.I., Jackson, A.A., Evans, D., Wingfield, R.T.R., Barnes, R.P. and Arthur, M.J. (1995), *United Kingdom Offshore Regional Report: The Geology of the Irish Sea*, British Geological Survey, HMSO, London.

Levison, A. (1988), 'The Geology of the Morecambe Gas Field', *Geology Today*, **4**(3), pp. 95–100.

Meadows, N.S. and Beach, A. (1993), 'Controls on Reservoir Quality in the Triassic Sherwood Sandstone of the Irish Sea', in J.R. Parker (ed.), *Petroleum Geology of Northwest Europe*, Proceedings of the 4th Conference, The Geological Society, **2**, pp. 823–33.

Meadows, N.S., Trueblood, S.P., Hardman, M. and Cowan, G. (eds) (1997), 'Petroleum Geology of the Irish Sea and Adjacent Areas', *Geological Society Special Publications*, **124**.

NWRA (2005), *North West Sustainable Energy Strategy Draft Version*, North West Regional Assembly, Wigan.

Obdam, A., Van Der Meer, L., May, F., Kervevan, C., Bech, N. and Wildenborg, A. (2003), 'Effective CO_2 Storage Capacity in Aquifers, Gas Fields, Oil Fields and Coal Fields', in J. Gale and J. Kaya (eds), *Proceedings of the 6th International Conference on Greenhouse Gas Control Technologies*, Pergamon, Oxford.

Stewart, T.J. and Scott, L. (1995), 'A Scenario Based Framework for Multi-Criteria Analysis in Water Resources Planning', *Water Resources Research*, **31**, pp. 2835–43.

Stirling, A. and Mayer, S. (2001), 'A Novel Approach to the Appraisal of Technological Risk: A Multi-Criteria Mapping of a Genetically Modified Crop', *Environment and Planning C: Government and Policy*, **19**, pp. 529–55.

Stoker, G. (2004), *Transforming Local Governance: From Thatcherism to New Labour*, Palgrave Macmillan, Basingstoke.

Stuart, I.A. and Cowan, G. (1991), 'The South Morecambe Field, Blocks 110/2a, 110/3a, 110/8a, UK Irish Sea', in I.L. Abbots (ed.), *United Kingdom Oil and Gas Fields*, 25 Years Commemorative Volume, Geological Society, London, Memoir, **14**, pp. 527–41.

Upham, P. and Shackley, S. (2006), 'The Case of a Proposed 21.5 MWe Biomass Gasifier in Winkleigh, Devon: Implications for Governance of Renewable Energy Planning', *Energy Policy*, **34**(15): 2161–2172.

Wilbanks, T. and Kates, R. (1999), 'Global Change in Local Places: How Scale Matters', *Climatic Change*, **43**(3), pp. 601–28.

Wildenborg, A., Gale, J., Hendriks, C., Holloway, S., Brandsma, R., Kreft, E. and Lokhorst, A. (2004), 'Cost Curves for CO_2 Storage, European sector', in E.S. Rubin, D.W. Keith, and C.F. Gilboy (eds), *Proceedings of 7th International Conference on Greenhouse Gas Control Technologies. Volume 1: Peer-Reviewed Papers and Plenary Presentations*, IEA Greenhouse Gas Programme, Cheltenham.

Yaliz, A. and McKim, N. (2003), 'The Douglas Oil Field, Block 110/13b, East Irish Sea', in J.G. Gluyas and H.M. Hichens (eds), *United Kingdom Oil and Gas Fields, Commemorative Millenium Volume*, Geological Society, London, Memoir, **20**, pp. 63–75.

Yaliz, A. and Chapman, T. (2003), 'The Lennox Oil and Gas Field, Block 110/15, East Irish Sea', in J.G. Gluyas and H.M. Hichens (eds), *United Kingdom Oil and Gas Fields, Commemorative Millenium Volume*, Geological Society, London, Memoir, **20**, pp. 63–75.

Yaliz, A. and Taylor, P. (2003), 'The Hamilton and Hamilton North Gas Fields, Block 110/13a, East Irish Sea', in J.G. Gluyas and H.M. Hichens (eds), *United Kingdom Oil and Gas Fields, Commemorative Millenium Volume*, Geological Society, London, Memoir, **20**, pp. 63–75.

Chapter 7

A Regional Integrated Assessment of Carbon Dioxide Capture and Storage: East Midlands, Yorkshire and Humberside Case Study

Clair Gough, Michelle Bentham, Simon Shackley and Sam Holloway

7.1   Introduction

This chapter describes a follow on study from the exploration of the potential for CCS in North West England, described in Chapter 6, adopting a similar approach for two the regions of the East Midlands and Yorkshire and Humberside combined (EMYH). Here, we are interested in the potential for CO_2 storage at locations in the southern North Sea basin, how this relates to potential storage requirements emanating from the region and the perceptions of a variety of stakeholders from the public and private sectors regarding the role of CCS within the region. As a follow on study, the research explores similar questions, namely:

1. What is the potential for CO_2 storage in gas fields and saline aquifers beneath the southern North Sea?
2. How does this potential relate to CO_2 captured from power stations within the region for different energy scenarios to 2050?
3. How do different stakeholders evaluate these different energy scenarios and their implications for the region?

Again we have conducted a geological assessment, scenario analysis and a Multi Criteria Assessment of future power generation scenarios to explore these questions. This area of the UK has been chosen for two reasons: i) it is adjacent to a large concentration of gas fields and saline aquifers in the North Sea representing a large potential for offshore geological CO_2 storage and ii) many of the UK's fossil fuelled power stations are located within the two regions. We chose to combine the two administrative regions of the East Midlands and Yorkshire and Humberside in order to include this large amount of generating capacity and hence gain greater insight into the relative potential for storage in relation to the potential for CO_2 capture.

This combined region includes almost 35 per cent of the UK's fossil fuel generating capacity, of which over 20 per cent is coal, and is adjacent to the main area of North Sea gas fields. In 2000, the region generated 101 TWh electricity – equivalent to 31 per cent of the UK's demand for electricity (final consumption) in that year (DTI, 2004). Any strategy for significant uptake of CCS in the UK will have a major impact on these regions.

The approach adopted here was similar to that in the NW case study region – in which we developed a set of scenarios describing electricity generation and potential carbon dioxide storage within the region to 2050, followed by stakeholder review of the scenarios using a Multi-Criteria Assessment (MCA) process. The main difference in approach was that we developed the scenarios in greater detail (although the final figures were presented in a similar format) and, based on responses from the stakeholders in the NW region, made slight alterations to the criteria set used.

This chapter begins with a review of the storage potential in the Southern North Sea. In Section 7.3 we present the scenarios to 2050 for the region, incorporating both alternative power generation and storage regimes. Section 7.4 describes the MCA process through which these scenarios were reviewed by stakeholders.

7.2   Geological Assessment of CO_2 Storage Potential in the Southern North Sea

This section discusses the potential for storing CO_2 in the closed structures (domes) in the Bunter Sandstone Formation (saline aquifer) and in gas fields in the Southern North Sea Basin of the UK. The estimated storage potential in the closed structures of the Bunter Sandstone Formation and gas fields is 14.3 Gt and 3.9 Gt of CO_2 respectively. Many of the Southern North Sea gas fields are produced by depletion drive (by natural depressurisation) with very little aquifer support (water influx into the reservoir) during production. This makes them particularly favourable for CO_2 storage; the reservoir pressure after production is low, making CO_2 injection less costly. The gas fields also have gas seals proven over geological timescales. Most of the closed structures in the Bunter Sandstone Formation have not stored gas and the injectivity of the Bunter Sandstone Formation is largely unknown, as a result storage in this aquifer carries more uncertainties than in the gas fields. It is important that before CO_2 injection takes place at any geological storage site a full site investigation, characterisation and testing should be carried out. The storage sites identified in this study were used to produce the scenarios, outlined here (Section 7.3). The purpose of the scenarios is to present stakeholders with a range of options for reducing CO_2 emissions using geological storage of CO_2.

The Southern North Sea Basin The southern North Sea Basin lies to the east of England (Figure 7.1). It contains three major reservoir rocks; the Leman Sandstone Formation, of early Permian age, the Bunter Sandstone Formation, of Triassic age, and Carboniferous sandstones of Silesian age.

The CO_2 storage potential of the Carboniferous sandstone aquifer has not been investigated in detail, as there is insufficient data available on the distribution and structure of these sandstones to make a meaningful analysis. It is the reservoir rock for several gas fields, which are discussed further.

The CO_2 storage potential of the Leman Sandstone Formation lies principally in its gas fields. There is good reason to suppose that all closed structures in the Leman Sandstone Formation were originally full of gas and thus its CO_2 storage capacity can be estimated with a high degree of confidence on the basis of its recoverable gas reserves.

The CO_2 storage potential of the Bunter Sandstone is mainly aquifer potential. It contains some very large dome-shaped structures, the majority of which do not contain natural gas. The pore spaces in these structures are filled with highly saline water that could be displaced by CO_2. There are a few gas fields in the Bunter Sandstone, but these do not represent the bulk of its potential storage capacity.

Location of the Bunter Sandstone Formation The Bunter Sandstone Formation is a major sedimentary rock formation which is widely distributed in northern and central England (Figure 7.1). It is continuously present beneath a large area that stretches from its outcrop (the places where it comes to the surface) between Nottingham and Teesside, beneath the East Midlands and eastwards without

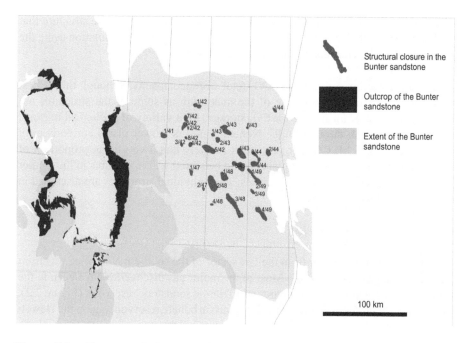

Figure 7.1   Extent and closures of the Bunter Sandstone Formation in the UK sector of the southern North Sea

interruption beneath the southern North Sea. This study focuses on the Bunter Sandstone in the UK sector of the southern North Sea. The Bunter Sandstone has many of the characteristics required for CO_2 storage, including large closed structures (domes), good average porosity and permeability, and a good seal in the overlying Haisborough Group, which consists of mudstones and evaporites (evaporites are rocks that have been formed by the evaporation of saline water, and those in the Haisborough Group include thick beds of rock salt). The Bunter Sandstone is a proven gas reservoir in the southern North Sea Basin and the Irish Sea Basin. Nonetheless, it is not possible to demonstrate conclusively that the large structures in the southern North Sea will not leak if filled with CO_2; many of the structures are cut by faults and the sealing efficiency of these faults is not known.

Closures within the Bunter Sandstone Formation with the potential for CO_2 storage were identified using a combination of existing maps, offshore well data and 2D seismic data (Brook *et al.*, 2003). The location of the closure sites is shown in Figure 7.1. The volumes of the closures and their CO_2 storage capacities were estimated using Equation 7.1:

Equation 7.1

$$CO_2 \text{ storage potential (tonnes)} = (\text{Area} \times \text{thickness} \times \text{porosity}$$
$$\times \text{density of } CO_2 \text{ at reservoir conditions}) \times 0.4$$

where 0.4 is the maximum estimated fraction of the pore space in the structure that could be filled with CO_2. This factor was derived from reservoir simulation using the Esmond field reservoir model (Obdam *et al.*, 2003).

This produces a figure of 14.3 Gt CO_2 total storage capacity, although this is best regarded as a theoretical estimate of the maximum as some of the structures are likely to prove to be unsuitable for CO_2 storage, for a variety of geological reasons, e.g. they might leak through faults. Nonetheless, UK power plants currently emit in the order of 176 Mt CO_2 per year, so it is likely that a very large proportion of CO_2 emissions from UK power plants could be stored within closures in the Bunter Sandstone Formation for several decades. The individual structures are discussed in more detail below.

Southern North Sea Gas Fields

The first gas to come ashore from the UK sector of the southern North Sea was from the West Sole gas field, in 1967. Most of the major gas discoveries have been in the Lower Permian, Upper Carboniferous and Triassic sandstone reservoirs (Figure 7.2). Gas has also been found in the Upper Permian carbonate reservoir, e.g. in the Hewett field (Cameron *et al.*, 1992). The major source of the gas in the southern North Sea is coal seams in the Upper Carboniferous Coal Measures. After the gas was generated it migrated to fill reservoirs in the Carboniferous, Permian and Triassic.

The Permian Leman Sandstone Formation contains the majority of the gas in the southern North Sea and as a result has the greatest potential for CO_2 storage. The total storage capacity of gas fields in this region is estimated at $3.9\,GtCO_2$.

CO_2 storage may be made more difficult due to various geological conditions within the reservoir discussed here. Compartmentalised fields such as Barque, Indefatigable, Schooner, Viking and Leman may require more wells to access all of the available storage in each compartment. Faults present within the reservoir that act as barriers to flow during production will make injection harder and more complicated. Fields in which the reservoir has been artificially fractured to allow increased production rates may also make CO_2 injection more problematic, examples of such fields are Clipper and Trent. Gas fields, which have more than one separate accumulation in different reservoirs, for example Trent, may prove more costly to inject CO_2 into, as more wells or deviated wells may have to be drilled to access each depleted reservoir.

Large amounts of water influx into the gas fields after production, for example South Sean field, will be a problem as CO_2 injected into the field will have to push the water back out of the pore spaces making injection more difficult. Fields without any evidence of water ingress back into the field after production (e.g. Clipper, Barque and Leman) would be a better choice for CO_2 storage because empty pore spaces at lower pressure than the initial reservoir would make CO_2 injection back into the reservoir much easier.

As fields are depleted and the pressure decreases the caprock may become damaged, allowing CO_2 migration out of the reservoir during re-injection. However, the ability of salt (which is a major component of the Zechstein cap rocks overlying the gas fields) to creep may counteract any cracking due to compression of the reservoir. As a general rule the initial reservoir pressure of the gas field should not be exceeded in the injection period, unless the seal is tested before injection.

The storage capacities of the southern North Sea gas fields were calculated in the GESTCO study, according to Equation 7.2 (GESTCO, 2003). The calculation assumes all the gas produced from the field can be replaced by CO_2.

Equation 7.2

$$VCO_2 = (V_{GAS}\,(stp)\,/\,Bg) \times \rho CO_2$$

Where:

$VCO_2 = CO_2$ storage capacity (10^6 tonnes)
Stp = standard temperature and pressure
$V_{GAS}\,(stp)$ = volume of ultimately recoverable gas at stp ($10^9\,m^3$)
Bg = gas expansion factor (from reservoir conditions to stp)
ρCO_2 = density of CO_2 at reservoir conditions ($kg\,m^{-3}$)

Water invasion into the reservoir after gas production will affect the amount of CO_2 that can be injected back into the gas field. The effect of this can be most accurately assessed by using reservoir simulation. But for this study no reservoir simulations were available. In the absence of reservoir simulations the following factors were used to adapt Equation 7.1 (from studies by Bachu and Shaw (2003) on oil and gas fields in Alberta).

1. In gas fields with depletion drive, i.e. those where the wells are opened up and the pressure in the gas field simply depletes, as it would if the gas were being produced from a sealed tank, it is assumed that 90 per cent of the pore space could be occupied by CO_2.
2. In gas fields with water drive, i.e. those where water encroaches into the pore space formerly occupied by the produced natural gas reserves, it is assumed that 65 per cent of the pore space could be occupied by CO_2.
3. In gas fields where the drive mechanism is both pressure depletion and water drive it has been assumed that each mechanism is acting equally on the reservoir, it is assumed that 77.5 per cent of the pore space could be occupied by CO_2.

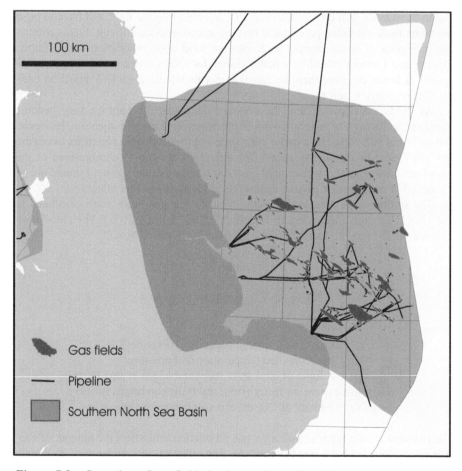

Figure 7.2   Location of gas fields in the southern North Sea

Where the drive mechanism is unknown, the following assumptions have been made. If the reservoir rock for the gas field is the Leman Sandstone the drive mechanism is depletion drive. This assumption has been made on the basis that most of the Leman Sandstone fields are depletion-produced fields. If the reservoir is in the Triassic or Carboniferous it has been conservatively assumed that the field is acting under water drive, as are most of the fields within these reservoirs.

7.3 CO_2 Capture and Storage Scenarios

For this Case Study, more detailed energy scenarios were considered to be necessary given the concentration of fossil fuelled power generation in the East Midlands and Yorkshire and Humberside (EMYH) and its role in the national electricity supply network. Detailed analysis of the implications of the scenarios in terms of the potential for CO_2 storage from the region at suitable offshore sites was also conducted.

The basic methodology in developing the scenarios for the EMYH region is as follows:

1. For the baseline year (2000), identify baseline data describing current generating capacity by fuel type, current electricity generation and demand within the region.
2. For each scenario identify plausible projections for each fuel type, such that total electricity supply is maintained at 2000 levels, based on assumed load factors. In the case of renewables, figures for 2050 are based on the accessible resource for the two regions (AEAT, 2002; EMRA, 2003).

The data used in generating the energy scenarios can be found in Gough *et al.*, (2006). As in the scenarios described in Chapter 6, we have not explored different views of energy demand explicitly in these scenarios. There is currently an active debate surrounding the evolution of electricity demand in the UK – whether it will continue to increase or whether demand side measures will take effect. For the purposes of this study this was considered to add an unnecessary level of complexity to the scenarios that would inevitably divert discussion away from the key issue of CCS. The aim was to consider the relative trade offs and issues associated with different electricity supply technologies; hence keeping electricity demand constant across the scenarios, while unrealistic, avoids diverting discussion away from specific concerns relating to supply technology onto demand side issues and enables us to restrict the number of alternative scenarios to five. The Tyndall Integrated Scenarios (Anderson *et al.* 2005) presents national energy scenarios which do explore alternative projections for energy demand and supply technologies on a national scale.

Power Generation Scenarios

Five scenarios were developed for EMYH as follows: Fossilwise; Renewable Generation; Nuclear Renaissance; Capture as a Bridge; Spreading the Load. The main difference from the NW scenarios (see Chapter 6) is the move to a single

Spreading the Load scenario; adopting two quite different scenarios as variations using the same name (with high and low exploitation of CCS) was felt by some respondents to be confusing. The fifth scenario therefore became Capture as a Bridge developed to explore the implications of adopting a more short term approach to CCS as the renewables capacity is built up.

Table 7.1   Key features of the scenarios

Scenario	Summary	Electricity balance (2050)	Dominant CCS technology (coal)
Fossilwise	Large scale exploitation of CCS	Increase in electricity generated within the region	IGCC with capture
Renewable Generation	Maximise use of renewables	Reduce electricity generated within the region	None
Nuclear Renaissance	Maintain or expand nuclear power generation	Reduce electricity generated within the region	None
Capture as a Bridge	CCS adopted in near term, to be phased out as renewables capacity is developed	Reduce electricity generated within the region	Ultra supercritical with capture
Spreading the Load	A broad mix including renewables, CCS and nuclear	Maintain electricity generated within the region	IGCC with capture

As already stated, the region generates considerably more electricity than is consumed within the region – predominantly due to the concentration of coal fired plant. It was thus assumed that this balance would vary across the scenarios with the relative importance of coal fired power generation. This implies that there will be a corresponding change in generating capacity in other regions (or that demand changes); since we wanted to focus the impacts to a particular region we have not specified the details of these implications to other regions, merely noting the necessary increase in capacity of the relevant type to make up any shortfall associated with a reduction in assumed fossil fuel capacity within EMYH. The key features of the scenarios are summarised in Table 7.1 and the fuel mix in Figure 7.3 and Table 7.2. In the case of Renewable Generation and Nuclear Renaissance it was assumed that CCS would not be adopted. Consequently, coal capacity is reduced with the assumption that other UK regions must increase electricity production from either renewables or nuclear to compensate; the generating capacity required to make up the shortfall is illustrated in Figure 7.3.

Electricity Supply

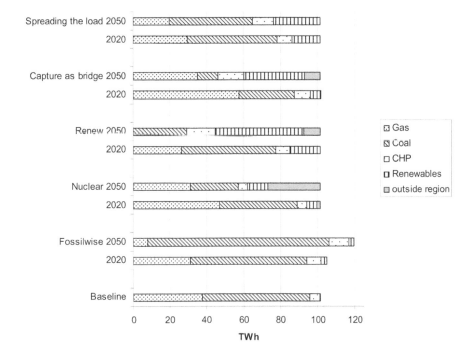

Figure 7.3   Summary of fuel mix in the five scenarios

There is currently no nuclear power plant within EMYH and it was considered that even in the Nuclear Renaissance scenario it would be unlikely that new plant would be constructed within the region, rather that any new build would be made at sites with existing nuclear capacity.

In addition to the summary shown in Figure 7.3 and Table 7.2, each scenario was accompanied by a visual summary of the generating mix, an example of which is shown for the Fossilwise scenario in Figure 7.4. Note also that in this Case Study we have not presented the scenarios in relation to the 60 per cent CO_2 reduction targets. This would require applying the scenario methodology to all demand sectors and adopting some means of allocating emission reductions across national regions. The level of CO_2 reduction is calculated for electricity generation in this case study region; this will have implications for measures required in other regions and sectors if a national target of 60 per cent reduction is to be met. In this way, the scenarios expose the relative contribution the supply technologies included in the scenarios make towards achieving that target.

Carbon Capture and its Storage

Table 7.2   **Quantitative summary of fuel mix and CO_2 storage in the five scenarios**

	CCGT	Coal	CHP	Renewables	Total
Baseline					
Generating Capacity (GW)	4.5	13.8	1.3	0.1	19.8
% (supply)	29%	66%	5%	0%	
Fossilwise					
2020 (GW)	3.8	12.8	2	0.8	19.4
% (supply)	23%	66%	8%	3%	
Capture (MtCO₂pa)					75.8
2050 (GW)	1.2	15	2.6	0.8	19.6
% (supply)	7%	82%	9%	2%	
Capture (MtCO₂pa)					55.1
Nuclear Renaissance					
2020 (GW)	6.4	8.7	1.3	1.5	17.9
% (supply)	44%	45%	5%	6%	
Capture (MtCO₂pa)					0
2050 (GW)	5	5	1.3	2.5	13.8
% (supply)	39%	39%	8%	14%	
Capture (MtCO₂pa)					0
Renewable Generation					
2020 (GW)	2.8	10.9	2	4.5	20.2
% (supply)	18%	58%	8%	16%	
Capture (MtCO₂pa)					0
2050 (GW)	0	5	4	11	20.0
% (supply)	0%	31%	17%	52%	
Capture (MtCO₂pa)					0
Capture as a Bridge					
2020 (GW)	7.4	8.8	2.2	1.4	19.8
% (supply)	45%	41%	8%	5%	
Capture (MtCO₂pa)					52.8
2050 (GW)	5.5	2.0	3.5	9.1	20.1
% (supply)	38%	12%	15%	36%	
Capture (MtCO₂pa)					18.6
Spreading the Load					
2020 (GW)	3.8	10.6	2.4	4.5	21.3
% (supply)	24%	50%	9%	16%	
Capture (MtCO₂pa)					59
2050 (GW)	3.0	7.5	2.8	6.5	19.8
% (supply)	20%	46%	11%	24%	
Capture (MtCO₂pa)					34

As noted earlier, a high proportion of the UK's coal fired electricity generation is located within this case study region; in order to explore the implications of large

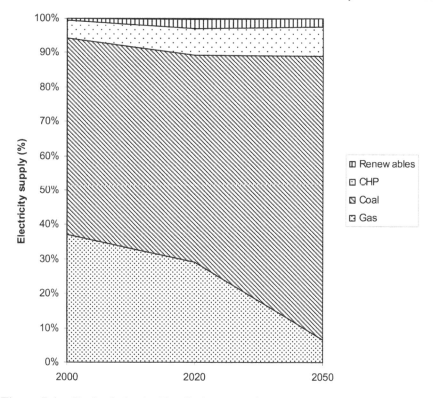

Figure 7.4   Fuel mix in the Fossilwise scenario

scale CCS with respect to potential storage sites it was necessary to estimate amounts of CO_2 to be captured in the different scenarios. This requires assumptions of the efficiency of different plant types to be made; as CCS is such a new technology, there is a large range of estimates for potential performance characteristics available in the literature. In preparing the scenarios we have employed different estimates according to the particular scenario. Past experience has shown that as new technologies develop, expertise and economies of scale lead to improvements in the cost and performance of those technologies, known as technology learning (Argote and Epple, 1990); there is a large literature in this area and methods for estimating learning rates for different technologies and how these may be influenced by policy have been developed (for example, Goulder and Mathai, 2000; McDonald and Schrattenholzer, 2003[1]). By using different studies from the literature and consulting experts within the industry we have implicitly allowed for different technology

---

1   Riahi *et al.* (2004) have explored alternative scenarios using different learning curves for CCS; their research demonstrates that different assumptions about the technology learning make a significant difference to the relative use of CCS in future mitigation approaches.

learning rates without developing our own detailed models; this was not considered to be necessary for the purposes of developing indicative scenarios and would have been beyond the scope of this study.

Geological Storage Scenarios

For the three scenarios that include CCS, an amount of CO_2, derived from power plants in EMYH, is made available for storage in geological formations beneath the southern North Sea. It has been assumed that the CO_2 from the power stations is collected and delivered via a pipeline to a single coastal gas terminal before being piped out to the storage site. The amount of CO_2 available for storage is detailed at the beginning of each scenario, the potential storage sites for these emissions are described in Section 7.2, i.e. the aquifer potential in the Bunter Sandstone Formation and the southern North Sea gas fields. Only gas fields which could store over 40 Mt where considered. These are listed in Table 7.3.

Table 7.3   Gas fields with the potential to store over 40 Mt of CO_2

Field Name	Estimated CO2 Storage capacity in million tonnes (Mt)
Leman	1203
Indefatigable	357
Viking	221
Ravenspurn N & S	146
West Sole	143
Galleon	137
Barque	108
Victor	70
Vulcan	63
Clipper	60
Audrey	53
Amethyst E & W	63
Sean N & S	45
Schooner	41
Total	**2710**

Based on calculations described in Section 7.2

All of the Bunter Sandstone Formation closures shown in Figure 7.1 were considered for the scenarios and a selection was made based on their estimated storage capacities and proximity to the gas fields. Closures were selected based on the proximity to the gas fields identified and depending on whether the scenario dictated use of an aquifer site.

It should be noted that insufficient geological data was available to properly characterise the individual potential CO_2 storage sites presented here. They were chosen solely because of their potential storage capacity and location, and their actual geological suitability is not known at this stage. Enhanced gas recovery and gas field abandonment dates were not considered in this study. Each of the scenarios used a different rationale for choosing storage sites, e.g. to use gas fields only or storage sites near to the Bacton gas terminal. The rationale is explained at the beginning of each scenario.

Fossilwise Total accumulated CO_2 available for storage in this scenario is 1.9 Gt. This scenario uses all of the available storage sites closest to existing onshore gas terminals, using one pipeline from the terminal to a cluster of storage sites where it branches to reach each site, as shown in Figure 7.5. The gas fields were preferentially filled up with CO_2 before moving on to the nearest aquifer sites, creating a storage hub. This is considered to be a 'cost averse' strategy in which a hub of reservoirs is chosen, in contrast to a 'risk averse' strategy in which a broader network of gas fields is used before saline aquifers are deployed. The total capacity of the reservoirs highlighted in Figure 7.5 is estimated at 2,445 Mt CO_2. It should be noted that the aquifer reservoirs are not as well understood as the gas fields, and due to a lack of data the presence of faults within the aquifers cannot be ruled out. The aquifers would require geological characterisation before they could be used as storage sites with a high degree of certainty.

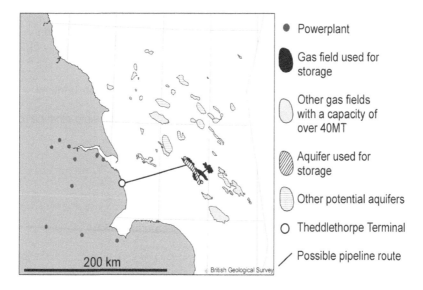

Figure 7.5   Map of the Fossilwise scenario (2050)

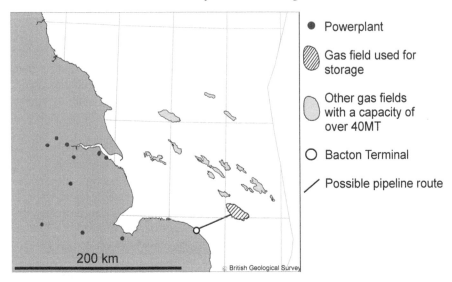

Figure 7.6   Map of the Capture as a Bridge scenario (2020)

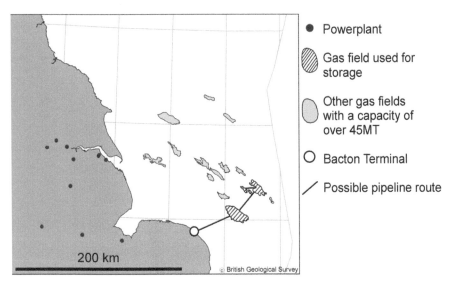

Figure 7.7   Map of the Capture as a Bridge scenario (2050)

Capture as a Bridge Total accumulated CO_2 available for storage is 212 Mt in 2020 and 1279 Mt in 2050. Storage has been considered in gas fields only. This is because there is enough storage in the gas fields between now and the projected end of the scenario in 2050, when CCS is phased out; commencing storage in gas fields is considered to be a risk averse strategy. The scenario was played out only from the Bacton terminal and aimed to use as few gas fields as possible, adding nearby gas fields onto the system when required. A snapshot of the scenario was taken in 2020 (Figure 7.6) and 2050 (Figure 7.7). The total capacity of the reservoirs highlighted in Figure 7.7 is estimated at 1,557 Mt CO_2.

Spreading the Load This scenario uses the same rationale as Fossilwise – a 'cost averse' strategy, using gas fields followed by nearby aquifers, shown in Figure 7.8. Total accumulated CO_2 available for storage is 1.4 GT. The total capacity of the reservoirs highlighted in Figure 7.8 is estimated at 2,441 Mt CO_2.

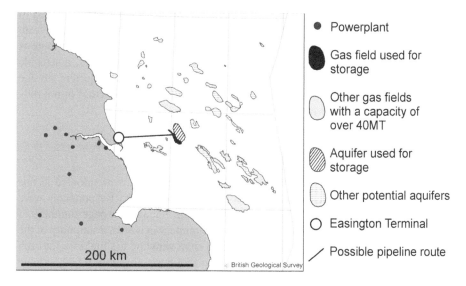

Figure 7.8   Map of the Spreading the Load Scenario (2050)

Summary

Five alternative scenarios have been defined – suggesting possible power generation futures for the EMYH region.

1. Fossilwise, in which large scale exploitation of CCS is adopted with an emphasis on coal fired generation. The region maintains its significance as an exporter of electricity. CO_2 is stored in both gas fields and saline aquifers.
2. Renewable Generation, in which the use of renewable energy is maximised, CCS is not adopted. The region generates slightly less electricity than in the present day.
3. Nuclear Renaissance, in which there is a revival of nuclear power in the UK, although new nuclear power plant are not constructed within EMYH, which

produces less electricity (although it is still a net exporter). CCS is not adopted in this scenario.

4. Capture as a Bridge, in which CCS is adopted in the near term, with storage in gas fields only, to be phased out as renewables capacity is developed.

5. Spreading the Load, which utilises a broad mix of supply technologies including renewables, nuclear and CCS using gas fields and saline aquifer storage sites.

There is a huge potential for CO_2 storage in the southern North Sea, in both depleted gas fields and the Bunter Sandstone Formation. However, any of the storage sites considered would have to undergo a rigorous geological site characterisation and risk assessment before use. Although the scenarios are theoretical and designed to provoke stakeholders' views of the different CO_2 reduction options, they do give an overview of how a CO_2 capture and storage scheme might work. Because there is greater geological uncertainty associated with the aquifer storage sites, a sensible strategy for those scenarios using a combination of gas fields and aquifers storage sites might be to test CO_2 injection into the aquifers (to find out whether or not they leak) whilst filling the gas fields.

7.4   Stakeholder Multi Criteria Assessment

The stakeholder interviews and multicriteria assessment framework adopted in this case study region was the same as that used in the NW study. Respondents were presented in advance with material described in section 7.3: a) a one page summary of each of the five scenarios, including a very short summary in words, a summary of the energy mix (see Figure 7.3 and Table 7.2) and a chart showing the fuel mix for each scenario (see for example, Figure 7.4); b) a map showing utilisation of CO_2 storage sites and offshore pipelines (Figures 7.5–7.8). It was assumed that all respondents were familiar with the concept of CCS and at the start of the interview, respondents were asked whether they wished to clarify any of the information provided or specific facts to do with CCS. Eight interviews were conducted as follows:

I:	Environmental Non-Governmental Organisation Campaigner
J:	Energy Business Director, Regional Development Agency
K:	Environmental Modeller, Electricity Supplier
L:	National Coal Mining Company
M:	Chair, Regional Sustainable Energy Forum / Politician
N:	Energy Coordinator, Regional Government
O:	Low Carbon Economy Advisor, Regional Development Agency
P:	Equipment Manufacturer

Respondents were then asked to score each of the scenarios against the nine criteria that the project team previously developed (Gough and Shackley, 2006). The wording of the criteria was adapted slightly to facilitate the scoring process by making it clearer that in all cases a high score reflected a positive performance against the criteria (for example, environmental impact was changed to environmental performance). The criteria and their meaning are described in Table 7.4. These criteria broadly match those used in the NW Case Study (Table 6.3) with the exceptions that the 'lifestyle' criterion has been removed and replaced by a criterion reflecting the relative contribution to achieving a 60 per cent national CO_2 reduction target. The lifestyle criterion proved difficult to score and it was felt that there was too much overlap with the public perceptions criterion. The CO_2 target criterion was added to account for the differing levels of emission reduction achieved in the EMYH scenarios; this criterion was intended to capture not just the perceived performance of the scenarios, in terms of CO_2 reductions, but also the extent to which the measures adopted in the scenarios in the power sector within this region facilitate achieving targets across the economy at a national level.

Stakeholders were also given the opportunity to add their own criteria, which five did (see Table 7.4). The scoring procedure is the same as that used in the North West case study, whereby respondents are asked to allocate 100 points across the five scenarios for each criterion allowing the relative performance of each scenario to be indicated. Thus, if all scenarios are considered to be equal for a particular criterion, each would score an average of 20. Respondents were then asked to weight the importance of the nine criteria, again having a total of 100 points to assign. The criteria scores for each scenario were then multiplied by the relevant weighting and summed to give an overall score. The scorings and weightings were entered directly into an Excel spreadsheet so allowing immediate ranking of the scenarios to be apparent to the respondent. Respondents were invited to revisit, and if necessary change, their scorings once they had seen the final output. Throughout the MCA process, respondents were invited to enter into discussion to explain their approach to the scoring and weighting. We recorded and transcribed all of the interviews. All the respondents performed the MCA exercise as we intended.

Criteria Weighting

Adopting a similar approach to the analysis of the results taken in Chapter 6, we begin by considering how the respondents weighted the criteria, as shown Figure 7.9. However, unlike in the NW Case Study it is not possible to identify the two clusters of business versus environmentally and socially focused criteria in the weighting. Overall the two criteria reliability of supply and security of supply emerge as clearly important to the respondents, all of whom assigned weights of 10 or higher for these criteria. As respondent M put it:

> I can tell you if the lights went off for 24 hours, my God there'd be a revolution.   (M)

Table 7.4   The criteria used in the assessment of the scenarios

Criterion	Explanation of the Criterion
Cost Effectiveness	The economic performance of the scenario
Infrastructure	The disruption and level of change in infrastructure required (as distinct from costs)
Security of energy resource	The security of the fuel inputs to the energy system (i.e. coal, gas, oil, nuclear fuels and renewables)
Quality of Environment	The environmental performance of the scenario excluding CO_2 (e.g. air, water quality, landscape etc)
Public perceptions	The public reaction to the scenario (from local to national)
Reliability of supply	The extent to which the scenario implies a challenge in delivering a constant and reliable electricity supply (e.g. 'keeping the lights on' when faced with problems of intermittency)
Resilience to major disaster	The protection against large-scale failure with adverse consequences for the environment and/or human health and safety
Avoidance of lock in	The extent to which decisions taken in the shorter term may come to limit the opportunities for changing those decisions in the longer term
Consistency with achieving 60% target	The extent to which the scenarios facilitate the achievement of a 60% CO_2 reduction in the UK
Additional Criteria	
Technical feasibility (deliverability)	The extent to which the scenario could be delivered in practice (proposed by J and M)
Political Feasibility	The extent to which the political decisions required to realise the scenario are likely to be taken (proposed by N)
Fit with an international effort to reduce CO_2	Synergistic effects of multilateral approaches to carbon reduction (in terms of technology) (proposed by L)
Consistency with reaching global targets	Transferability of technology to other countries (proposed by P)
Benefits to UK commerce and industry	The relative economic opportunities to the UK (proposed by L)

The cost effectiveness criterion also receives a middle to high weight by all respondents and is given the highest weight of any criteria by K. This respondent is in the electricity supply business and initially gave cost a weight of 40 (although subsequently reduced this to 20 during the process of weighting the remaining criteria) on the grounds that it would be cost that would drive whether or not a technology would be adopted and whether or not government would support it. The quality of environment criterion received the narrowest range of weights, with all respondents assigning a mid-weight.

Public perceptions and avoidance of lock-in both generated a broad range of weights – public reactions were seen as very important by K and M and as relatively unimportant by I and P both of whom considered that any negative public reactions to a technology may quickly reverse as a technology becomes more familiar (I) or in the event of disruptions to the power supply. Avoidance of lock-in was awarded a particularly low weight by L, who couldn't envisage anything new coming in quickly enough to make potential lock-in a problem, and P (who weighted it zero). Avoiding lock in was considered to be important by I because, according to I, once you start down a particular energy pathway you are committed to that technology for 20 or 30 years.

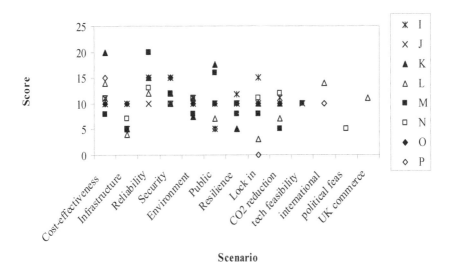

Figure 7.9   Chart showing criteria weighting

Resilience to major disasters was not generally weighted highly, with the lowest weighting given by K and L – both business sector respondents. It appears that the other criteria were simply seen as being more important rather than that this criterion is seen as being unimportant; it was also considered that there is a low probability of a major disaster occurring (K).

The compatibility with existing infrastructure criterion was assigned a weight of 10 or less by all correspondents indicating that this was widely viewed as not being a significant issue across the scenarios – generally respondents considered that over the timescale of the scenarios, to 2050, significant infrastructural changes would occur irrespective of the mode of power generation and that problems encountered could be overcome.

    Compatibility with CO_2 targets received a pattern of weighting somewhat similar to resilience to disaster, only L and M assigning particularly low weights. N proposed the additional criterion 'political feasibility' but gave it a low weight on the grounds that she did not believe political expedience was a good basis on which decisions about energy should be made.

    Respondent J opted to weight all criteria the same, on the grounds that they were all, effectively, equally important criteria and that to some extent they are all interrelated – for example, if something is not reliable and is incompatible with existing infrastructure it will not be cost effective and similarly an option performing poorly on environmental quality will be more subject to public opposition. None of the respondents raised this issue and it appears that the relationship between variables was not perceived in this way by other respondents. Given the complex nature of the electricity supply system, it is to be expected that there will be diverse viewpoints about how the criteria are related.

Scoring the Scenarios

In the following section we describe how the scenarios were scored against the nine criteria. As explained earlier, each respondent was asked to allocate 100 points across the five scenarios for each criterion, thus if all scenarios are considered to be equivalent for a particular criterion (i.e. all are thought to perform equally well/ poorly) then an 'average' score of 20 points would be given to each scenario.

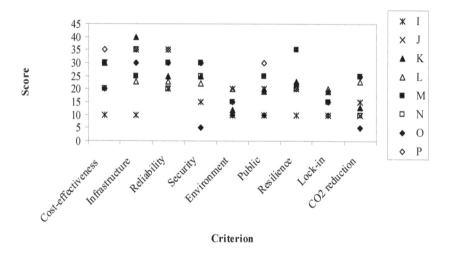

Figure 7.10 Scoring of the Fossilwise scenario

Fossilwise Figure 7.10 shows that generally this scenario performs better under economic and business criteria (costs, infrastructure, reliability) than for the environmentally or socially focused criteria (environment, public perceptions, resilience, lock-in and compatibility with CO_2 targets). A notable exception to this is respondent I (the environmental campaigner) who generally scored this scenario low for all criteria but particularly costs and infrastructure due to the dominance of CCS in this scenario. Despite performing less well overall against environmentally and socially focused criteria, there does not appear to be any clustering of respondents.

Reasons cited for the relatively poor environmental performance of this scenario are generally concerned with emissions (CO_2 emissions are not included in this criterion), such as so-called 'rare earth metals', waste products associated with burning coal and from the capture process but also the environmental impact of coal extraction (I, L, O). Despite being the scenario that delivers one of the highest levels of CO_2 reductions (for this region and sector), Fossilwise did not receive a particularly high score for the CO_2 reduction criterion from any of the respondents – only L, M and P gave it a score slightly above the average despite some strong statements that CCS is essential in achieving CO_2 targets. Other respondents considered that CCS does not advance the necessary reduction in fossil fuel (and other energy) use (K, N), that it does not genuinely address the CO_2 problem (O) or that leakage rates might negate any early benefits (I). This first point is also reflected in a poor score for avoidance of lock-in awarded by all respondents.

Table 7.5   Clusters of scoring for Fossilwise and Nuclear Renaissance scenarios

	Fossilwise		Nuclear Renaissance	
	Higher performance	Lower performance	Higher performance	Lower performance
Costs	J, K, L, M, P	I	P	I, J, K,L, M, N, O
Infrastructure	J, K, L, M, N, O, P	I	I, J, O, P	M, N
Reliability	J, K, L, M, O, P		K, L, N, O, P	I, M
Security	K, L, M, N, P	I, J, O	I, J, K,	M,N, O, P
Environment		I, K, M, N, O, P	P	I, J, M, N, O
Public	M, N, P	I, K		I, K, L, M, N, O, P
Resilience	K, L, M, G	I		I, J, K, M, N, O, P
Lock in		I, J, K, M, N, O, P		I, J, K, M, N, O, P
CO_2 targets	L, M, P	I, J, K, N, O	K, P	I, J, M, N, O

There is a fair spread of opinion about the extent to which Fossilwise provides a secure energy supply. Respondents K, L, M, N, P all gave a positive score, citing large global coal reserves, generally from stable countries, and the option for exploiting domestic coal reserves if necessary. Respondent O gave Fossilwise a very low score for security because of its reliance on a single fuel; whilst I was concerned about global demand for coal. There was also a range of views concerning how Fossilwise performs against public opinion – some considered that as a broadly 'business as usual approach' there would be little public reaction – reflecting a view that CCS will not generate a large public response (unless something goes wrong (N)). Those anticipating a more negative public response attributed this to either the perceived risks to humans imposed by the onshore components of CCS, such as pipelines, (respondent I) or a negative response to the sheer scale of coal use in this scenario with respect to the low uptake of renewables (which respondent K thought would be viewed more favourably by the general public).

Nuclear Renaissance The scoring for the Nuclear Renaissance scenario is illustrated in Figure 7.11. This scenario generally scored poorly for all criteria except compatibility with infrastructure, reliability and security of supply, reflecting a level of scepticism towards this scenario, particularly on the part of I, M and N. The socially and environmentally focused criteria and, in particular, costs were all given a low (or at best average) score by all respondents except P who thought that nuclear will be cheaper than renewables (in his view, the most expensive option) and that it will have the best environmental performance of all the scenarios, not being personally concerned about nuclear waste (although he did consider this to be a major problem for public perceptions). Nuclear was widely viewed as an uneconomic option, only kept open, in L's opinion, by effective lobbying from within the industry. Several respondents (e.g. J and M) made a point of distancing themselves from campaigning opponents of nuclear power whilst still giving it a low score. The following statement from respondent M characterizes this viewpoint; respondent M did not score the nuclear scenario higher than 15 for any of the criteria.

> I'm not against nuclear, I've never been phobic about nuclear, I just think we've constantly subsidised the nuclear industry, the amount of money we've poured into that industry … why do you have to turn it into a whole generating industry that produces enormous amounts of power as baseload at excessive cost?   (M)

The scoring for security of supply is fairly spread around the centre for this scenario – the lower scores related to the significant amount of gas assumed in the scenario. Should the gas component have been converted to coal in the scenario it would have been ranked highest for this criterion by H instead of the low score given by this respondent.

Only K and P give this scenario a positive score for its compatibility with reaching CO_2 reduction targets. Because this scenario still deploys significant fossil fuel power generation (assumed to be without CCS), much of which is located within the study region, CO_2 emissions remain relatively high in the region. Respondent J and L thought this made it harder to reach the broader targets:

if that's what [the Nuclear Renaissance scenario] implies, that by doing things with nuclear, (…) we've got do an awful lot in transport and energy efficiency then I just don't see that [targets being reached]. (J)

any scenario with lots of CCS is consistent with that because nuclear doesn't take you through the transport, unless you produce hydrogen from nuclear which is still a quite distant prospect … but you have to crack transport emissions and if you're still reliant on fossil fuels then that really does point to CCS at some point in the supply chain. (L)

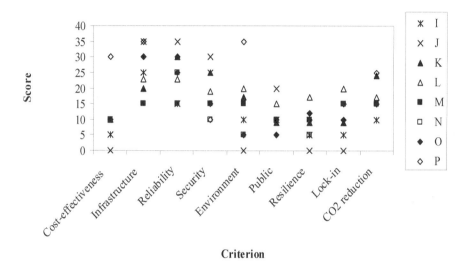

Figure 7.11   Scoring of the Nuclear Renaissance scenario

Scepticism about the scope for establishing a hydrogen economy, even by 2050, was also voiced by respondent O.

Renewable Generation In contrast to Fossilwise and Nuclear Renaissance the Renewable Generation scenario performs poorly against economic and business focused criteria and well against the environmentally and socially focused criteria – the pattern illustrated in Figure 7.12 for this scenario is almost the mirror of the scoring for Nuclear Renaissance shown in Figure 7.11.

Although there appears to be a fairly high degree of consensus in the pattern of scoring outlined above it is not so easy to identify clusters of individuals in the scoring. The strongest advocates of renewables are I and N, with I (an environmental campaigner) consistently awarding higher scores to this scenario. Despite this positive scoring, N considered the scenario to fare poorly against public perception

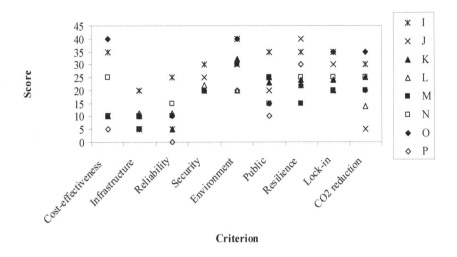

Figure 7.12   Scoring for the Renewable Generation scenario

– mainly on the grounds of visual intrusion, which was expected to be an issue particularly for large wind farms off the East Midlands coast. The split scoring for public perceptions seems to reflect the idea that the general public are in favour of renewables *in principle* but frequently opposed *in practice*.

J and L appear to be the most sceptical about renewables being adopted on a large scale, with J specifying an additional 'technical feasibility' criterion for which renewable generation is scored zero. Respondent L was very sceptical about renewables and felt that the technology was surrounded by a myth widely perpetuated (particularly in schools):

> the solution is presented as being renewable energy, everything's green … they're being sold a dream which isn't ever going to happen but those seeds are sown and therefore we have a population which is going to be largely anti-nuclear and anti-fossil.   (L)

There is a large spread of opinion over the costs of renewables across our respondents – with I and O both giving a particularly high score. Given the general acknowledgment that all energy technologies have an environmental impact in some way or another, the key impact of renewable energy (in particular wind) on the landscape was not felt to be important by any of the respondents; N went as far as to comment on the potential positive environmental effects that offshore wind structures might have by acting as an artificial reef.

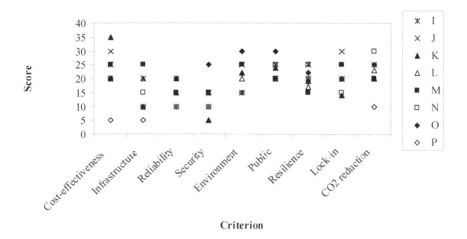

Figure 7.13 Scoring for the Capture as a Bridge scenario

Capture as a Bridge The scoring of this scenario (Figure 7.13) tends to reflect the two phase structure of this scenario – the first half being similar to the Fossilwise scenario (although using different boiler technology for coal firing) while the end phase becomes more like the renewables scenario; consequently the overall pattern most resembles the renewables scoring. This transition caused respondent N concern over potential risks of using 'one technology to push for another', introducing a 'pinch point' during the transition phase, when coal generation is in decline and the renewables component expanding. If the risk paid off, however, respondent N thought that this scenario could bring about the necessary step change and this is reflected in the scoring against compatibility with CO_2 targets criterion. In fact, with the exception of I (who doesn't like CCS) and P (who doesn't like renewables), all respondents considered this scenario to be more compatible with achieving the UK's CO_2 targets than the Renewable Generation scenario.

The majority of respondents saw this scenario as being more cost-effective than the Renewables Generation scenario; for example, K thought it would be cheaper to implement the upgrading to supercritical technology that is deployed on coal plant in this scenario than the new IGCC technology adopted under Fossilwise. The two respondents that thought renewables to be the most cost effective option (I and O) however, did not agree. Respondent P scored the costs criterion solely on the basis of significant renewables being expensive and maintained the low score for this scenario.

Table 7.6   Pattern of scoring for Renewable Generation, Capture as a Bridge and Spreading the Load scenarios

	Renewable Generation		Capture as a Bridge		Spreading the load	
	Higher performance	Lower performance	Higher performance	Lower performance	Higher performance	Lower performance
Costs	I, N, O	J, K, L, M, P	I, J, K, M	P	I, J, L, M, P	J, O
Infrastructure		J, K, L, M, N, O, P	M	J, K, N, O, P	I, M	J
Reliability	I,	J, K, L, M, N, O, P		J, K, M, P	K, M, P	J, O
Security	I, J		O	I, J, K, L, M, N, P	K, N, O, P	I
Environment	I, J, K, M, N, O		J, M, N, O	I, P	I, J, N, O, P	K, M, O, P
Public	I, L, M,	N, O, P	I, N, O		K, N, O, P	
Resilience	I, J, N, P	M	I, N,O	M	I, K, M, N	
Lock in	I, J, N, O, P		J, M	K, N	I, J, K, N	
CO$_2$ targets	I, N, O	J, L	I,J, N, O	N, P	J	

Spreading the Load The scoring for this scenario (Figure 7.14) tends to be somewhat 'flat' around the central average – because it has a broad mix of supply options, it features both respondents' favoured and disliked technologies. No patterns appear between the business or environmental/social criteria – although public perceptions, resilience to disaster and lock-in receive no scores below the average. Scores are generally positive for all criteria, although moderate – around 25 – with only very high scores from O for security and public perception. The key strength of this scenario was seen to be its diverse fuel mix – that no single supply option dominates, which was widely seen as beneficial with respect to both security and public perceptions.

Aggregating Scores across the Scenarios

As in the NW England Case Study we have carried out a linear additive approach to combining the criteria scores and weights across the scenarios. Unlike the first Case Study the weighting made no difference to overall order of the scenario ranking for individual respondents (i.e. each respondent's final ranking was in the same final order with or without the weightings). Including weights had the effect of increasing the range of the final scores, i.e. made the scoring slightly more extreme, and had a small effect on the relative magnitude of the totals for the different respondents.

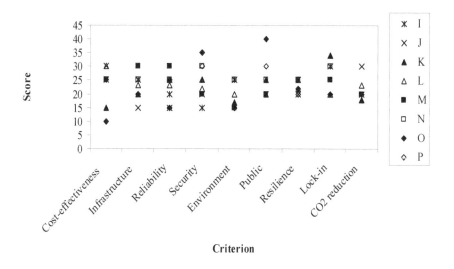

Figure 7.14 Scoring for the Spreading the Load scenario

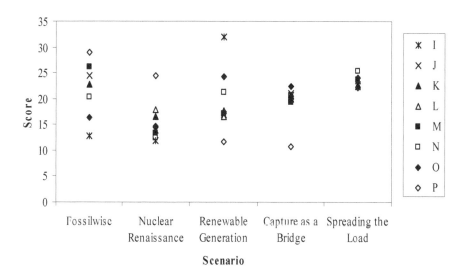

Figure 7.15 Final ranking of scenarios

Here we illustrate (Figure 7.15) and describe the overall ranking taking account of the respondents' criteria weighting.

The greatest range in scoring is associated with the Renewable Generation scenario and the smallest range with Spreading the Load which achieves a relatively high score from all respondents.

Respondent I:	clearly favoured the Renewable Generation scenario, gave Spreading the Load then Capture as a Bridge mid-range scores and very low scores to Fossilwise and Nuclear Renaissance.
Respondent J:	awarded highest scores to Fossilwise, followed by Spreading the Load and Capture as a Bridge. Renewable Generation then Nuclear Renaissance received the lowest scores.
Respondent K:	favoured Fossilwise and Spreading the Load followed by Capture as a Bridge (the three scenarios in which CCS is adopted) with Renewable Generation then Nuclear Renaissance both less favoured.
Respondent L:	showed similar final scores to respondent K, with the exception that Nuclear was very slightly preferred over Renewable Generation.
Respondent M:	clearly favoured Fossilwise and disliked Nuclear Renaissance; Spreading the Load and Capture as a Bridge then Renewable Generation lie between these two.
Respondent N:	preferred the Spreading the Load scenario, with Renewable Generation, Capture as a Bridge and Fossilwise also faring well. This respondent was clearly very sceptical of the Nuclear Renaissance scenario.
Respondent O:	favoured Renewable Generation then Capture as a Bridge and Spreading the Load and was also very sceptical of Fossilwise and Nuclear Renaissance
Respondent P:	showed the most extreme response, clearly favouring Fossilwise but also supportive of Nuclear Renaissance and Spreading the Load and deeply sceptical of Capture as a Bridge and Renewable Generation.

In general, those that preferred the Renewables Generation scenario tended to be sceptical about the Fossilwise and Nuclear Renaissance scenarios and vice versa.

Approaches to Scoring

The MCA process in the NW Case Study region revealed strategic and explorative approaches to the scenario scoring; a similar response was observed in the EMYH Case Study. A strategic approach to scoring was manifested as a bias either against a particular scenario (negative strategy) or in favour of a particular scenario (positive strategy) or component thereof, or in some cases both. For example, respondent P clearly demonstrated a negative strategy, with a deep scepticism of renewable energy dictating the scoring. This respondent gave the Renewable Generation scenario a score of 0 or 5 for costs, compatibility with existing infrastructure and reliability; the scoring of these criteria for the other scenarios was made according to the amount of renewables in the scenario. This respondent did, however, give a high score to renewables against the avoidance of lock-in criterion but then went on to weight the criterion at zero, effectively removing it from the assessment. Although to a lesser extent than for renewables, this respondent also demonstrated negative strategic scoring against gas fired generation. Similarly, for security of supply he made it clear that the low score for Nuclear Renaissance was due to the gas component of the scenarios and that if gas were to be replaced by coal this would be his preferred scenario. Respondent P works for a manufacturer of fossil and nuclear power generation and offshore oil and gas equipment.

In contrast to P, respondent I was clearly implementing a positive strategic scoring approach in favour of renewable energy combined with negative scoring against nuclear and CCS. For example, I, an environmental campaigner, awarded a high score of 30 or more to the renewable generation scenario for all criteria, except infrastructure and reliability to which he gave 20 and 25 respectively – no other respondent scored renewable generation above 15 or 11 for these two criteria. In fact, this respondent was alone in considering that the Renewable Generation scenario would be the most reliable of the five when all the other respondents considered it to be the least reliable:

> I think if you've got them well spread out then you're virtually always going to have somewhere which is windy.   (I)

This view is in contrast with the same respondent's idea of nuclear which suggests a somewhat selective view of the relative performance of the two technologies:

> I'm still a bit concerned about how often they do have to not operate at full capacity it does seem to happen quite often.   (I)

The highest score that I managed for Fossilwise was 20 for reliability of supply (considering it to be less reliable than wind energy) – although discussion focused on the relative reliability of renewables and nuclear for this criterion, the respondent seems to have struggled with the implications of capture processes on reliability. Generally, as each criterion was scored, this respondent focused on negative attributes of CCS for each criterion. As this respondent is an environmental campaigner it would be easy to simply attribute this approach to scoring as a 'knee jerk' reaction to the technology but it should be pointed out that I had read widely on CCS and was well-informed about the technology, leading him to feel 'less confident about CCS' than the official position of his organization.

Both I and N gave consistently low scores to Nuclear Renaissance for all criteria except Infrastructure (A: 25), Reliability (F: 25) and Security (A: 25), which were still only slightly above average. Respondent N explicitly refers to the influence of her scepticism of nuclear power by saying 'I'm loathed to score it too highly' despite considering it to perform well against, in this case, the reliability criterion.

The Capture as a Bridge scenario brought out some of the strategic approaches to scoring; since it begins with a large proportion of fossil fuel with CO_2 capture which is subsequently replaced by renewable power it contains elements of the same respondents' preferred and less favoured options. In the case of the strategic scorers this appears to have had the effect that scoring was focused on the component of the scenario that was disliked rather than on a view of the scenario as whole. For example, despite the final fuel mix comprising a large proportion of I's favoured renewables, this respondent did not give this scenario anywhere near the high scores seen for the Renewable Generation scenario. Likewise, P gave consistently low scores for this scenario despite it deploying significant CCS in its first phase, disliking this scenario on the grounds that he couldn't envisage a reason for moving away from CCS in the second phase.

Respondent J displayed the clearest example of explorative scoring with some extreme scoring, both positive and negative for particular scenarios. For example, this respondent gave Nuclear Renaissance a score of zero for cost effectiveness, environmental performance and resilience to disaster and avoidance of lock in whilst giving the same scenario a score of 30 or more for compatibility with existing infrastructure, reliability and security of supply. Similarly, J gave Renewable Generation very low scores for cost effectiveness, compatibility with existing infrastructure, reliability and contribution to achieving CO_2 targets and high scores for environmental performance, resilience and avoidance of lock in. Respondent O demonstrated an explorative approach to scoring by awarding a broad spread of scores to several of the scenarios depending on the criterion. For example, Renewable Generation, which comes out as this respondent's preferred scenario in the final ranking is given a score of 35 or more for cost effectiveness, environmental performance, avoidance of lock-in and contribution to achieving CO_2 targets, while receiving a score of 15 or less for compatibility with existing infrastructure, reliability and public perceptions.

Implications of CCS

During the course of the MCA interviews, certain recurring topics were raised that relate to the implications of introducing CCS in the UK.

IGCC or Supercritical?   Two quite different techniques for coal fired power generation are described in the scenarios after 2020 – Ultra Super Critical (USC) with capture and Integrated Gasification Combined Cycle (IGCC) (these technologies are described in Chapter 3). It is currently not clear which of these is likely to become the dominant technology for coal fired power generation in the UK. There are many, largely unknown, factors governing the development and diffusion of a new technology (technical, economic, political, social, cultural, etc.) (see for example, Geels, 2005). USC technology allows for incremental upgrades on existing plant whereas IGCC would entail a total redesign and redevelopment of plant. Even though there are several demonstration projects, IGCC is not yet proven on a commercial scale; however, most of the coal fired power stations in the UK are approaching their end of life.

Respondent P was sceptical of IGCC being suitable for large scale integration to the power generation network because it would be better suited for baseload generation and not sufficiently flexible for load following. The remainder of those that expressed opinions on this subject all considered that, while in the short to medium term upgrading to USC would be the most straightforward and cost-effective option, IGCC would appear increasingly over the longer term:

> its [retrofit] not a long term solution because you've got to get an IGCC plant up ... but what they'll do is go to supercritical boilers because its cheaper, quicker and easier. (M)

> the first phase of that [Capture as a Bridge] should be one of the least expensive options ... first of all you upgrade to supercritical and then you probably make a decision at some point about whether you're going to go onto capture so you have a situation where you have a more efficient plant and you weren't investing totally straight away in the new technology.   (K)

The importance of the global context was also raised by respondent L in relation to these technologies, notably what happens in China. Although they are not currently pursuing IGCC technology for power generation, the rate at which the Chinese are developing new coal fired capacity suggests that the next 10–20 years could see significant changes.

British Coal The security of supply criterion raised the issue of the potential for domestic coal reserves. Currently 59 per cent of the UK's coal use is imported (DTI, 2005) and large global supplies are available from a wide variety of sources (in contrast to gas supplies). However, Respondent L (a representative of the coal industry in the UK) noted that should the current trend for increased coal prices continue or should the supply of imports be challenged in some way, the UK does still have significant reserves of coal that could be exploited. Accessing these reserves may involve developing a new mine to access coal seams previously mined elsewhere (where the mines have been closed and cannot be reopened).

A corollary to this was voiced by J – that although the region has a strong identity with coal, there is a distinction between coal for power generation and coal mining; although there is a long history of coal mining in the region:

> Even with communities like Wakefield, the Coalfield Community Campaign etc,...its actually not all that long before they get to coal mining is a horrible industry. [...] In that sense is it coal or is it gas – at the economic level is not seen as a big issue.   (J)

The Concept of CCS Beyond the responses to specific criteria most of the respondents made clear statements about their opinion of the principle of using CCS as an approach to climate change mitigation. Since all the scenarios incorporate different combinations of power generation technology and three incorporate CCS, we have separated these comments from the scenario analysis – in order to highlight particular opinions relating to the concept of CCS. Similar to the participants of the Citizens Panels (described in Chapter 5), the stakeholder respondents fit into one of three groups:

1. Resistant to CCS – due to deep seated opposition to the principle of CCS, because of scepticism about the risks associated with the technology or both;
2. Ambivalent – somewhat sceptical of the technology but see that it may have a role to play;
3. CCS advocates – believe it to be an essential element of any climate policy, that it will be impossible to meet CO_2 targets without it.

1. Resistant to CCS – three respondents, I, N, and O, expressed fundamental concerns about the use of CCS. I held the strongest views of the two and was concerned about the legacy to future generations, clearly not identifying CCS with long term benefits in climate change terms:

> So if you're basically producing something that has to be monitored for thousands of years, I'm not sure that's a good idea, that we should be passing that sort of thing on to future generations where they're getting the cost and no benefit. [...] That's what I don't like about it that its basically forcing your descendents to monitor an area virtually for ever potentially.   (I)

In addition I, along with N and O expressed doubts about the long term storage security of reservoirs and the integrity of the cap rock. Although in the final scoring respondent N ranked the three scenarios that incorporated CCS relatively highly, this appears to have been a function of her strong opposition to nuclear rather than support for CCS. This respondent made various statements that revealed a fairly deep scepticism against CCS:

> its not really going to deal with the major issue, its using CCS as an easy way out – in the longer term you're not restructuring the whole thing. [...] for whatever reason its not going to cause a change, its not going to make us think in a different way about how we use the resources that we've got.   (N)

2. Ambivalent – respondents C and K identified CCS as being part of our future energy policy, although K saw it as a relatively short term solution, in the absence of any better alternative currently available:

> I'm not seeing it as a long term issue around the UK ... over 50 years you could do it and by then hopefully you've got some other technology.   (K)

> ... reflecting the inability of the other technology in my view to fill the gap, [with the] best will in the world its not credible to have so much renewable or nuclear is unlikely to significantly increase.   (K)

Respondent J on the other hand was simply sceptical about whether CCS could be made to work on a larger scale:

> I am predisposed towards clean coal as a good way of doing it, however I do think the idea of piping CO_2 offshore and sticking it in the wells is ridiculous. It seems to me what we're very good at doing as human beings is building plants with Process Integration ... the idea of carbon abatement where you're doing it (process engineering) it seems to me we have a track record, the moment you start bringing lots of links in the chain and pipelines and moving things off site ... somehow as human beings we're not ever so good at doing that.   (J)

3. CCS Advocates – M, L and P were all strongly in favour of CCS as a necessary technological approach to CO_2 reduction in the context of the challenges to achieving reductions in other sectors and through demand and efficiency measures:

> its got to happen it's the only way to do it – to run a modern industrial economy, you cannot transform the economy to a low energy economy to the extent it's required, you cannot reduce your energy demand far enough you've got to have CO_2 capture.   (M)

> any scenario with lots of CCS is consistent (with achieving national CO_2 targets) [ ] CCS ... gives other sectors of the economy a hook on which they can attach their emissions as well.   (L)

Respondent M (a politician) even identified CCS as a personal goal:

> If we can get to CO_2 capture and IGCC by 2050 I will die a happy man – with carbon capture in the North Sea we've effectively turned around the power industry in the right direction.   (M)

Although all respondents were presented with maps (Figures 7.5–7.8) showing potential storage locations adopted in each of the scenarios there was a general reluctance to engage in discussion over the relative merits of different types of storage reservoir. Since the majority of respondents were drawn from the energy sector and hence would not be familiar with the geological debate in this area, this is perhaps not surprising. The two exceptions to this are I, our strongest opponent of CCS who was very sceptical about the long term storage security of aquifers, and M, our strongest supporter of CCS, who considered the large capacity of aquifers to be a great asset to the region.

7.5   Summary

The MCA process implemented here with seven regional stakeholders (and one more from outside the region), has highlighted some key points of disagreement (and agreement) concerning future power generation technologies. Whilst the number of respondents is too small to draw general conclusions about stakeholder opinion within the region (which was not the intention of the study) it has been successful in exposing some of the different opinions that might be expected from a variety of perspectives.

Historically, the Yorkshire and Humberside and East Midlands regions have been closely associated with coal, both mining and for power generation. There is thus a perception of significant economic benefit to the region in maintaining a fossil fuel economy in general and coal firing in particular. This is reflected in the popularity of the Fossilwise scenario in most cases – the exceptions being the environmental campaigner and the two of the respondents with a regional sustainable energy remit. Although the combined region has a large coastline there is generally less enthusiasm for the large scale deployment of renewables from the group as a whole. The Renewable Generation scenario incorporated a very high penetration of renewables (just over 50 per cent of generation) and many respondents were highly sceptical about the feasibility of this. One aspect of renewable energy that did yield some highly contrasting views was cost, with three of the respondents considering it to be the cheapest of the scenarios and the remaining five considering it to be the most expensive. One factor considered by one respondent to make renewables a relatively cheaper option was the rising price of fuels as global energy demand increases.

Nuclear Renaissance was the lowest ranked of any of the scenarios by six of the respondents, next to lowest by one, leaving just one respondent (manufacturer of nuclear equipment) giving it a second place ranking. Several respondents made a point of stating that they were not opposed to nuclear power *in principle* but that it was widely felt to be an expensive option with high environmental risks that would be unpopular with the public and vulnerable to potential major disasters. The fact that there are no nuclear power stations located within the region in any of the scenarios may have removed the effect of any potential employment or regional economic benefits leaving respondents to focus on the negatives. The two more mixed scenarios (Capture as a Bridge and Spreading the Load) produced a high degree of consensus amongst the respondents. Spreading the Load was the highest ranked scenario for two of the respondents and second highest for the remainder; the key strength of this scenario was widely viewed as being its diversity of supply.

7.6   References

AEAT (2002), *Development of a Renewable Energy Assessment and Targets for Yorkshire and the Humber*, Final Report to Government Office Yorkshire and the Humber, July 2002 AEAT, Didcot, Oxon.

Argote, L., Epple, D. (1990), 'Learning Curves in Manufacturing', *Science*, **241**, pp. 920–924.

Anderson, K., Shackley, S., Mander, S. and Bows, A. (2005), *Decarbonising the UK: Energy for a Climate Conscious Future*, Tyndall Centre Technical Report 33, Tyndall Centre, Manchester.

Bachu, S.. and Shaw, J. (2003), 'Evaluation of the CO_2 Sequestration Capacity in Alberta's Oil and Gas Reservoirs at Depletion and the Effect of Underlying Aquifers', *Energy Conservation and Management*, **42**, pp. 51–61.

Brook, M., Holloway, S., Shaw, K.L., and Vincent, C. J. (2003), 'GESTCO Case Study 2a-1. Storage Potential of the Bunter Sandstone Formation in the UK Sector of the Southern North Sea and the Adjacent area of Eastern England', *Commissioned Report* CR/03/154, British Geological Survey, Nottingham.

Cameron, D.J., Crosby, A., Balson, P.S., Jeffery, D.H., Lott, G.K., Bulat, J. and Harrison, D.J. (1992), *United Kingdom Offshore Regional Report: The Geology of the Southern North Sea*, HMSO for the British Geological Survey, London.

DTI (2004), *Digest of UK Energy Statistics*, Department of Trade and Industry, London.

DTI (2005), *UK Energy Sector Indicators 2005*, Department of Trade and Industry, London.

EMRA (2003), *Towards a Regional Energy Strategy*, East Midlands Regional Assembly, Melton Mowbray.

Geels, F.W. (2005), *Technological Transitions and Systems Innovations*, Edward Elgar, Cheltenham.

GESTCO (2003), *Geological Storage of CO_2 from Combustion of Fossil Fuels*, Final Report, European Union Fifth Framework Programme for Research and Development, Project No. ENK6-CT-1999-00010, Brussels.

Gough, C. and Shackley, S. (2006), 'Towards a Multi-Criteria Methodology for Assessment of Geological Carbon Storage Options', *Climatic Change*, **74**(1–3), available on-line 24 February 2006.

Gough, C., Shackley, S., Holloway, S., Cockerill, T., Bentham, M., Bulatov, I., Klemeš, J., McLachlan, C., Kirk, K. and Angel, M. (2006), *An Integrated Assessment of Carbon Dioxide Capture and Storage in the United Kindgom*, Tyndall Centre Technical Report, Tyndall Centre, Manchester.

Goulder, L. and Mathai, K. (2000), 'Optimal CO_2 Abatement in the Presence of Induced Technological Change', *Journal of Environmental Economics and Management*, **39**, pp. 1–38.

McDonald, A. and Schrattenholzer, L. (2003), 'Learning Curves and Technology Assessment', *Special Issue of the International Journal of Technology Management*, **23**(7–8), pp. 718–745.

Obdam, A., Van Der Meer, L., May, F., Kervevan, C., Bech, N. and Wildenborg, A. (2003), 'Effective CO_2 Storage Capacity in Aquifers, Gas Fields, Oil Fields and Coal Fields', in J. Gale and J. Kaya (eds), *Proceedings of the 6th International Conference on Greenhouse Gas Control Technologies*, Pergamon, Oxford.

Riahi, K., Rubin, E., Taylor, M., Schrattenholzer, L. and Hounshell, D. (2004), 'Technological Learning for Carbon Capture and Sequestration Technologies', *Energy Economics*, **26,** pp. 539–564.

Chapter 8

The Implementation of Carbon Capture and Storage in the UK and Comparison with Nuclear Power

Simon Shackley

8.1   Introduction

In this chapter we will assess the extent to which CCS may be implemented in the United Kingdom over the next few decades. The analysis draws upon the previous chapters, but also upon other published and grey literature. Potential deployment of CCS needs to be assessed in the context of the wider on-going changes in the energy system; these drivers (long-term target for a 60 per cent reduction in CO_2 emissions, market liberalization and competition, technological innovation, price of CO_2 on the EU Emissions Trading Scheme, volatile oil and gas prices and increasing global demand for fuels, etc.) will have a profound influence upon whether, and if so how, CCS is deployed in the UK and other countries. Much of this Chapter will consist of a sustainability appraisal of fossil CCS versus nuclear power. This is because the extent of CCS deployment in the UK will depend to a large extent on whether the base load is being met through nuclear power or fossil CCS, given that: a) renewables are, by themselves, probably insufficient to provide reliable power to 2050 and b) there may be insufficient funds to support major investment in *both* nuclear and fossil CCS.

Because of the retirement of existing nuclear capacity and, indeed, of a probable 50 per cent of the existing coal plant, decisions on replacement of the existing large-scale supply technology are required soon given the time that it takes for planning, construction and addressing technical and operational teething troubles. It has been estimated that by 2020, it will be necessary to replace or refurbish 50GW of generation capacity, 2/3rds of existing capacity (Deloitte, 2006). This is equivalent to: 55 new CCGTs, 30 new nuclear power stations, 95,000 onshore wind turbines or 40,000 offshore wind turbines (ibid.). Deloitte's (2006) analysis suggests that in a 'business as usual' scenario, 1/2 of the new build will be CCGTs, but if this should happen the power generation sector emissions would be between 120 and 140 MtC/pa, against a target (assuming a long-term 60 per cent CO_2 reduction by 2050) of 105 MtC/per annum. Deloitte's considers that there are risks associated with such a reliance upon CCGT due to: increasingly volatile gas prices, difficulty of meeting CO_2 reduction targets and security of fuel supplies. The performance of the replacement technology with respect to CO_2 emissions is likely to be a major part of the replacement decision making. A key advantage of nuclear power has been its low CO_2 emissions compared to traditional fossil fuel generation; the emergence of CCS challenges this advantage and brings coal based power generation back into the debate as a major low carbon option.

8.2   Early Opportunities for Deployment of CCS in the UK

Given the additional costs of CO_2 capture and storage, early opportunities for CCS are somewhat limited because there is currently no clear long-term market price for a tonne of CO_2 abatement. The nearest we come to this is the market value provided by the EU Emissions Trading Scheme (EU ETS) and the present value (at the time of writing) of approximately €25 is perhaps half the current cost of CCS (see Chapter 3 for a more detailed discussion of CCS costs). The EU ETS has no guaranteed existence beyond 2012, however, and its future may well depend upon what is decided in commitments made under the United Nations Framework Convention on Climate Change (UNFCCC) beyond the Kyoto phase. Post-Kyoto commitments are highly uncertain at the present time and all that we know for certain is that the Parties to the UNFCCC will discuss the nature of such commitments in due course (as agreed at the 11th Meeting of the Conference of the Parties in Montreal in December 2005). Industry has expressed the opinion that it needs certainty regarding a policy mechanism to price the value of CO_2 reduction for '15 years beyond 2012' (HoC, 2006, p. 56).

A further problem with the EU ETS is that there is as yet no established protocol for inclusion of CCS as a way of mitigating carbon emissions (e.g. with respect to monitoring, reporting and verification) and that the 'grandfathering' scheme used by the UK Government to allocate emissions to generators (on the basis of their past emissions) is only guaranteed for three years. This makes it difficult for companies that are contemplating investment in CCS to know whether their investment is secure and will make an adequate return. In short, at present, the EU ETS seems shrouded in too much uncertainty for it to be a strong policy signal for CO_2 producers who may be contemplating CCS deployment. As BP has put it: '... the specific European system [EU ETS] is currently insufficient [with respect to CCS], even if the rules were to be clarified, because it fails to provide a framework of sufficient duration and the current (and indeed, forecast) level of carbon price is inadequate to encourage business to invest the very large sums required' (BP, 2005). Industry has indicated that it would need a minimum price of about £20 to £40 (€30 to 60) per tonne CO_2 over the next twenty years to justify investment in CCS projects now (HoC, 2006). BP has suggested that £40/tCO_2 would be at the lower end of what would be required for the company to break-even on CCS projects (ibid.).

One of the niche opportunities for CCS is in Enhanced Oil Recovery (EOR) in the North Sea. The UK government estimates that the UK North Sea contains about

1.5 billion barrels of oil which could be recovered using CO_2-EOR (DTI, 2003). One estimate is that this would allow storage of up to 700 million tones of CO_2 (Haszeldine, 2005). EOR would allow the lifetime extension of some oil fields by up to twenty years, hence securing income, jobs and maintaining and developing the skills base. Decommissioning of the oil and gas fields in the North Sea has already started and will pick-up speed in the next decade. This means that EOR needs to be applied in the next decade in order to extract the remaining resource and avoiding stranding a valuable asset. The capital cost of installing the North Sea infrastructure has been vast, estimated at £170 billion over the past 40 years and including 11,000 kms of pipeline costing £11 billion (HoC, 2006). As BP has put it: 'recycling the North Sea pipeline infrastructure could play an important part in enabling cost effective access to these reservoirs ... The UK's window of opportunity to gain material benefit from CCS technology will close as that infrastructure is removed' (HoC, 2006, p. 47). Clearly the opportunity of using infrastructure already in place reduces the capital outlay required, though whether existing pipelines are suitable for transporting CO_2 depends upon whether they have been designed to cope with natural gas containing CO_2. Furthermore, Statoil's Snohvit EOR project in the Norwegian sector of the North Sea is going ahead despite the fact that existing equipment and pipelines will not be re-used (HoC, 2006).

From a climate change perspective, CO_2-EOR might seem some what perverse in that it results in production of more oil with subsequent release of CO_2, emissions which are unlikely to be captured. Hence, it may be asked how effective is CO_2-EOR in actually reducing CO_2 emissions into the atmosphere and might it actually increase those emissions? Against this perspective, however, it has to be noted that the very high demand for oil means that alternative technologies for EOR would probably be used in any case and that production of additional oil from the North Sea in any case replaces demand for that quantity of oil from another field.

The advantages of CO_2-EOR in the North Sea include the availability of CO_2 where the gas is already being removed from natural gas and its physical proximity to potential geological storage reservoirs so that transportation costs are minimized. Estimates of the costs of EOR expressed in US\$ and £ per tonne of CO_2 avoided, range from US\$9 to 44 (£5-25) (for pulverized coal power plant), 19 to 68 US\$ (£11 to -39) (Combined Cycle Gas Turbine) and -7 to -31 US\$ (£-4 to -18) (Integrated Gasification Combined Cycle with coal) (i.e. a net gain, arising from the economic value of oil recovery using EOR) (IPCC, 2005). The costs of CCS with EOR are typically one third to one half of the costs of CCS without EOR (IPCC, 2005).

It is no surprise, therefore, that the UK's first major CCS project, due to come on line in 2009 (provided that the commercial partners agree on the viability of the business case), involves using CO_2 stripped from natural gas for use in EOR in the North Sea Miller Field. The methane is reformed, breaking it down into H_2 and CO_2. The hydrogen is used to power a 350 MW CCGT power plant, whilst the CO_2 is captured and sent to the Miller Field for EOR. The pipelines to the Miller Field have been designed to be CO_2-tolerant, hence it is highly suitable infrastructure to re-use for CO_2-EOR. It is anticipated that 1.3 million tones of CO_2 will be stored annually and that the EOR will extend the life of the oil field by approximately 20 years and extract an additional 40 million barrels of oil from the field (DTI Oil & Gas, 2005). The scheme, known as the Decarbonised Fuels Project (DFI), is being led by BP and will entail: the largest CO_2 EOR project in the North Sea; the first carbon dioxide pipeline in the North Sea; the largest hydrogen-fired power generation facility in the world; and the largest Auto Thermal reformer for generating hydrogen in the world (BP, 2005). One advantage of the Miller field scheme is that there are a group of other oilfields located nearby which would also be suitable for EOR and suitable incentives could allow extension of the CO_2 pipeline infrastructure to these fields (Haszeldine, 2005). Haszeldine has noted that:

> Miller is a crucial CCS opportunity for the UK, and it hard to over-emphasise the unique opportunity provided by the combination in sequence of: oilfield, pipeline, equipment, power station, willing companies, and timing. If this opportunity is missed it is hard to see another such combination on the UKCS. Miller can act as a crucial full-scale demonstration of CCS suitable for EOR, as a bridge to add-on EOR in neighbouring fields, and as learning for aquifer storage.   (Haszeldine, 2005, p. 81).

The economic viability of EOR projects depends to a large extent on the oil price. The Norwegian Petroleum Directorate examined the feasibility of EOR in the Ekofisk and Gullfaks fields and concluded that the break-even oil price was between $22 and $33 per barrel. The projected oil price used in the study was not given, but industry estimates at the time (the early 2000s) were giving a range from between $27-30, hence it probably was not considered a viable option to pursue (DTI Oil & Gas, 2005). In the period 2004-5, however, the oil price has risen to $62 per barrel, falling some what at the end of 2005 to $58 a barrel, but still anticipated to be between $55-58 a barrel in the near term future due to high demand and unexpected political and environmental events, though some industry experts expect the price to stabilise at around $40 a barrel in the medium term. Under these conditions EOR becomes a far more attractive option. Furthermore, there are other reasons why CCS with EOR may be more expensive in Norway than in the UK, namely the long transport distances required for transporting CO_2 to oil fields, and the sparse distribution of gas power stations.

It may be for this reason that BP has stated that: 'It is not fanciful to expect existing CCS technology to be in operation within five years, provided that stable market conditions and the necessary policy mechanisms are in place' (BP, 2005, p.85). The company argues that: 'Over a decade, there will be even greater scope to achieve significant improvements in the technology's cost performance, although the policy framework will always be important since it will always cost more to decarbonise fossil fuels than to burn them without decarbonisation' (BP, 2005, p.85). BP has argued that the UK's Climate Change Policy should not seek to 'pick' low- or zero-carbon winners, as it currently does for renewables through the ROCs scheme. Hence, the company would like to see the introduction of a support mechanism such as 'decarbonised electricity certificates' (DECs) for rewarding zero- and low-carbon energy technologies which would in effect create a 'level playing field' for CCS and renewable energy.

In summary, the opportunity for EOR is especially important to address because of the imminent decommissioning of the gas and oil infrastructure in the UK Continental Shelf. BP notes that the Miller Field was a candidate for decommissioning prior to the possibility of the DF1 EOR project. The gas fields in the Southern North Sea field are also due to be decommissioned in the next decade, yet they offer excellent CO_2 storage possibilities.

8.3   The DTI's 2005 Deployment Study

The UK Government has recently explored the prospects for CCS in the move towards a carbon-constrained energy system (Marsh *et al.*, 2005). The study used MARKAL, a bottom-up cost optimization model, with three different scenarios: Baseline (business as usual, GDP growth of 2.25 per cent per annum); World Markets (more individualism and globalization, with GDP growth of 3 per cent pa); and Global Sustainability (transition to a more sustainable society, GDP growth 2.25 per cent per annum). In each case a constraint is imposed on the model, namely a 60 per cent reduction in CO_2 emissions from electricity generation in 2050 relative to 1990, with appropriate intermediate targets. The study found that the timing and extent of CCS deployment depended upon the interaction of three factors.

1. The rate of improvement in energy efficiency in the economy (since energy efficiency is a more cost-effective way of reducing CO_2 emissions than the addition of supply-side capacity).
2. The rate of economic growth (since higher growth results in a higher demand for energy, the precise relationship depending on above changes in energy efficiency).
3. The deployment of other carbon abatement options, especially nuclear power.

In most scenarios CCS for electricity generation was deployed between 2010 and 2020, though deployment from 2010 was necessary in the high energy demand World Markets scenario. CCS for large-scale hydrogen production is utilized from 2040 in most scenarios, though a decade earlier in World Markets (ibid.). The overall level of CCS deployment increased over time from 0 to 25 Mt CO_2 per annum in 2010–2020 to between 50 and 180 Mt CO_2 pa by 2050.

Since our work did not assess the national scale, we do not have any comparable numbers to those in the DTI study by Marsh *et al.* (2005). Nevertheless, it is interesting to note that in a number of our scenarios, deployment occurs much more rapidly than in Marsh *et al.* (2005). Hence, annual CO_2 storage in 2020 in three of our scenarios for the EMYH region alone is between 53 and 76 Mt CO_2 per annum, considerably higher than the 25 Mt CO_2 per annum figure of Marsh *et al.* for the UK as a whole. Furthermore, the quantity of CO_2 stored actually comes down in our scenarios between 2020 and 2050, i.e. the same three scenarios for the EMYH region have a range of 19 to 55 Mt CO_2 per annum by the year 2050. This comparison raises two questions: why does deployment occur much more rapidly in some of our scenarios than in Marsh *et al.* (2005) scenarios? And why does annual CO_2 storage decrease between 2020 and 2050 in our scenarios but increase considerably in Marsh *et al.*?

The first point to note is the different philosophies and purpose behind the scenario exercise in our work and in that of Marsh *et al.* for the UK government. We attempted to create distinctive potential images of the future that, at the same time, were not implausible in order to: a) provide a challenging set of alternative pathways to provoke stakeholder opinion about the different energy supply options; and b), in the case of the southern north sea scenarios in particular, to gain a better understanding of the potential volume of CO_2 storage in relation to reservoir capacity. Our scenarios achieve fairly rapid deployment because we designed three of our scenarios such that a significant quantity of CO_2 would be available for storage, i.e. we deliberately took an optimistic or 'bullish' perspective on the prospects for CCS. Our scenarios are designed to provide some obvious choices regarding energy futures for the stakeholders to consider. Hence, we decided to have a Nuclear Renaissance as one scenario, but did not combine expansion of nuclear with deployment of fossil CCS (as in MARKAL). Our scenarios were also simpler than those of Marsh *et al.* (2005) in that we kept energy demand at a constant level (which will account in part for the reduction in CO_2 storage levels in our scenarios). Furthermore we kept the exploration of technological options relatively simple compared to Marsh *et al.*, e.g. we kept retrofitting for CO_2 capture of existing coal plant with advanced boilers (as in the Capture as a Bridge and Spreading the Load scenarios) separate from the introduction of IGCC with CO_2 capture (as in the Fossilwise scenario). Furthermore we did not consider the costs of different CO_2 capture technologies or of different carbon abatement technologies more generally in designing the scenarios. We assumed that it is difficult to make any sensible statements regarding the costs for the main competing carbon abatement technologies (renewables, fossil CCS and nuclear) for beyond a decade or so into the future, especially since present estimates suggest that the cost ranges for the main alternatives are rather similar and overlapping.

In contrast to our approach, Marsh *et al.* (2005) used a cost optimization model, MARKAL. If one technology is cheaper than another the model will favour that technology irrespective of other non-economic factors (sustainability, security, acceptability, political risks, etc.). Because the costs of base load nuclear power in MARKAL (3 p/kWh) are set at a lower level than fossil CCS (3.2 to 3.8 p/kWh), nuclear comes to dominate electricity generation by 2050 (accounting for 40 per cent of generation). Fossil CCS plant is in place from 2020 at very low levels, expanding somewhat to 2050, but still contributing less to electricity production than renewables. All the nuclear plant is operated at base load, whilst the load factor of fossil CCS plant varies between 25 and 55 per cent. The reason for this is that coal plant is more flexible in responding to changing demand than nuclear, hence it is cheaper to operate coal at lower load levels although the level of flexibility is likely to vary between different coal technologies. Fossil CCS becomes a cost-effective back-up to the intermittent renewable energy capacity (Marsh *et al.*, 2005).

Marsh *et al.* (2005) re-ran MARKAL but without permitting addition of any new nuclear capacity and found that in this case fossil CCS deployment reached 50 per cent of total generation by 2040; there was also an expansion in renewable energy. In this case CCS provided both base load and back-up to the intermittent renewables. In the baseline non-nuclear scenario, CCS begins to be deployed in 2020, whilst in the World Markets scenario CCS deployment begins at a low level from 2010 (involving retrofitting to existing coal plant then to gas plant and finally construction of entirely new coal plant with CCS between 2020 and 2030).

The MARKAL model assumed a rate of change in energy efficiency in the economy of 2.7 per cent per annum, but historically this has been difficult to achieve, with a reduction in energy intensity in the UK of 2.1 per cent per annum over the past 30 years. A high rate of change in energy efficiency will limit growth in demand for energy services, which could in turn influence the demand for carbon abatement technologies such as CCS. A model run was therefore conducted in which change in the energy intensity was limited to 2.1 per cent per annum. It was found that the overall demand for electricity in this model run was 30 per cent higher by 2050 (Marsh *et al.* 2005). The deployment of CCS in this run is double that using a 2.7 per cent per annum energy intensity improvement, though the deployment date (after 2020) does not change. If the constraint on new nuclear capacity is also imposed, the contribution of CCS becomes dominant, more so than when the 2.7 per cent per annum energy intensity change is applied with no new nuclear, and deployment begins in earnest after 2010.

Changes in energy intensity of the economy and of subsequent impacts on overall energy demand were outside of the remit of our scenarios. Likewise we did not look at the growth rate of the economy as a whole. Our scenarios are based on an intuitive view of how the energy supply system might evolve within the study regions; the quantification is introduced in order to 'bench mark' these broad alternatives in contrast to the quantitative modelling approach adopted by Marsh *et al.* Our 'high CCS' scenarios were designed to include a large quantity of CO_2 capture and this estimate is not constrained as in MARKAL by consideration of the costs of competing CO_2 abatement technologies. In our scenarios the amount of CO_2 being captured reduces over time as a consequence of changes in electricity generation technologies which result in substantial improvement in their overall efficiency, i.e. less fuel is consumed (and less CO_2 produced) to produce a given quantity of power.

Clearly, MARKAL is a more complete and complex tool for creating and assessing energy scenarios than the more 'ad hoc' approach taken in our regional case-studies. On the other hand, it could be argued that the information requirements for MARKAL are too onerous given current knowledge and data uncertainties. This limitation is reflected well in the issue of the relative position of fossil CCS and nuclear power in providing low-carbon electricity, especially for base-load generation. As Marsh *et al.* (2005) themselves acknowledge, uncertainties as to the 'true' future costs of nuclear generation versus fossil CCS, including the future costs of fuel inputs, mean that it is not possible to conclude whether new nuclear or coal CCS will become dominant in a carbon constrained energy system.

The uncertainty is well illustrated by the use of fossil fuel cost estimates in MARKAL. The original fossil fuel costs estimates used in 2002 for the preparation of the Energy White Paper to 2050 appear to be overly conservative, even in the World Markets and Global Sustainability scenarios. For example, the price of oil fell in the baseline and Global Sustainability scenarios by c. 10 per cent and nearly 50 per cent respectively, whilst it increased in World Markets by c. 25 per cent (Marsh *et al.*, 2005). Yet, given that many knowledgeable commentators believe that oil will have run out well before 2050 (e.g. Deffeyes, 2005) it seems highly conservative to limit price increase to just 25 per cent. Indeed, we know that oil and gas prices have been highly volatile over the past few years: the price of oil in the UK doubled during 2005, and gas prices doubled in the space of a few weeks during late 2005. Marsh *et al.* applied changes in fuel prices of about 30 per cent from the Baseline scenario values for 2020, and found that the price of gas and coal is important in the relative performance of gas and coal plant with CCS. At low (-30 per cent) and central (baseline) prices for both gas and coal new build gas CCS plant is cheaper than new coal CCS plant. However, with high (+30 per cent) gas prices and low or central coal prices, new coal CCS plant is as cheap and even cheaper than gas CCS plant.

Marsh et al. (2005) conclude that: '... the main conclusion to be drawn from this study is that CCS has the potential to make a major contribution to UK abatement targets. The current database does not provide a firm basis for choosing between technologies'. It appears that deployment of CCS could be desirable from 2010 onwards, depending on economic growth, demand for energy and fuel prices. Marsh *et al.* have also used MARKAL to explore the impacts of the EU ETS upon deployment of CCS. They have found that the effect of an EU-ETS price of €20/ tCO_2 is to reduce the amount of coal generation because of its high carbon intensity, it being replaced by gas. Retrofitting of existing coal plant with advanced boilers occurs, but there is no new build of new coal plant with a €20/tCO_2 price, unlike in the baseline scenario with no EU ETS: i.e. the EU ETS at a relatively low CO_2 price tends to reduce consumption of coal compared to the situation with no EU ETS scheme. A higher CO_2 price (and/or a higher gas price) would be required before the EU ETS would encourage deployment of coal CCS. In summary, the Government's analysis implies an important role for CCS in achieving a 60 per cent reduction in CO_2 emissions from electricity generation but the details of deployment, e.g. which technology, when it would be implemented, how much CCS would be developed, etc., are not possible to define because of intractable uncertainty.

8.4   Comparison of CCS and Nuclear Power

In the UK, as is likely to be the case in many other national contexts, a key debate is already emerging between the advocates of nuclear power and fossil generation with CCS. Whilst in theory there is sufficient potential for renewable energy to supply the UK's energy entire needs, in practice there are problems arising from the scale and costs of deployment which would thereby be necessary. The Government estimates that it is likely to need at least 30 per cent to 40 per cent of electricity to be provided by renewables by 2050 to meet the minus 60 per cent target (DTI, 2003a, pp .4-5). No robust estimate of the potential upper limit of electricity supply from renewables in the UK has yet been determined (Nick Jenkins, pers. Comm., 15.3.06). Nevertheless, there is a good chance that base- and peak-load will be supplied from either new nuclear capacity or from new fossil fuel generation with CCS (as is also the implication of the Government's own modelling work as reported in Section 8.3).

It is uncertain whether there would be sufficient capital available to support both nuclear *and* a fossil fuel CCS programme. Some experts have suggested that a nuclear programme on the scale of that anticipated by nuclear advocates (c. 10 GW capacity) would inevitably limit the capital available for investment in other energy technologies (MacKerron, 2005). Other experts do not agree with this and consider that there are a range of investment options which could be explored and which could, conceivably, support both a nuclear and fossil fuel power plant build programme. Indeed, MacKerron *et al.* have concluded in a more recent publication that: 'given the international interest in financing electricity generation it is perhaps unlikely that the existence or otherwise of a UK nuclear programme would have any material impact on the availability of finance for other sources' (SDC, 2006a). Whilst MacKerron *et al.* were referring specifically to the impact of investment in new nuclear capacity upon renewables, the same point would seem to apply equally to investment in fossil fuel CCS.

Given that some guarantee would have to be provided by government to limit the investment risk attached to new nuclear, and that some type of incentives scheme is almost certainly required to support application of CCS, the main constraint on deployment may well be whether Government is itself prepared to develop support schemes for both technologies. In this section we compare the sustainability of fossil fuel with CCS and nuclear power, i.e. in terms of their economic, environmental and social impacts and repercussions. It is not possible for this to be more than a preliminary evaluation, since a huge range of issues is involved and we cannot do justice to their complexity: this would take a book in its own right to achieve.

Economic Dimensions of Sustainability

Estimating the costs of new energy technologies is notoriously complex and fraught with uncertainty. There is no single indicator of costs and therefore we will consider three different ways of comparing the costs of CCS energy technologies: namely the cost of abating a tonne of CO_2, the capital costs of plant construction, and the overall construction, operation and maintenance costs.

The UK Government's Performance and Innovation Unit (PIU) compared the carbon abatement costs of the major mitigation options, estimated for 2020 (i.e. taking technological change into consideration) (PIU, 2002). In making these calculations it is assumed that Combined Cycle Gas Turbines (CCGT) are replaced through adoption of the specified technology (for the other detailed assumptions see Table 6.1, PIU, 2002). For nuclear, the range (minimum to maximum value) in £ per tonne of CO_2 avoided was £19 to £55. The quoted range for CCS was £22 to £76. These figures were updated in the DTI's Carbon Abatement Technologies report (DTI, 2005), though estimated for 2010 rather than for 2020, and the new range for nuclear was quoted as £14 to £26 and that for CCS at £15 to £88 tCO_2 avoided (though the high figures refers only to constructing a brand new coal plant). It can be seen that the lower cost estimates have come down such that there is considerable overlap at the lower end of the cost range between nuclear and fossil CCS. It can also be seen, however, that the carbon abatement costs for nuclear at the top end of the range appears to have come down, whereas the costs for CCS at the top of the range have been increased some what. The IPCC's (2005) more recent cost values for CCS have a lower range than the DTI (2005) estimates at $30 to $90, or £17 to £52 tCO_2 avoided. The Science and Technology Committee provided a similar range, the cost for CO_2 emissions avoided being from £17/tCO_2 for coal to £40/tCO_2 for gas (HoC, 2006). Given the high uncertainty attached to such cost estimates of two technologies which have yet to be implemented we suggest that, to all intents and purposes, the carbon abatement costs of nuclear compared to fossil CCS are therefore indistinguishable.

Turning to plant construction or capital costs the IEA (2004) estimates that a new fossil CCS plant would cost between $500 million and $1000 million (£300 million to £600 million) of which about half represents the additional costs required for CCS, but the IEA do not specify the plant capacity. The IPCC (2005) has estimated the costs of alternative CCS options with current technologies as follows: a new supercritical pulverized coal plant with post-combustion CO_2 capture is between £1100 and £1500 per kW (reference plant, i.e. with no capture, £650–850/kW); an IGCC with pre-combustion decarbonisation is between £800 and £1300 per kW (reference plant £670–900/kW); and a natural gas CCGT plant with post-combustion CO_2 capture is between £520 and £720 per kW (reference plant £300–420/kW). In terms of capital costs, CCS applied to CCGT is therefore the cheapest option with post-combustion PF coal and pre-combustion IGCC being more or less equally expensive. Note, however, the very large uncertainty (20 to 50 per cent) associated with these estimates. In addition to lack of experience in building such plants, uncertainty also arises from unknown exchange rates, commodity costs, labour costs and so on. Capital cost is only part of the real costs of electricity generation, however, since it does not take account of the fuel input and other operational costs, to which we now turn.

The third way of evaluating economic performance is through comparing the costs per kilowatt hour at the power station for the different generation options under current market conditions. CCGTs have, in recent years, been the cheapest option, generating at between 2.0 p/kWh (low gas prices) and 2.5 p/kWh (high gas prices) (British Energy, 2001), though the OECD has extended this range somewhat to 2.4 to 3.8 p/kWh for a 10 per cent discount rate (OECD, 2005). The DTI estimates the costs of fossil CCS to range from 3.2p/kWh for retrofitting CCS to existing coal fired capacity, 3.6 p/kWh for new gas fired plant with CCS, and 3.8p/kWh for new coal fired plant with CCS (Marsh et al., 2005). The UK Committee on Science and Technology has selected a higher baseline figure of 3.4p/kWh for gas-fuelled generation reflecting recent higher gas prices (£4/GJ) (HoC, 2006); the Committee's estimates of the costs of electricity generation from gas and coal with and without CCS, based upon analysis of five independent sources, are provided in Table 8.1. The reason why the studies disagreed on the costs of electricity generation from gas with CCS are not evident (HoC, 2006). If a lower gas price is assumed (£3/GJ) then a lower figure for the cost of generation with CCS should be assumed (i.e. 4.3p/kWh). Also included in Table 8.1 are earlier estimates for the 2003 Energy White Paper (recorded in HMT, 2006) which extend the range with respect to coal generation with CCS. The IPCC Special Report concurs with these figures, providing a range of generation costs for CCS of between 4.3 cents/kWh and 9.9 cents/kWh or 2.5 p/kWh to 5.7 p/kWh.

Table 8.1   Costs of power generation with and without CCS

	Without CCS	With CCS	Difference
Coal (pre- or post-combustion) (HoC)	2.6 p/kWh	3.7 p/kWh	1.1 p/kWh
Coal (2000) (HMT)	3.6 to 3.9 p/kWh	5.7–6.1 p/kWh	1.8–2.5 p/kWh
Gas (£4/GJ) (HoC)	3.4 p/kWh	4.3–5.7 p/kWh	0.9–2.3 p/kWh
Gas (2000) (HMT)	2.2–2.4 p/kWh	3.5–3.7 p/kWh	1.1–1.5 p/kWh

Source: HoC, 2006, p. 51; HMT, 2006, p. 14

In summary, the additional cost of CCS appears to be 1–2 p/kWh (HoC, 2006). The evidence does not currently allow us to select a clearly cheaper technology-fuel choice for CCS, i.e. we cannot presently discriminate between gas and coal, and between pre- and post-combustion capture routes for coal.

Nuclear power operators believe that they will be able to generate at between 2.2 and 3p/kwh (British Energy, 2001) with the new designs that would be constructed in a new nuclear build programme. OECD (2005) has provided a more optimistic view of nuclear generation costs, with a range from 1.3 to 1.9 p/kWh (5 per cent discount rate) (i.e. cheaper than CCGT implying that nuclear would become the marginal electricity generation technology of choice) to 1.8 to 3.0 p/kWh (10 per cent discount rate). In its review of the evidence, however, the PIU report concludes that: 'the central inter-quartile range of nuclear costs is 3 to 4 p/kWh, with both lower (industry) and higher outcomes possible' (PIU, 2002, p. 196). Comparing seven studies of the costs of nuclear generation, three fall within the PIU range, whilst the other four studies all estimate generation costs at less than 3 p/kWh (Howarth, pers. comm., 2005). There is clearly significant uncertainty attached to nuclear generation costs. The company E.ON UK plc provided the following cost estimates for different generation options:

Coal using CCS: 3.9 – 5.1 p/kWh
Nuclear: 2.5 – 4.0 p/kWh
Onshore wind: 4.2 – 5.2 p/kWh
Offshore wind: 6.2 – 8.4 p/kWh

One reading of the figures would appear to indicate that nuclear has some cost advantages over CCS, though there is once again too much overlap in the two ranges to make this claim with any reliability. Furthermore, to put these figures into a meaningful context it is necessary to examine the economic and commercial conditions which govern private sector investment in a liberalized energy market.

The Commercial Context into which CCS is being introduced

The reality in the UK is that CCGTs have been the only technology in which private sector investment has been evident since privatisation (accounting for 22 GW out of the 25 GW capacity which has come on line in the UK since 1990). Investment in renewables has taken place because of large public subsidies which have made it economically attractive to invest in them. There is general agreement amongst analysts that renewables such as wind, micro-hydro, and solar are all more expensive than coal, gas or nuclear (ranging from 50 per cent more to 2 to 3 times more costly) (OECD, 2005). MacKerron (2004) has accounted for the advantages of CCGTs as follows: '[they] were also (apparently) a low risk option, well-suited to the new market conditions of liberalization. In these new conditions, wholesale power prices were in principle outside the control of generators, so that unexpected cost increase could no longer be automatically passed through to consumers' (MacKerron, 2004, p. 1959). The reasons why CCGT were more suited to the new conditions are as follows (after MacKerron, 2004, see also Economist, 2001 and Winksell, 2002):

- they could be planned and built without opposition in about 2 years, allowing investors to begin to recoup capital investment rapidly;
- since CCGTs are modular, they can be built in a range of sizes from 300 MW upwards. CCGTs at 300 MW are as efficient as plants at higher capacity, hence there is no economy of scale which dictates an optimal plant size;
- the total capital outlay for CCGTs is relatively small, accounting for 25 to 30 per cent of the generating costs (their principal costs being fuel);
- as demand for CCGTs grew, competition between plant and equipment manufactures developed and it was possible for purchasers to agree fixed cost construction terms with plant suppliers;
- financing costs were therefore minimized because of the low cost of capital required.

MacKerron proceeds to analyse nuclear power in the same terms. His conclusions are as follows:

1. The planning and construction time for nuclear is lengthy, meaning that the earliest a reactor could generate electricity is perhaps 2020. The quoted costs of nuclear power generation (2.2 to 3 p/kWh or even lower) are not based on real operating plants (whereas the quoted costs of CCGT are based on actual plant), and the last nuclear reactor built in the UK produces electricity at about 6 p/kWh (PIU, 2002). The quoted figures for cost are based upon the construction of new reactors such as the CANDU or AP1000 designs, but no such reactors have yet been built. Hence, there is a risk in assuming that costs will be as low as quoted. As the PIU note: '[the 3 to 4 p/kWh range] still represents a major decrease in costs compared to all previous nuclear construction in the UK, including Sizewell B' (PIU, 2002, p. 196).

2. Nuclear power plants are very capital intensive, with capital costs accounting for 70 per cent of the total cost. This means that generating costs are highly sensitive to any increase in cost or time required for plant construction, or to a lower plant performance than anticipated. MacKerron notes that: 'For all these reasons, nuclear vendors are unlikely to match the cost or performance guarantees now commonplace for CCGTs and the required rate of return for nuclear power is likely to be higher than for CCGTs, inflating total capital costs significantly' (2004, p. 1960).

3. There are important economies of scale entailed in building nuclear power plants, at least to the designs being envisaged in a new build programme. Current reactors tend to be a minimum size of 1000 MW. Furthermore, the lower costs quoted are predicated upon a series of nuclear reactors being commissioned. Ten or so reactors (10 GW capacity) would need to be built before the 2.2 p/kWh cost would be realized. As MacKerron comments: 'The problem here is inflexibility: private markets are unwilling to commit easily to 1000 MW of new capacity of a new and untried technology, and attempting to commit at one moment to 10,000 MW is virtually unimaginable' (2004, p. 1960). An alternative perspective is that new reactor designs such as the EPR and AP1000 are only adaptations of designs, some of them over 25 years old, that are delivering 90 per cent availability in several countries (Butler, pers.comm., March 2006). It has also been pointed out that investors would not be required to put up the finance for 10 new plants at once, but staged over a 10 to 15 year period of re-build, i.e. one every 12 to 18 months (Bull, pers. comm., 16.2.06).

4. The long time scales involved in constructing nuclear power plants means that the cost comparisons with other energy generation technologies may be some what misleading. By 2020 the costs of renewable sources of energy and of other potential energy technologies should have been reduced through unit cost reduction as the level of production increases and technological and managerial innovation.

MacKerron's framework is very helpful, though it only considers the situation from the perspective of the private investor, not from society or the economy more generally. What is a desirable level of flexibility from the perspective of a private investor is not necessarily the desired-for flexibility from a societal-wide perspective. This limitation can be illuminated by considering the issue of fuel input prices. World oil prices trebled between mid-1999 and mid-2000 and nearly doubled again between mid-2004 and mid-2005 (Grimston, 2005). There have been major changes in fuel prices paid by UK power producers in the period 2004 to 2006. The cost of oil rose from about 1.2–1.3 p/kWh in the period 2003 to the last quarter of 2004 to 2.5 p/kWh in the middle of 2005. Gas prices have also risen dramatically in the last quarter of 2005, doubling in price in the course of a single week! Even the price of coal has increased slightly, though it has typically been much less volatile than gas or oil. If the cost of fuel inputs increases unexpectedly, then a power plant will have to produce at a high cost and pass on the costs to its customers. If an energy system is over-reliant on this type of fuel, then the costs to society and the economy as a whole will be large. In late 2005, the sharp increase in gas prices, and the importance of natural gas to the UK energy supply system (constituting 40 per cent of primary energy demand) has led to rapidly escalating energy bills for companies and other organizations. It appears that part of the explanation for the increase in energy costs arises from the lack of a free market in other European countries, which is not some thing which is readily within the control of the UK government or UK businesses to change. There is, therefore, value in having diversity in the energy supply system, including in the fuel types used.

Whilst a private investor may be primarily interested in a rapid return on capital investment, the wider societal perspective might put more emphasis on the running and operational costs. At a time of high fuel input prices, there is benefit in technologies such as renewables, nuclear and coal power plants, since the price of their input fuels is either zero or has conventionally been less volatile than for gas and oil based technologies. This is, in part, a function of the sheer availability and dispersion of these fuel types compared to gas and oil, which tend to be far more spatially concentrated and hence more subject to control politically and economically by a few countries (Deffeyes, 2005). Uranium supplies are regarded as being relatively under-explored, hence it is assumed that if the price should increase, then new supplies would be identified (SDC, 2006).

MacKerron (2004) uses his economic-based framework to assess renewable energy. The costs of different renewables range widely, but on-shore wind can sometimes compete in price with fossil fuel generation. Renewables tend to be capital-intensive investments, the fuel often being free (e.g. wind or wave power). They tend not to be subject to large economies of scale, however, and tend to be economic at a small scale. Renewables can be planned, constructed and in operation quickly, though planning problems have frequently thwarted rapid deployment of technologies such as onshore wind. It is possible to make a return on investment in renewables incrementally and fairly quickly, i.e. units can be constructed and operational within a short period of time, and successive units added over time.

MacKerron (2004) argues that this feature of renewables also means that technological learning can occur rapidly, since the costs per unit are relatively low and the design life relatively short. Innovation and learning can therefore be incorporated effectively and rapidly into new designs, whereas larger plant has a much longer life-time and there are a smaller number of actual units from which learning can occur. If necessary, renewable technologies can be dismantled and removed without incurring a high cost because of the typically low unit cost and this is another feature which makes them more flexible than large plant.

How then would fossil fuels with CCS compare with CCGT, nuclear and renewables with respect to the economic criteria set out above?

1. Construction and planning timescale: The timescale is clearly project specific depending on the choice of site, technology, CO_2 sink and any potential public or stakeholder concerns or objections. It should nevertheless be possible to construct and operate a fossil fuel CCS system in 4 to 6 years from the decision to proceed (E.ON, pers. comm., 2005). Two projects in the UK have been announced, the BP Peterhead/Miller project and the Progressive Energy Teeside IGCC and EOR project, both of which have proposed a 2009 start, implying 4 years development time. In both these cases, the use of EOR means that the legal situation with respect to the London and OSPAR Conventions is clear, i.e. they are permitted operations. If CO_2 storage were to be planned without EOR then there would be a higher risk arising from the legal uncertainty regarding CO_2 storage. Fossil CCS is therefore somewhere between CCGT/renewables on the one hand and nuclear on the other hand in terms of construction timescale. It does appear to perform better in this regard than nuclear power though, as with new nuclear designs, no one has yet constructed a large integrated CCS power plant and pipeline facility.

2. Modularity and costs: Whether fossil CCS is a modular technology or not depends upon whether the fossil fuel is gas or coal. If gas is used, as in the proposed BP Peterhead/Miller scheme, then the benefits of modularity can be enjoyed since a CCGT type technology is employed, the unit size being 350 MW. If coal is the fuel of choice, on the other hand, a larger plant is required for economy of scale, such as the Teeside IGCC, which is 800 MW. As with the new nuclear reactors, the real costs of fossil CCS are not known, since no large plants have been constructed. There is, therefore, a real risk to investors that the costs may be higher than anticipated, but they might also be lower and reduce over time through innovation.

3. Capital costs of fossil CCS: these will again depend upon whether gas or coal is the proposed fuel. Nevertheless, it is likely that the additional CO_2 capture, transportation and storage technologies and infrastructure would add considerably to the overall capital costs. Fossil CCS is therefore likely to be a somewhat 'lumpy' investment and there is also the problem that a sufficient number of plants and pipelines need to be constructed before the costs of CO_2 transport and storage decrease and stabilise. Freund (2005) estimates that the

lower transportation costs will only be realized when 10Mt/y of CO_2 are being piped through the system, yet this is likely to be a larger volume than that generated by a single point source (which may typically range from less than 1Mt/y for a gas scrubbing plant or chemical plant to 5 or 6 Mt/y for a 1GW coal power station). Fossil CCS therefore demonstrates some of the limitations of nuclear with respect to 'lumpiness' and inflexibility, but probably to a lesser extent than nuclear.

4. MacKerron estimates that a new nuclear plant might cost approximately £1,500 million for a 1000 MW plant, or £1500/kW (MacKerron, 2005). On the other hand, the OECD (2005) has estimated the 'overnight construction' cost of new nuclear at between £600 and £1200 per kW. One of the difficulties is that original sources of data tend to be reactor manufacturers whose tendency is towards appraisal optimism, and the lack of experience of building new nuclear power plants means that the validity of estimates is difficult to assess independently (MacKerron, pers.comm., 19.3.06). The new Finnish reactor is estimated to cost approximately £1,300/kW (ibid.). If £1300 to £1500/kW is closer to the real costs of new nuclear, and if the costings are comparable, then a fossil CCS plant appears to be less capital intensive and less lumpy than nuclear, with the exception of IGCC with CCS (which would cost about the same). However, if the OECD figures are employed, then it is not possible to make a definitive conclusion on whether fossil CCS or nuclear is the more 'lumpy' type of investment, since the range of estimates for CCS (£600 to £1300/kW) fall within the range of that for new nuclear.

5. As noted above, in terms of costs per kWh, it is not possible to distinguish reliably between new nuclear and fossil plant with CCS.

6. Future fuel prices: We cannot distinguish readily between coal, nuclear and renewable technologies with respect to potential future fuel costs, since all have favourable and reasonably stable fuel input costs (compared to gas and oil), at the same time as they all have high capital costs.

In its review of the role of nuclear power in a low carbon economy, the Sustainable Development Commission (2006) decided that it was impossible to put a reliable figure on the cost of new nuclear and pointed out three problem areas.

1. The waste and decommissioning aspects of the cost calculation for nuclear are highly uncertain, in large part because they occur so far into the future: 'With UK nuclear waste policy still undecided, there are no certain estimates of the total cost of waste disposal for new-build plant' (SDC, 2006, p. 10). Putting aside funds to cover these future costs may turn out to be insufficient due to changing requirements for waste storage, whilst revenue from the plant may be less than anticipated. To some extent the same argument could be leveled at long-term CO_2 storage. The decommissioning costs for new nuclear reactors alone are estimated at between £220 million and £440 million per GW of

capacity (SDC, 2006). The costs of nuclear waste disposal from new build reactors in the UK has been estimated at between £200 and £300 million for a 60 year plant life time (Chapman and McCombie, 2006). This cost estimate assumes, however, that ten new nuclear reactors would be constructed. If only one plant were commissioned, the disposal costs would, presumably, be significantly higher. Thus waste and decommissioning for new nuclear plant is estimated to cost between £440 and £770 million for a 1GW plant over 60 years. The decommissioning costs for an equivalent sized coal power plant are likely to be an order of magnitude smaller and could even be negative, depending upon the value of scrap metal and the extent of necessary land remediation. The disposal costs for CO_2 are approximately £8/tCO_2 (£2/tCO_2 for transportation assuming movement of the CO_2 over a distance of 250km in quantities of 10 Mt/y; £5/tCO_2 for the costs of injection; and £1/tCO_2 for the costs of monitoring over the life time of the project) (Freund, 2005; IPCC, 2005). For a 1 GW plant, this might translate to a total yearly disposal cost of between £25 and £50 million depending upon the load factor of the plant. Over a 30 year life time, these costs would amount to £750 to £1500 million. It is, therefore, difficult to distinguish between the waste and decommissioning costs for a new build coal-fired plant with CCS and a new build nuclear reactor, especially since the costs for nuclear assumes a serial-ordering of ten new plants.

2. A 'moral hazard' exists in relation to the nuclear industry, whereby, '… investors and companies who decide to take part in nuclear projects, both directly and indirectly, may be willing to take on higher levels of risk than otherwise under the expectation that the Government would be unwilling or unable to let the project or enterprise fail … This will tend to depress balance sheet costs, but could in the long run lead to large costs for the taxpayer, in effect acting as a form of subsidy that is virtually impossible to avoid' (SDC, 2006, p. 10). Based on the experience of the UK's privatized market in constructing and operating gas-fired power plants, it is unlikely that such a moral hazard would attach itself to fossil CCS plant.

3. There are additional external costs associated with nuclear power that are not usually included in standard cost calculations, including: safety and security arrangements, limited liability guarantees and the cost of possible foreign policy interventions in securing access to uranium (SDC, 2006). The fossil fuel supply chain is unlikely to incur such costs to the same extent as nuclear.

In summary, we cannot make any definite conclusions in an economic comparison between new nuclear build and fossil fuel plants with CCS. Both types of investment are rather lumpy and suffer from inflexibility compared to CCGT, but the influence of changing fuel and CO_2 prices is becoming ever more important, as analysed in IPCC (2005). Where the price of both gas and CO_2 are low then natural gas will be used in CCGT. As the gas price rises, coal will come to displace gas. This transition to coal occurs at a higher gas price as the carbon price increases from zero because of the higher carbon intensity of coal compared to gas. At a certain price of CO_2 abatement CCS is introduced. Gas with CCS is deployed at low gas prices, but coal with CCS is the preferred fuel and technology as the gas price rises.

Table 8.2   Deloitte UK power generation scenarios for 2020 and associated generation and investment costs

Scenario	Power Generation Mix (%)						Average generation costs (p/kWh)	Total capital expenditure (in £ billions)
	Gas	Existing Coal	Nuclear	Coal CCS	Renewables	CHP and fuel cells		
Business-as-usual low gas price	70%	10%	5%	0%	15%	0%	3.1 p/kWh	£22
Business-as-usual high gas price	50%	10%	5%	20%	15%	0%	4 p/kWh	£32
Diversified Portfolio	30%	0%	15%	20%	20%	15%	4.1 p/kWh	£51
Low Carbon	30%	0%	30%	25%	15%	0%	4 p /kWh	£50
Variant: Low Carbon (no new nuclear)	30%	0%	5%	45%	20%	0%	N/A	£52

Source: Deloitte (2006)

Deloitte's (2006) has undertaken a scenario modeling study for the UK electricity generation sector in which it has explicitly addressed the issue of the investment costs of the alternative scenarios: Business-as-usual, low gas price; Business-as-usual, high gas price; Diversified Portfolio; Low Carbon. The latter two scenario achieve much lower CO_2 emissions (at 77 MtC/pa and 57 MtC/pa respectively) than the two Business-as-usual scenarios (147 MtC/pa and 122 MtC/pa respectively). The four scenarios are compared in Table 8.2. Diversified Portfolio and Low Carbon achieve large CO_2 reductions through uptake of IGCC with CCS, renewables and CHP.

It can be seen from Table 8.2 that the generation costs of the four scenarios are all very similar, with the exception of the low-gas price Business-as-Usual scenario. The capital costs, on the other hand, are very different, with Diversified Portfolio and Low Carbon costing approximately twice the investment required for Business-as-Usual scenarios. A 'no new nuclear' variant of the Low Carbon scenario was explored in which the nuclear component was reduced to 5 per cent and the generation re-allocated to CCS (+20 per cent) and renewables (+5 per cent). The overall capital costs are similar to that for the nuclear version of Low Carbon, though CO_2 emissions are slightly higher at 66 MtC/pa (below the likely UK emissions target for the power sector according to Deloitte, 2006).

Deloitte has also explored the sensitivity of the average generation cost to variations in the fuel price for gas and coal and to the CO_2 abatement price. The Business-as-Usual low gas price scenario is most sensitive of the scenarios to fuel and CO_2 price fluctuations. 'Electricity markets may actually be more tolerant of the Low Carbon and Diversified Portfolio scenarios, as the overall volatility in electricity prices may be lower and generation prices more predictable' (Deloitte 2006, p. 24). The operational risks would be reduced relative to Business-as-usual by reducing imported fuel dependence. On the other hand, these two lower CO_2 emissions scenarios have delivery and affordability risks, associated with the need for technological and logistical availability, and the requirement for the Government to support nuclear new build, coal with CCS and renewables.

In conclusion, Deloitte's analysis supports the argument that we cannot make a clear distinction in economic terms between new nuclear and new coal with CCS. Both incur rather similar capital cost outlays and generate electricity at similar prices. Both tackle problems of fuel security and both bring down the power sector's contribution to CO_2 emissions to substantially below what the 2020 target could be anticipated at. The two technologies are effectively interchangeable in Deloitte's analysis, leading it to conclude that: '...there may be low carbon generation alternatives to nuclear [i.e. coal CCS] which may be economically viable in the generation mix' (Deloitte, 2006, p. 24). The study also confirms that the price of gas, and its volatility, has an important impact upon the desirability or otherwise of new nuclear and new coal CCS plant. In addition to CO_2 emissions reduction, the issues of fuel security and operational cost of generation are increasingly of importance to investment decision-making on what replaces the existing generation mix over the next several decades. A limitation of existing studies is a lack of analysis of the sensitivity of the cost estimates for nuclear and CCS to relative changes in the costs of the constituent technologies, e.g. percentage changes in capital costs, fuel costs, waste disposal costs, CO_2 reduction credits, etc. (Butler, pers. comm., March 2006).

Research, Development and Demonstration Expenditure

Before we address the issue of research spend, we should first ask why any public money should be spent on energy RD&D at all? Why not leave it entirely to the market to determine appropriate allocation of resources? The conventional argument that has been made for putting public money into energy R&D is the existence of potential positive externalities. The argument here is that because the benefits of R&D cannot always be fully appropriated by the private sector, there will be a tendency for under-investment by the private sector in such R&D. Hence, there is a role for government expenditure in R&D until such point as low carbon technologies are potentially attractive to private developers (MacKerron, 2004).

There has been very little expenditure in the UK to date on CCS RD&D. Public expenditure from 2000 up to the end of 2005 has been approximately £2.5 million (based upon data in HoC, 2006). However, this level is likely to increase quiet dramatically over the next few years. Already, £4.4 million has been committed under approved R&D initiatives lasting over the next 3 to 5 years. More significantly, the government is running the Carbon Abatement Technologies programme, which potentially has £35 million to spend over 4 years on CCS, in addition to the Cleaner Fossil Fuels programme which received £20 million over three years (HoC, 2006). The government's own advisory board on CCS had recommended that the government needed to provide £100 million for demonstrating carbon abatement technologies (ibid.), 3 times more than has been provided. The cost of the BP Peterhead/Miller field scheme is estimated at approximately £330 million. E.ON has announced that it is considering the construction of a new 450MW coal power station with CCS and estimates the cost at around £550 million.

To put these numbers into perspective, the UK government has allocated over £500 million to emerging renewables and low carbon technologies over the period 2002 to 2008 via R&D grants and capital grants, most of which represents expenditure on renewables (HoC, 2006). Historically a large proportion of public RD&D on energy was spent on nuclear fission, and reduction in this budget largely accounts for the dramatic decrease in public R&D on energy in the UK, from several hundred million pounds in the 1970s and 1980s to around £50 million a year or less in recent years (less than 10 per cent of the budget in the late 1980s). Even today, the largest single energy R&D expenditure by far is on nuclear fusion, at £38 million per year, as the UK's contribution to the European fusion research activity, equal in 2001 to the amount of money spent by the DTI's Energy Group on all other forms of energy RD&D (Watson and Scott 2001). This is despite the fact that since the mid-twentieth century, nuclear fusion has consistently been claimed to be 50 years away from commercialization!

At the global scale, the same trend of a reduction in energy R&D budgets has been evident, with expenditure in 2002 half of what it was in 1980. The budget for fossil fuels-based research has been in particular decline, falling from a peak of $2.7 billion in 1981 to just $0.7 billion in 2002, a decrease of 73 per cent (IEA 2004). The IEA has estimated that $159 billion (in year 2000 $) was spent globally on nuclear research between 1974 and 1998 (IEA, 2001). The current budget for CCS R&D is estimated by the IEA to be approximately $100 million – not even 1 per cent of the historic nuclear R&D budget in the 1970s and 1980s.

Excluding the proposed large projects, the amounts of money spent on RD&D to date are not large compared to other expenditure on RD&D and for this reason, and also because CO_2 capture is a new technical challenge, many commentators believe that there is a significant cost reduction potential to be achieved through appropriate RD&D investment. Given the large amounts of money which have been spent on other forms of energy R&D it could be argued that it is now time to spend a lot more on CCS, and concurrently a lot less on 'tried and failed' technologies. On the other hand, MacKerron (2004) argues that the past failures and excesses surrounding nuclear RD&D do not necessarily count against future investment in nuclear RD&D on the grounds of its positive externalities, i.e. the potential benefits for energy security, reliability, and environmental protection. He points out that nuclear technology is diverse and that future designs will need to embrace 'once-through' cycles rather than requiring reprocessing of spent fuel, as well as 'passive' (rather than engineered) safety features. There is also the potential for development of smaller-scale reactors, which would help to increase the modularity and reduce the capital costs of nuclear power.

It could be argued that the urgency of the agenda for decarbonising energy systems is such that funding for CCS RD&D should not be in competition with that for nuclear. I.e. energy RD&D should not at the present time be perceived as a zero-sum game. The IEA (2004) has estimated that a 30 per cent increase in the current global R&D budget for fossil fuels and power & storage technologies is required in order to provide sufficient funds to develop CCS. The UK government's allocation of £35 million to CCS RD&D goes some way towards increasing spend on cleaner fossil fuels. However, as the Science and Technology Committee has recommended, government investment in RD&D probably needs to increase by an order of magnitude in order to 'pump prime' the initial demonstration projects (HoC, 2006).

Environmental Dimensions of Sustainability

In terms of CO_2 reduction per kWh electricity generated, nuclear performs better than fossil CCS. Life-cycle assessment suggests that the CO_2 emissions arising from nuclear power production (i.e. including plant construction, fuel extraction and preparation but not decommissioning or waste treatment) are 16 kg CO_2/MWh (SDC, 2006, p. 5) which is considerably lower than the CO_2 emissions arising from gas-fired power generation with CCS (52 kg CO_2/MWh) and PF coal-fired generation (112 kg CO_2/MWh) (and the figures for fossil fuels are in reality higher since the numbers quoted here do not include upstream life cycle emissions) (see Table 3.3). Nevertheless, the difference in CO_2 emissions per kWh between nuclear and gas or coal with CCS is only 10 per cent of the CO_2 emissions arising from gas or coal without CCS. The Sustainable Development Commission has argued that the contribution of a new nuclear build programme of 10GW would displace 25 MtCO$_2$, a cut of 4 per cent on overall UK CO_2 emissions (from 1990 levels), whilst a 20GW new build programme would reduce emissions by 8 per cent (SDC, 2006). The Commission note therefore that nuclear by itself could only contribute an element of a 60 per cent reduction in CO_2 emissions and that it could not in reality contribute until after 2020. However, more or less the same arguments can be applied to fossil fuel with CCS, since there is a similar scale of potential replacement of old coal-fired plant with new incorporating CCS.

There are major differences between nuclear waste and CO_2. Nuclear waste is highly toxic and remains so for thousands to tens of thousands of years because of the very long half-life of the nuclear elements which are decaying. Plutonium 239 has a half-life of 24,100 years. By contrast, CO_2 is not toxic, though it can cause death by asphyxiation and, if it should leak, could cause localized problems as discussed in Chapter 2. The sheer scale of the nuclear waste problem in financial terms is illuminated by the statistic that the UK government has found it necessary to put aside £57 billion for nuclear waste treatment, remediation and decommissioning (though this is not all for power-related facilities). On the other hand, much of the past cost arises from poor planning and a new-build nuclear programme of 10 GW would add less than 10 per cent to the total UK nuclear waste inventory (SDC, 2006).

It is probably robust to conclude that there are greater uncertainties regarding the risks associated with nuclear waste disposal than there are with respect to CO_2 storage: these risks arise from the longevity and toxicity of nuclear waste compared to CO_2. On the other hand, the small quantities of waste produced by nuclear power means that a given per cent increase in the waste disposal costs will have a considerably smaller impact that a similar percentage increase in the costs of CO_2 storage (Butler, pers. comm., March 2006). Technical solutions to nuclear waste storage are probably available and the most difficult issue appears to be that of achieving a social and political consensus on storage options (MacKerron, 2005). The Sustainable Development Commission expresses its concerns on nuclear waste in the following terms:

> Even if a policy for long-term nuclear waste is developed and implemented, the timescales involved (many thousands of years) lead to uncertainties over the level to which safety can be assured. We are also concerned that a new nuclear programme could impose unanticipated costs on future generations without commensurate benefits.
>
> (SDC, 2006, p. 19).

Some of these same concerns could also be directed at CO_2 storage, however, given that it would take place over thousands of years and is also shrouded in considerable uncertainty as to the longer-term fate of the stored CO_2. Both coal and uranium mining, processing and transportation raise environmental issues and the impacts would need to be evaluated in each country where mining takes place (SDC, 2006). The combustion of coal itself results in the release of uranium and thorium which emit low-level radiation. Gabbard from the Oakridge National Laboratory in the USA (no date) claims that:

> Naturally occurring radioactive species released by coal combustion are accumulating in the environment along with minerals such as mercury, arsenic, silicon, calcium, chlorine and lead, sodium, as well as metals such as aluminium, iron, lead, magnesium, titanium,

boron, chromium, and others that are continually dispersed in millions of tons of coal combustion by-products … coal-fired power plants are allowed to release quantities of radioactive material that would provoke enormous public outcry if such amounts were released from nuclear facilities.   (Gabbard, no date).

Gabbard also argues that if operators of coal-fired power stations had to take responsibility for the release of radioactive materials from coal then their economic advantage relative to nuclear power stations would be reduced. This opinion appears to be a minority viewpoint, however. Other experts point out that uranium is also found in normal soil dust and in sea salt which gets distributed in the coastal zone as spray (Gibbins, pers. comm., 2006). Whilst there are radioactive substances in ash from coal plants, they are not considered to be at levels likely to be harmful to the workforce or the general population (although there may be other health effects associated with the ash and the way it is discharged to the environment, especially in developing countries) (Bull, pers. comm., 2006). These issues clearly need to be further analysed and included in a comprehensive assessment, but they are beyond the scope of the present evaluation.

Social Dimensions of Sustainability

Stakeholder and citizen preferences The social aspects of sustainability are defined here as public and stakeholder preferences and opinions and the socio-political risks associated with deployment of alternative energy options given such perceptions. In Chapter 5 we provide evidence that the public sample consulted in this research tended to support fossil CCS over nuclear as a way of achieving large cuts in CO_2 emissions. In Chapters 6 and 7 we discovered that CCS is accepted as a legitimate component of decarbonisation scenarios for the two regions studied. None of the respondents reacted against fossil CCS in the way that many did with respect to nuclear generation. In one of the regions (EMYH), there was some positive support for fossil CCS from respondents, whilst in the NW it was accepted as a legitimate component, though a Renewable Generation scenario without CCS was preferred by most respondents. For just about all respondents there was acceptance of the benefits of fossil CCS in terms of infrastructure, costs, public opposition and reliability, though opinion was more mixed regarding environmental impacts, energy security and potential for disaster. It seems a reasonable conclusion to draw that the public and stakeholder samples consulted accept fossil CCS as a legitimate component of sustainable energy scenarios for 2050 and would, in the majority, have some preference for fossil fuel CCS compared to nuclear. However, much more research is required to confirm this preliminary finding.

    Within the literature on risk perceptions, radioactive waste and nuclear reactor accidents have tended to be regarded as a technology which is both 'unknown' and 'dreaded' (compared to most other technologies); characteristics which have tended to increase the negative perception of risk associated with nuclear power (Slovic, 2000). Slovic's work suggests that coal mining accidents are also regarded with 'dread', but are considered to be reasonably well 'known' risks. Yet, whilst risk perceptions may emerge over time and be relatively static, and certain technologies such as nuclear power 'stigmatised' (Slovic, 2000), some surveys of public perceptions of nuclear power in the UK have shown that there has been some movement in opinion over the past few years towards a more favourable perspective. Hence, one survey repeated the following question over the past few years: 'To what extent would you support or oppose the building of new nuclear power stations in Britain to replace those which are being phased out over the next few years? This would ensure the same proportion of nuclear energy is retained.' In 2001, 60 per cent were opposed to replacement and 20 per cent were in favour, but by the end of 2004 support and opposition for replacement were about equal (35 per cent supported, 30 per cent opposed) (Grimston, 2005). At the end of 2005, the supporters were at 41 per cent compared to 28 per cent against (MORI, 2005). Meanwhile, when asked 'How favourable or unfavourable are your overall opinions or impressions of the nuclear industry/nuclear energy', the percentage of 'favourable' opinion crossed over 'unfavourable' opinion in December 2004 and in late 2005 stood at 33 per cent favourable, and 27 per cent unfavourable (MORI, 2005).

It is interesting to note that there was a large increase in respondents replying unfavourably to the above question from early 2000 to late 2002 and this is likely to be due to the controversy surrounding the MOX data falsification scandal that broke in September 1999, resulting in the resignation of the Chief Executive in January 2000 (Bull, pers. comm., 2006). This suggests that public opinion on nuclear power is highly sensitive to perceived accidents and scandals, a finding which is consistent with other research on the public response to technical risks. By the same reasoning, it is plausible to suggest that public perceptions of CCS might also be highly sensitive to perceived accidents, incidents and scandals.

Grove-White *et al.* (2006) have pointed out that other recent surveys do not entirely confirm the MORI results. For example, an ICM poll for the BBC in May 2005 found that only 21 per cent of respondents supported the proposition that nuclear represented 'the most feasible way of meeting the UK's future energy needs while reducing carbon dioxide'. 52 per cent considered it wrong 'for the government to consider nuclear power as an energy source for the future', compared to 39 per cent who considered it right. A further poll in July 2005 found that only 18 per cent of respondents supported building new nuclear power stations as a way of reducing the energy gap following the phase-out of nuclear, whilst 79 per cent of respondents supported 'investing in (other) energy generating plants, particularly renewables'. 59 per cent of the sample felt closest to the statement that: 'nuclear waste problems remain and it would be irresponsible to build new nuclear power stations' (Grove-White *et al.*, 2006). The problem of nuclear waste appears to be especially important in accounting for negative perceptions of the nuclear option. The some what conflicting nature of the results reported above point to the need for much more research on public perceptions of energy options and what influences them, especially in the comparative context of the major decarbonisation options.

Part of the reason for the possible shift in perceptions of nuclear power may be the rising importance of climate change on the political agenda. Climate change is regarded by many as posing greater risks than nuclear power. For example Poortinga *et al.* (2006) note that in their survey 41 per cent of respondents agreed with the statement that it is better to accept nuclear power than to live with the consequences of climate change, whilst 54 per cent were willing to accept the building of new nuclear power stations if this would help to tackle climate change. Some well known environmental scientists and commentators have argued that nuclear power might well be needed as a key technology to deploy in response to climate change, most notably the famous inventor of 'Gaia theory' James Lovelock (2006) but also Sir Crispin Tickell (Tickell, 2004). Certainly the MORI 2005 poll showed a huge increase in public concern over climate change. When asked 'What issues to do with the environment and conservation, if any, concern you most these days?', by far the most frequent response was climate change at 34 per cent, representing an increase in 19 per cent from 2004 (MORI, 2005). The respondents also prioritized 'effect on the environment' and 'effect on global warming' as the two most important factors to take into account when deciding which methods of energy/electricity production should be used in the future. The three main benefits identified by respondents arising from the use of nuclear as a source of electricity were: 'ensure reliable supply of electricity', 'no carbon dioxide produced/little impact on global warming', and 'clean air' (MORI, 2005).

The survey results do not permit us to conclude that public perceptions towards nuclear energy have become more favourable over recent years *because* of growing concern about climate change, though there is good reason to relate the two processes. One survey (Poortinga *et al.*, 2006) has concluded that a reasonably high level of acceptance of nuclear power by the British public is related to its contribution to climate change mitigation, though it also notes that it is a 'reluctant acceptance' akin to the lukewarm acceptance of CCS that we observed in our focus group work (Chapter 5). Existing surveys also seem to confirm that the respondents' expressed a strong preference for renewables and demand reduction rather than nuclear power, if such a choice were made available to the public (Chapter 5; Curry *et al.*, 2005; Poortinga *et al.*, 2006).

Another survey has illuminated that Members of Parliament have a perception of much greater public opposition to nuclear power than appears to be the case, e.g. the survey identified that MPs thought that 84 per cent of the public would be unfavourably inclined to the nuclear industry, whereas only 25 per cent of the public sample actually expressed this sentiment. Likewise, the MPs massively underestimated the percentage of the public who were generally favourably inclined towards the nuclear industry (Grimston, 2005).

Grove-White *et al.* (2006) have argued that any shift in favour of nuclear power would be highly conditional on the political, industrial and institutional framework in which nuclear were to be developed. Based upon an analysis of the development of the nuclear industry in the UK from the 1950s to the late 1980s, Grove-White *et al.* point to the unusually centralizing, inflexible and security-sensitive character of nuclear energy technology. For example, the highly complex and extended nuclear fuel cycle required long-term and expensive commitment from Government, whilst the possibility of catastrophic accidents led to unique and privileged liability terms. Particularly onerous security arrangements were required because of the properties of radioactive materials. The notion of 'series ordering' of power plants to achieve economies of scale was promoted in 1973 and 1979, but if accepted, would have reduced the flexibility available to policy makers to modify policy decisions as circumstances changed. Grove-White *et al.* (2006) point to the tendency of the nuclear industry and its advocates to attempt to re-create these centralizing and inflexible properties of a new nuclear build programme, e.g. through proposed changes to the land use planning system, streamlining of generic licensing procedures for new reactor designs and guaranteed series-ordering of nuclear plants. They conclude that such advocacy by the nuclear industry could back-fire because such proposals could readily ignite opposition from a wide range of influential public and policy constituencies based upon the perceived undesirability of highly centralized inflexible technologies. Grove-White *et al.* do not argue that a new nuclear programme is entirely infeasible, however. Rather:

> Greatly increased openness and transparency about the economics of the industry, a convincing long-term solution to the nuclear waste issue, and the adoption of manifestly 'fail-safe' new reactor types without the associations of past ventures ... could conceivably alter the picture to bring about shifts in public responsiveness.
>
> (Grove-White *et al.*, 2006, p. 12)

There are further socio-political risks associated with development of nuclear power and fossil fuel with CCS which need to be incorporated into any sustainability appraisal. These include the issue of waste (considered briefly above), but also health and safety issues and the potential for disaster.

Accident record and health and safety issues   The risk of accidents in the nuclear industry have to be taken extremely seriously, given their potential to become major disasters. It is not possible here to review comprehensively the accident records of nuclear installations, but the name Chernobyl has become synonymous with major industrial disaster. The Reporting of Injuries, Diseases and Dangerous Occurrences Regulations (RIDDOR) of 1995 describe the requirements for employers to report certain occurrences relating to health and safety in the following categories: deaths, major injury, over-three day injury and reportable injury (a category which includes both major injury and over-three-day injury). There have been no fatalities between 1995 and 2005 in the following organizations, which constitute the bulk of the 'nuclear sector' in the UK: British Nuclear Fuels (BNFL), British Energy (BE), Atomic Weapons Establishment (AWE) and UK Atomic Energy Authority (UKAEA). Most other sectors in the UK have less impressive fatality rates, the worst being Agriculture, Hunting, Forestry and Fishing, where there is 1 death per 11,000 workers per year averaged over 1995 to 2005. In terms of injury rates, the nuclear sector also compares favourably with other sectors, with approximately 328 injuries per 100,000 employees per year (inj/100k/y); this compares with 487 inj/ 100k/y in the service industries, 614 inj/100k/y in all of industry, 1130 inj/100k/y in manufacturing and 1350 inj/100k/y in extractive and utility industries (Health and Safety Statistical Highlights, 2002/3). Other types of incidents related to nuclear operations in the UK (i.e. not related to health or injury) have to be recorded and these range from 0 to 7 per annum over the last 5 or so years (2001:7, 2002:5, 2003:0, 2004:2) (HSE website).

Comparing the health and safety risks of nuclear power with coal is problematic in that it is not comparing like-with-like. For coal mining the risks are associated with occupational injury and death from accidents, mostly underground and often involving a few workers. The major coal-related diseases are those induced by dust such as pneumoconiosis and emphysema, noise-induced hearing loss and musculoskeletal disorder (HSE website). There have been few if any direct risks to non-employees in the UK arising from use of coal in the past few decades, though in the past many thousands of people are thought to have died prematurely as a consequence of smog in urban areas arising from the combustion of coal without proper cleaning of the emissions. One infamous smog incident in London in 1952 is thought to have resulted in an additional 3,500 to 4,000 more deaths over and above that expected for the period (Farmer, 1997).

Whilst the nuclear industry does of course encounter occupational health and safety hazards, the wider risks of greatest concern are those arising from potential explosions and contamination to employees, local residents, ecosystems and indeed members of the public and ecosystems in other countries. The Chernobyl accident had very far-reaching consequences, for example, with restrictions on the sale of agricultural produce in North West England still in force nearly 20 years later due to retention of radioactive elements in soils.

Table 8.3   Health and safety statistics for the UK coal industry

Year	Fatalities	Major injuries	Over 3 day injuries	Total injuries	Per 100,000 workshifts	Diseases
1997/98	4	163	931	1098		6
1998/99	4	109	879	992		28
1999/2000	0	95	752	847		122
2000/2001	0	89	747	836		199
2001/2002	1	80	690	771		328
2002/2003	0	62	542	604		408
2003/2004	0	48	444	492	31.61	237

*Source: Health and Safety Executive, Coal Mines – Incident statistical digest (provisional),
www.hse.gov.uk/mining/accident/coalisd3.htm*

We will look first at the occupational health and safety risks in the coal industry. Because approximately 60 per cent of the UK's coal is imported, it would be desirable to examine the health and safety record of the major countries from which coal is imported: Australia, the USA, Colombia, Poland, Russia and South Africa. Note that in the future, coal imports could occur from different countries. Unfortunately there is no comparable data set on coal-related deaths, accidents and disease in these countries (Vannieuwenhuyse, pers. comm., 2005). Table 8.3 summarises the number of fatalities, injuries and illnesses suffered within the UK coal mining industry from 1997/8 to 2003/4.

The Uranium Information Centre Ltd (UIC) in Australia has compared energy-related accidents between 1977 and 2004 with serious reactor accidents (www.uic. com.au/nip14app.htm). The UIC lists 45 large-scale energy-related accidents of which 27 are related to coal-mining and account in total for 3113 deaths, an average of 115 deaths per accident. Over half of these accidents occurred in China, followed by the Ukraine (5) and Russia (3). In the last few years there have been approximately 6000 deaths per year from coal mines in China. (The number is larger than the previously quoted figure because many of the fatalities arising from Chinese coal mining occur in small mines and involve one or a few deaths). The yearly death toll in Ukranian coal mines has varied from 270 (in 1999) to 459 (in 1992). Even in Australia, with one of the safest coal mining industries in the world, there have been 112 deaths since 1979. The UIC claims that coal mining deaths are 7 per million tonnes of coal mined in Ukraine, 5/Mt coal in China, 0.034/Mt in the USA and 0.009/Mt in Australia. Mining has become safer in countries which have shifted away from underground deep-shaft mining to open-cast or strip mining. China relies heavily upon deep-shaft mining, whilst safety equipment is also often rudimentary in Chinese mines. The massive range in the death rate from coal mining in different parts of the world suggests that the power generation sector in the UK can to some extent limit the health and safety impacts arising from coal mining through only sourcing imported coal from those countries with the best health and safety records.

The UIC also present data on damage or malfunction of the reactor core of nuclear power stations between 1952 and 1989. Ten such incidents are listed, of which no impact is recorded with respect to four. Two incidents are described as 'very minor radioactive release' (SL-1, USA in 1961, Lucens, Switzerland, 1969) and one more as 'minor radioactive release' (Saint Lauren-A2, France). That leaves three incidents which are, in chronological order:

- Windscale-1, UK, 1957: widespread contamination, farms affected (1.5×10^{15} Bq released);
- Three-Mile Island-2, USA, 1979: minor short-term radiation dose (within ICRP limits) to public. Delayed release of 2×10^{14} Bq of Kr-85';
- Chernobyl-4, Ukraine, 1986: major radiation release across Eastern and Northern Europe and Scandinavia (11×10^{18} Bq).

In terms of deaths, three operatives died from the SL-1, USA accident at an experimental military reactor in 1961. According to the UIC, no other deaths have occurred from nuclear accidents except at Chernobyl. 28 people died during the acute phase of Chernobyl in 1986 and a further 23 people died between 1987 and 2002, i.e. 51 in total (Berkovski, 2004). A further 171 patients are undergoing medical examination, whilst a total of 2,608,354 people have been included on the Ukrainian National Register as Chernobyl 'sufferers', meaning person actually and potentially affected and who have therefore been included in the group for long-term health monitoring (ibid.). Clearly, the long-term legacy of Chernobyl has yet to be fully comprehended. According to Berkovski (2004) the 3rd International Conference on Health Effects of the Chernobyl Accident in Kiev in 2001 concluded that.

1. There has been an increase in the incidence of thyroid cancer in children who were aged from newborn to 18 years old at the time of the accident and that this is a direct consequence of the accident (ibid.).
2. An increase in the number of cases of thyroid cancer are also expected amongst the workers who were involved in the clean-up of the facility post-accident. There is also a trend of increasing incidence of leukaemia amongst the clean-up workers on site in 1986 and 1987 who were exposed to significant radiation.
3. There has been a significant increase of leukaemia in adults and children living in the contaminated territories of the three affected countries.

A report by the World Health Organisation, UN and IAEA concluded that there would be 4000 eventual deaths arising from Chernobyl, and this figure has been widely accepted and quoted (IAEA, 2005).

The Uranium Information Centre data shows that in terms of sheer numbers, coal mining has accounted for a much larger number of deaths and injuries than nuclear power. On the other hand, coal is a more important fuel at the global scale than nuclear power, so a comparative measure of deaths or injuries per unit of electricity produced would be a helpful comparison. There have been a few attempts to compare accident records arising from different energy supply chains. The most comprehensive is a report by Hirschberg *et al.* (1998) at the Paul Scherrer Institute in Switzerland, which attempted to compare the incidence of 'severe accidents' (defined as 5 or more fatalities) and other impacts across different energy sector chains between 1969 and 1996. The study looked at accidents not only at the generation stage, but also during the fuel extraction, processing, preparation, transportation and waste disposal stages. Expressed as deaths per Terrawatt Year, nuclear power comes out as safest at 8 deaths/TWy, followed by gas at 85 deaths/TWy, coal at 342 deaths/TWy, oil at 418 deaths/TWy and hydro at 884 deaths/TWy. This data suggests that in a comparative sense workers in the nuclear power sector have, over the past few decades, been less prone to death than those working in the fossil fuel industries. Note that Hirschberg *et al.* (1998) use a probabilistic method to assess the risks arising from a nuclear power plant in Switzerland rather than relying upon data from Chernobyl, which is not regarded as representative of the risk in the OECD countries.

Hirschberg *et al*. (1998) also compared injury and evacuee levels per TWy and in this case coal tends to do better than oil, natural gas, hydro and nuclear (which has a particularly high evacuee level). Also, when compared in terms of the economic losses incurred by energy-related severe accidents, the nuclear sector incurs very high costs, much higher than those in the other energy supply chains. Coal has the lowest economic costs arising from accidents of all the energy chains compared.

Clearly, (western style) nuclear power plants outperform coal in terms of severe accidents per TW/y across the fuel chain. Furthermore, the definition of 'severe accidents' as five fatalities means that a large proportion of the deaths in coal mines in countries such as China are not being accounted in the database complied by Hirschberg *et al*. (1998). On the other hand, when accidents do occur in the nuclear industry, they have far greater repercussions for evacuation of residents and are far more expensive than accidents in the coal supply chain.

The types of risks arising from the two technologies are, in many ways, simply different. The fact that we still do not have a complete understanding of the full risks emanating from the Chernobyl accident twenty years after the event, underlines the uncertainty attached to the risks from nuclear power, however unlikely experts regard a Chernobyl-style nuclear accident in OECD countries. The risks from coal mining are much better understood and are also to a large extent quite manageable. Australian coal mining is 555 times safer than Chinese coal mining, suggesting that it would be possible to dramatically improve the safety record of coal mining in those countries which presently have a poor record. The potential for accident is unlikely to be eliminated entirely from any complex technological system (Perrow, 1984). The consequences of accidents from coal mining and nuclear power are in their own way both serious, but those from nuclear may well be less controllable and the repercussions of accidents more wide-ranging and uncertain. It is probably for this reason that the risks of terrorist attack upon nuclear power stations have been widely mentioned in the wake of the 9/11 attacks on the Twin Towers in New York.

Nuclear Proliferation

Nuclear proliferation is defined as:

> the spread of nuclear weapons production technology and knowledge to nations which do not already have such capabilities. ... [The] primary focus of anti-proliferation efforts is to maintain control over the specialized materials necessary to build such devices because this is the most difficult and expensive part of a nuclear weapons program. The main materials whose generation and distribution is controlled are highly enriched uranium and plutonium. Other than the acquisition of these special materials, the scientific and technical means for weapons construction, although non-trivial, to develop rudimentary, but working, nuclear devices are considered to be within the reach of most nations.
>
> (http://en.wikipedia.org/wiki/Nuclear_proliferation)

The British nuclear industry was founded in order to produce plutonium for nuclear weapons (Walker, 1999). Nuclear weapons also require highly enriched uranium and tritium. Nuclear power reactors use either highly enriched uranium or plutonium. Therefore the basic technologies for producing nuclear fuels for power generation are the same as those required for producing nuclear weapons. This creates a risk that nuclear fuels can be diverted to military purposes (Durie and Edwards, 1982). The 1970 Treaty on the Non-Proliferation of Nuclear Weapons (NPT) aims to prevent civil nuclear power being used for military purposes. The NPT has been signed by 188 countries, including the five declared Nuclear Weapons States: the USA, Russia, China, France and the UK. Israel, India and Pakistan have not signed the NPT although they are known to have nuclear weapons, whilst North Korea has withdrawn from the NPT. The International Atomic Energy Agency (IAEA), which implements the provisions of the NPT, believes that without the NPT there might have been 30 to 40 Nuclear Weapons States (SDC, 2006). Despite these successes, there are reasons for questioning whether the NPT will be sufficient in the future to ensure the separation of civil and military applications of nuclear technology. In particular:

- the difficulties of enforcing international treaty obligations;
- proliferation risks associated with the widespread use of nuclear technologies in countries with very diverse systems of governance;
- the capacity and resources available to enforce international obligations in a potentially growing number of states with a nuclear capacity; and
- how to deal with states that withdraw from treaties or develop nuclear capability outside of them (SDC, 2006, p16).

In summary the risk of proliferation can be controlled in some countries, but there is no guarantee that this level of control could be extended to all countries which may be interested in pursuing the nuclear power route. If existing nuclear powers expand their civil nuclear programmes, will this not potentially increase the demand in non-nuclear states for equivalent nuclear power programmes?

It could be argued that expanding nuclear power in countries which already have nuclear weapons capability will not have any bearing upon proliferation in other countries. The technologies and skills are already within that country, hence expansion domestically will not necessarily increase the extension of such skills to non-nuclear weapons states. Furthermore, existing nuclear powers could support expansion of civil nuclear power plants into (what are currently) non-nuclear states by offering the service of nuclear fuels preparation (as Russia did in 2006 vis-à-vis Iran), thereby overcoming the problem of proliferation. On the other hand, expansion of nuclear even in existing controlled nuclear states could be argued to generate greater demand for, and supply of, nuclear technology, skills and services which directly or indirectly will stimulate the nuclear sector in other non-controlled countries or otherwise support proliferation where there is a political will for nuclear weapons development.

The argument is made even more complex by the fact that the nuclear power and nuclear weapons sectors are so intertwined; it is hard to imagine de-proliferation of nuclear power from existing nuclear sates without nuclear weapons disarmament also occurring. If a nuclear weapons state such as the UK is not prepared to contemplate nuclear disarmament, then it could be argued that expansion of the nuclear power sector is more or less incidental. Or in other words, that avoiding proliferation is inescapably concerned with the whole issue of nuclear weapons, which is a huge and complex political debate in its own right, and will not be entered into further here.

8.5   Conclusions

This chapter has considered the prospects for the implementation of CCS in the UK and has highlighted the importance of CCS with Enhanced Oil Recovery (EOR), given the currently favourable economics arising from the high oil price, the presence of highly suitable gas and oil fields, suitable off-shore infrastructure and closeness to other potential EOR and geological storage sites. However, there is a time-limited window of opportunity for utilizing the infrastructure before decommissioning of off-shore structures commences. Decisions would have to be taken in the next 5 years, and it may be necessary for the policy framework to provide sufficient economic incentive for companies to undertake the necessary investment, an issue to which we return in Chapter 9.

The chapter then went on to compare nuclear power and fossil CCS from a sustainability perspective. On balance, it is impossible to claim a clear 'winner'. One other attempt at comparing the two technologies (Pooley, 2005) has likewise not clearly picked a winner, though it has presented a somewhat more favourable evaluation of nuclear power than we have. In terms of responding to the need for rapid CO_2 emissions reduction, whilst providing new capacity to replace that which is being retired, CCS plant can probably offer a solution that is five or so years ahead of new nuclear. Companies such as BP and E.ON believe that they could have a CCS plant operational by 2010, compared to new nuclear construction, which could be anticipated to be operational by 2015 at the earliest.

From an economic and financial perspective, fossil CCS may have some advantages over nuclear. Whilst it shares some of the same characteristics of being 'lumpy' capital-intensive investment, it is not so dependent as nuclear upon multiple orders for the economics to work. From a social perspective, fossil CCS also has advantages over nuclear in terms of public acceptability and perceptions of the relative risks entailed. Existing fossil fuel power generation technologies are not associated with the inflexibility, centralization and high levels of security and secrecy which appear to have rendered the British nuclear industry so unpopular in the past. On the other hand, there is (inconclusive) evidence that the perception of the risks of nuclear power might be undergoing some change, apparently in response to the heightened risk perceptions of climate change and the consequent need for large carbon reduction. In terms of health and safety and accidents, we ended up attempting to compare 'apples' and 'oranges'. The coal fuel chain most certainly has a worse safety record than the nuclear chain in terms of fatalities and injuries, though the costs of a nuclear accident and the number of evacuees tend to be much greater. There has, however, been only a single nuclear accident (Chernobyl) in the time period over which safety was evaluated, upon which the risks from nuclear have had to be extrapolated. There seem to be reasonable prospects for reducing the risks arising from coal mining as witnessed by the huge variation in death and accident rates in different countries. Furthermore, electricity producers in the UK can choose to purchase coal only from countries with the best health and safety records. Nuclear advocates argue that safety features will be greatly improved in the next generation of reactors, e.g. through passive design features, though there is still the risk arising from terrorism and no complex technological system has yet been designed that is one hundred percent 'fail-safe'.

Nuclear power also encounters the particularly tricky problem of nuclear proliferation and only through enforcement of international treaties, and through exertion of other economic, political and even potentially military pressure (prior to a nation acquiring nuclear weapons capability) can nuclear proliferation be conclusively tackled. On the other hand, it is questionable whether expansion of domestic nuclear power in an existing nuclear state, such as the UK, has any real impact on the risk of nuclear proliferation. It was difficult to distinguish between nuclear and coal CCS with respect to fuel security, since both global uranium and coal supplies look to be reasonably plentiful.

Finally from an environmental perspective, both nuclear and fossil CCS raise issues arising from the impacts of mining, though the volume of coal required for a unit electricity output means that the scale of coal mining is much larger than uranium extraction. Again, neither type of environmental impact is a 'show-stopper' but rather requires stringent environmental impact assessment and restoration of mined landscapes. The management of both nuclear waste and CO_2 is a technical and managerial challenge, but it can probably in both cases be adequately tackled if there is sufficient social and political agreement on the underlying framework and process for selecting and evaluating storage sites. The challenge in identifying an acceptable solution for nuclear waste storage is probably greater than that for CO_2 storage, though we have as yet very little experience of the risk assessment issues facing CO_2 storage. Both are long-term problems, which require new ways of devising long-term and secure systems of governance, including monitoring and remediation if necessary.

The most important conclusion arising from this chapter is not that fossil CCS is 'better' than nuclear power or vice versa. The more important conclusion is that fossil CCS is a major CO_2 abatement option which has to be considered alongside nuclear power as a potential source of low-carbon base load as well as peak-following capacity. Many discussions about energy policy in the media and in policy circles still refer to nuclear power as if it is the only available and credible low-carbon large-scale supply option for electricity (and hydrogen) generation. This chapter, if nothing else, hopefully establishes that whenever we include nuclear power in energy policy debates we also need to include fossil fuels with CO_2 capture and storage in that debate.

6.6   References

Anderson, K., Shackley, S., Mander, S. and Bows, A. (2005), *Decarbonising the UK: Energy for a Climate Conscious Future*, Tyndall Centre Technical Report 33, Tyndall Centre, Manchester.

Berkovski, V. (2004), 'Position Statement', in *Nuclear Power: Global Warming Escape ... or Unnecessary Risk?*, an edited transcript of a debate at the Royal Institution, London, on May 20th, 2003, Tyndall Centre, Manchester, pp. 17-23.

BP (2005), *Memorandum from BP*, Appendix 22, Select Committee on Science and Technology, UK Parliament, London, pp. 83–87, www.publications.uk/pa/cm200506/cmsctech/578/578m26.htm.

British Energy (2001), *Replace Nuclear with Nuclear: Submission to the Government's Review of Energy Policy*, British Energy, Edinburgh.

Chapman, N. and McCombie, C. (2006), *Cost Estimates for Disposal of Spent Fuel from New Build Reactors in the UK*, MCM International, Switzerland, www.mcm-international.ch.

Curry, T., Reiner, D., *et al.* (2005), 'A Survey of Public Attitudes Towards Energy and Environment in Great Britain', Laboratory for Energy and the Environment, MIT, Cambridge, MA, http://lfee.mit.edu/publications.

Deffeyes, K. (2005), *Beyond Oil: The View from Hubbert's Peak*, Hill and Wang, New York.

Deloitte (2006), *2020 Vision, the Next Generation: Meeting UK Power Generation Objectives in 2020, a Strategic Insight*, Deloitte, London.

DTI (2003), *Review of the Feasibility of Carbon Dioxide Capture and Storage in the UK*, DTI Report URN 03/1261, September 2003, Department of Trade and Industry, London.

DTI (2003a), *Our Energy Future – Creating a Low Carbon Economy*, The Stationery Office for the Department for Trade and Industry, Norwich.

DTI (2005), *Carbon Abatement Technologies*, Department of Trade and Industry, London.

DTI Oil & Gas (2005), 'BP and Partners Announce Engineering Studies on CO_2 EOR for Miller Field', DTI Oil and Gas – Maximising Recovery Programme, http://ior.rml.co.uk/issue11/page-people/.

Durie, S. and Edwards, R. (1982), *Fuelling the Nuclear Arms Race: The Links Between Nuclear Power and Nuclear Weapons*, Pluto Press, London.

Economist (2001), 'A Renaissance That May Not Come', Special Report Nuclear Power, *The Economist* 19 May 2001, pp. 27–31.

Farmer, A. (1997), *Managing Environmental Pollution*, Routledge, London.

Freund, P. (2005), Submission to the Select Committee on Science and Technology, House of Commons, Appendix 2, UK Parliament, UK, www.publications.uk/pa/cm200506/cmsctech/578/578m26.htm.

Gabbard, A. 'Coal Combustion: Nuclear Resource or Danger?', http://www.ornl.gov/info/ornlreview/rev26-34/text/colmain.html.

Grimston, M. (2005), *The Importance of Politics to Nuclear New Build*, International Institute for International Affairs, Chatham House, London, http://www.chathamhouse.org.uk/pdf/research/sdp/Dec05nuclear.pdf.

Grove-White, R., Kearnes, M., Macnaghten, P. and Wynne, B. (2006), *Nuclear Paper 7: Public Perceptions and Community Issues*, Sustainable Development Commission, London.

Haszeldine, S. (2005), Memorandum to the HoC Science and Technology Committee, Appendix 21, UK Parliament, London, pp. 79–83, www.publications.uk/pa/cm200506/cmsctech/578/578m26.htm.

Hirschberg, S., Spiekerman, G. and Dones, R. (1998), *Severe Accidents in the Energy Sector*, PSI Bericht Nr. 98–16, ISSN-1019-0643.

HMT (2006), *Carbon Capture and Storage: A Consultation on Barriers to Commercial Deployment*, HM Treasury, London.

HoC (2006), *Meeting the UK's Energy and Climate Needs: The Role of Carbon Capture and Storage, Volume 1, Report together with Formal Minutes*, House of Commons Science and Technology Committee, HC 578-1, UK Parliament, London.

IAEA (2005), http://www.iaea.org/NewsCenter/PressReleases/2005/prn200512.html.

IEA (2001), *Nuclear Power in the OECD*, International Energy Agency, Paris.

IEA (2004), *Prospects for CO_2 Capture and Storage*, International Energy Agency, Paris.

IPCC (2005), *IPCC Special Report on Carbon Dioxide Capture and Storage*, prepared by Working Group III of the Intergovernmental Panel on Climate Change, B. Metz, O. Davidson, H.C. de Coninck, M. Loos and L.A. Meyer (eds), Cambridge University Press, Cambridge.

Lovelock, J. (2006), *The Revenge of Gaia*, Allen Lane, London.

MacKerron, G. (2004), 'Nuclear Power and the Characteristics of "Ordinariness" – the Case of UK Energy Policy', *Energy Policy*, **32**, pp. 1957–65.

MacKerron, G. (2005), 'Who Puts Up the Cash?', *The Observer*, 4 December, p. 4.

Marsh, G., Pye, S. and Taylor, P. (2005), *The Role of Fossil Fuel Carbon Abatement Technologies (CATs) in a Low Carbon Energy System – A Report on Analysis Undertaken to Advise the DTI's CAT Strategy*, Report No. COAL R301, DTI/Pub URN 05/1894, Department of Trade and Industry, London.

MORI (2005), *Attitudes to Nuclear Energy November 2005: Report Prepared for Nuclear Industry Association*, http://www.niauk.org/pdf/NIAMORINOV06full.pdf.

OECD, (2005), *Projected Costs of Generating Electricity: 2005 Update*, OECD, Paris.

Performance and Innovation Unit (PIU) (2002), *The Energy Review*, The Cabinet Office, UK Government, London.

Perrow, C. (1984), *Normal Accidents: Living with High-Risk Technologies*, Basic

Books, New York.

Pooley, D. (2005), 'Contemplating Future Energy Options', *Nuclear Future*, **1**(5), pp. 192–197.

Poortinga, W., Pidgeon, N. and Lorenzoni, I. (2006), *Public Perceptions of Nuclear Power, Climate Change and Energy Options in Britain: Summary Findings of a Survey Conducted Between October and November 2005*, Understanding Risk Working Paper 06-02, University of East Anglia, Norwich.

Slovic, P. (2000), *The Perception of Risk*, Earthscan, London.

Sustainable Development Commission (2006), *The Role of Nuclear Power in a Low Carbon Economy*, UK Government, London.

Sustainable Development Commission (2006a), *Nuclear Paper 4: The Economics of Nuclear Power*, An evidence-based report for the Sustainable Development Commission by Science and Technology Policy Research (SRRU) and NERA Associates, UK Government, London.

Tickell, C. (2004), 'Foreword' in *Nuclear Power: Global Warming Escape ... Or Unnecessary Risk?*, an edited transcript of a debate at the Royal Institution, London, on 20 May 2003, Tyndall Centre, Manchester, pp. 4–5.

Vannieuwenhuyse, F. (2005), personal communication, International Federation of Chemical, Energy, Mine and General Workers' Unions.

Walker, W. (1999), *Nuclear Entrapment: Thorp and the Politics of Commitment*, Institute for Public Policy Research, London.

Watson, J. and Scott, A. (2001), *An Audit of UK Energy R&D: Options to Tackle Climate Change*, Tyndall Centre Briefing Note Number 3, Tyndall Centre, Norwich.

Winksell, M. (2002), 'When Systems are Overthrown: The Dash for Gas in the British Electricity Supply Chain', *Social Studies of Science*, **32**(4), pp. 563–598.

Chapter 9

Conclusions and Recommendations

Simon Shackley and Clair Gough

In this book we have undertaken an assessment of the potential role of CO_2 Capture and Storage (CCS) in reducing CO_2 emissions from the UK's electricity sector. We have sought to address several key questions concerning the viability of CCS in the UK context and in this chapter we review and summarise the answers to these questions. We then examine some of the key future research questions arising the project, as well as making recommendations to the UK and other governments regarding policy changes and developments which may be necessary to take CCS forward in a responsible and effective fashion.

9.1   Is CCS Viable from a Geological Perspective?

The first question in assessing CCS is to ask whether there is a strong and solid geological case for the safe and secure long-term storage of CO_2 in depleted oil and gas reservoirs and in saline aquifers. Unless this question can be answered in the affirmative, the case for CCS is greatly weakened. The underlying geological arguments for CO_2 storage are presented in Chapter 2, with more specific examples of geological reservoirs in the UK context provided in Chapters 6 and 7. It is argued in Chapter 2 that there are sound underlying reasons why secure storage should, in principle, be possible. For example, many oil and gas reservoirs have stored natural gas (sometimes themselves containing significant volumes of CO_2) for thousands to millions of years, whilst natural storage reservoirs for CO_2 are also known which have been there for similarly long periods of time. However, Chapters 2, 6 and 7 are adamant that the suitability of each potential geological reservoir needs to be explored on a case-by-case basis because of the high degree of heterogeneity of reservoirs. There are no generic assumptions that can be made regarding the suitability for CO_2 storage at particular sites.

9.2   What is the Capacity for CO_2 Storage in UK Geological Reservoirs?

Having established the 'in principle' viability of CCS, the next question concerns the capacity for storage in geological reservoirs. Chapter 2 presents data which suggests that the CO_2 storage capacity of the UK's oil and gas fields is approximately 6.2 Gt, sufficient to store about 35 years worth of CO_2 emissions from UK power stations (at current emission levels). The UK's oil and gas fields are well understood geologically because of the extensive exploitation of such reserves and the associated accumulation of scientific data, knowledge and tools (such as reservoir simulation models). Much greater uncertainty is associated with the storage capacity of aquifers, because these structures are much less well characterized scientifically. For instance, a major uncertainty concerns the proportion of the pore space that will be occupied by supercritical CO_2 that is injected via a bore hole. Depending on what assumption is made about this proportion (from 0 to nearly 100 per cent), a massive range of values for the storage capacity of aquifers can be generated.

For example, earlier estimates by Holloway and Baily (1996) for geological storage in the UK have been quoted by the UK Department of Trade and Industry (DTI, 2003). The DTI report quotes the storage capacity of oil and gas reservoirs at 7.5 Gt CO_2, slightly larger than the value provided in Chapter 2. The real difference concerns the storage capacity of aquifers, however, with a 'high' storage estimate of 8.6 Gt CO_2 for 'trapped' or confined parts of aquifers and a massive 240 Gt CO_2 for open, or non-confined, parts of aquifers (Holloway and Baily, 1996). If such a large figure were to be used, then the CO_2 emissions from the UK's power stations (at current levels) could be stored for well over 1000 years.

In Chapter 2 of this book, Holloway *et al.* have argued that their earlier estimates quoted by the DTI (2003) are misleading because they are based on a very optimistic assumption, namely that a particular fraction of the total pore volume of all potential reservoir formations would be available for CO_2 storage. Holloway *et al.* consider that the earlier methodology is clearly inadequate, albeit the best available at the time. He and his colleagues (Chapter 2 and Holloway *et al.*, in press) have now studied the Southern North Sea basin in much greater detail, including consulting maps and extensive geological data. This method led to the estimate for the CO_2 storage capacity of aquifers in the Southern North Sea Basin of up to 14.26 Gt CO_2, sufficient to store about 89 years worth of CO_2 emissions from UK power stations (at current emission levels). Even this figure is regarded as being only a very rough estimate by Holloway *et al.* and does not include, for example, considerations of leakage. There is no widely accepted methodology for calculating the storage capacity of aquifers in the absence of detailed geological data, and such detailed data is rarely already available for non-hydrocarbon aquifers. We are therefore faced at the current time with endemic uncertainty, and this can cut both ways. Capacity may be less than 14 Gt CO_2 and it could conceivably be larger than 240 Gt CO_2 – we simply do not know. It is likely that we will only get a better understanding of actual capacity through learning-by-doing, i.e. by initiating CO_2 storage projects and devoting sufficient R&D activity to learning more about reservoir capacities and how reservoirs respond to storage of increasing volumes of CO_2.

9.3   Changes in the Energy System and in Energy Demand

The calculations of CO_2 storage years above, and in Chapters 6 and 7 with respect to storage in the North West and East Midlands, Yorkshire and the Humber, assume *current* levels of CO_2 emissions from the electricity sector. The scenarios developed in this project did not allow for change in the production of transport fuels such as hydrogen from natural gas or coal, with consequent CO_2 emissions and their potential capture and storage. Nor did we explore changes in overall energy demand levels. Other projects have, however, developed wider-ranging scenarios which depict changes in the energy system on both the supply and demand sides. The Tyndall Centre's *Decarbonising the UK* (DUK) scenarios (Anderson *et al.*, 2005) explored a range of future energy demand levels from 90 million tonnes of oil equivalent (mtoe) to 330 mtoe, compared to today's level of 170 mtoe. All the scenarios were designed to reach a 60 per cent reduction in CO_2 emissions by 2050.

There are two variants of the high energy demand (330 mtoe) scenario, the Pink and the Purple Scenarios (Anderson *et al.*, 2005). On the supply side (and expressed as contribution to the primary energy demand in 2050) the Purple Scenario is made up of a mixture of nuclear (43 per cent), renewables (29 per cent), oil (16 per cent), biofuels (10 per cent) and gas (2 per cent), i.e. it achieves a low-carbon electricity supply through a mixture of nuclear and renewables. The Pink Scenario, on the other hand, has the following primary energy supply-side mix: gas (29 per cent), coal (24 per cent), renewables (15 per cent), oil (12 per cent), nuclear (10 per cent) and biofuels (10 per cent). Both the coal and gas components of the electricity supply (constituting 55 per cent of electricity supply) are associated with CCS, though the Pink Scenario does not include any hydrogen production (unlike the Purple Scenario, where H_2 is generated from both nuclear power and renewables, becoming an important transport fuel).

In the Pink Scenario, the total amount of CO_2 which needs to be stored in the year 2050 is 538 Mt CO_2 (assuming a capture efficiency of 90 per cent). If a linear increase in storage is assumed between now and 2050, then the accumulative CO_2 requiring storage is 13.45 Gt CO_2. Using the storage estimates in Chapter 2, there would be sufficient capacity to store all the CO_2 implied by the high demand Pink Scenario, at least to 2050. Note, however, that about 70 per cent of this storage would occur in aquifers, which are less well known and therefore (at the present time) a some what higher-risk option than storage in depleted oil and gas fields. Furthermore, the storage capacity in the Pink Scenario would be exceeded in a decade or so after 2050 should the scenario continue in a similar fashion after 2050. We have not, however, included the potential storage capacity in aquifers in the Northern North Sea basin in this calculation.

The Pink Scenario is not the 'highest CCS' scenario that could be envisaged in the framework of the DUK Scenarios. We therefore modified the Pink Scenario to replace the nuclear component by coal with CCS. We calculated that there would be 868 Mt CO_2 to capture and store in the year 2050, or an accumulated total of 20.65 Gt CO_2 between now and 2050. Using the estimates of storage capacity in Chapter 2, it would still be possible to store this quantity of CO_2 in the storage reservoirs identified (still using aquifers in the Southern North Sea). After 2050, however, capacity in these reservoirs would have been exceeded. Note that it would be possible to envisage a version of the Pink Scenario in which there were even larger quantities of CO_2 being captured, e.g. replacement of the renewables proportion of primary energy supply with coal-derived fuels and electricity with CCS.

One interesting conclusion from the above analysis is that the CO_2 storage reservoirs identified in Chapters 2 and 7 are probably of sufficient capacity (c. 20 Gt CO_2) to store all of the CO_2 emitted to 2050 even in a high fossil-fuel, high energy demand scenario. This is the case even when the captured (and stored) CO_2 emissions are 3 to 5 times as large as current CO_2 emissions from the power generation sector. In figure 9.1 the storage capacity is shown on the far left, whilst the CO_2 storage demand under a number of scenarios described above is also illustrated.

CO2 storage capacity and demand

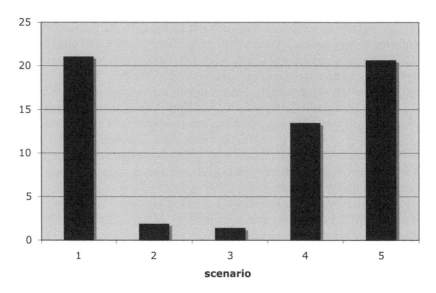

Figure 9.1 **Scenarios of cumulative CO_2 storage to 2050 (in Gt) compared to estimated capacity**

Entry 1: CO_2 storage capacity

Entry 2: East Midlands Yorkshire & Humber, Fossilwise scenario;

Entry 3: East Midlands Yorkshire & Humber, Spreading the Load scenario;

Entry 4: Decarbonising the UK 'Pink' (high energy demand) scenario;

Entry 5: Decarbonising the UK 'Modified Pink' scenario (replacement of nuclear by coal with CCS).

Beyond 2050, the reservoir capacity would be exceeded (using the estimates in Chapter 2). If the DTI's (2003) high estimates of storage capacity in geological reservoirs are used instead then it would be feasible to store hundreds of years worth of CO_2, even under a high fossil-fuel, high energy demand scenario (though we do not advocate the continued use of this earlier capacity estimate). Given the large uncertainty associated with estimates of storage capacity, too much should not be read into the above analysis and its associated numbers. Much greater certainty is required before we can begin to have confidence in these capacity figures. As noted above, it is likely that such certainty will only emerge from learning-by-doing as experience accumulates of actually using reservoirs over the next few years and decades.

An important further proviso to the above analysis is that the UK may wish to provide opportunities for other countries to store their CO_2 emissions. This could happen, for example, through the inclusion of CCS within the remit of the EU Emissions Trading Scheme (EU ETS); thereby the UK could offer to provide a CO_2 storage facility to other EU companies in the UK sector of the North Sea. The rationale is that the UK potentially has an abundance of suitable reservoir capacity (depending on how suitable aquifers turn out to be as we gain more understanding). If this provision of storage facilities to other EU countries was to occur, then clearly the demand for storage of CO_2 within the UK sector could potentially be far greater than that arising from the UK's own CO_2 emissions. Meeting such increased storage demand would obviously limit the number of years worth of CO_2 storage. Such a scheme could take place if CCS represented a cost-effective way of abating CO_2 relative to other possible carbon mitigation options and if there was agreement under the EU ETS that such an approach was acceptable.

9.4   What are the Risks and Potential Impacts of Leakage?

The key risk with respect to CO_2 storage is that of possible leakage. CO_2 might leak out of storage reservoirs for a number of reasons, as explained in Chapter 2. No generic assumptions about the risks of leakage are possible. Instead, it is necessary to evaluate the risks of leakage on a case-by-case basis.

Why does it matter if CO_2 does leak out of reservoirs? Leakage is important for two reasons. Firstly, leakage can reduce the effectiveness of meeting the objective of CCS, namely storing CO_2 to ensure that it does not enter the atmosphere thereby contributing to anthropogenic climate change. Secondly, leakage could pose potential health, safety and environmental risks to humans, other organisms and ecosystems.

Health, Safety and Local Environmental Impacts

The risks to human health and safety and to other organisms and ecosystems that might arise from the leakage of CO_2 from storage reservoirs have been covered in Chapter 2, the general conclusion being that such risks need to be assessed on a case-by-case basis but are by no means 'show stoppers'. Since we have only looked at CO_2 storage in off-shore reservoirs however, we have not considered in any detail the risks of leakage from an on-shore reservoir. The risks arising from the physical infrastructure associated with the collection, compression and piping of CO_2 on land to the storage facility were analysed by Det Norsk Veritas (DNV) for the UK Department of Trade and Industry (Vendrig *et al.*, 2003). DNV utilised a conventional risk assessment methodology used in the chemical industry known as HAZOP to analyse and quantify the risks associated with the operation of a CCS plant and pipeline to transport CO_2 to storage reservoirs under the seabed. Where specific numbers were not available because of a lack of experience, a Delphi process was employed with external experts providing estimates of the relevant numbers through several iterations. DNV concluded that the risks of explosions and the associated risks to human operatives were equivalent to those arising from other large industrial plant handling compressed gases at low temperature. DNV also noted that the risks could be adequately managed by following best practice guidelines in industries dealing with power plants and high pressure gases.

The Impact of Leakage upon the Effectiveness of CCS

If too much CO_2 leaks out from reservoirs, and/or this leakage occurs too quickly, then CCS could become pointless as a mitigation option. Let us consider some extreme scenarios to illustrate the problem. If a large CO_2 reservoir were to leak entirely within say 10 years (a rate of 10 per cent leakage per year), there would be no benefit at all in terms of reducing atmospheric CO_2 concentration. In fact, things would be worse because of the CCS 'energy penalty', meaning that more CO_2 would have been emitted from the reservoir than if a conventional power station with no CO_2 capture had been operating over the 10 year period. On the other hand, the idea of zero-leakage in perpetuity, as initially advocated by some environmental NGOs with respect to CO_2 storage, is not geologically credible and very unlikely to be necessary. Imagine a large CO_2 reservoir which was to leak at the rate of, say, 0.0001 per cent per year. It would then take 100,000 years for all the CO_2 to leak out. It is widely assumed that we cannot say any thing sensible about society, the economy or the energy system 100,000 years from now. In many ways, we have to assume that generations of humans so far into the future will have developed zero- and low-carbon technologies, or alternatively that the ravages of global climate change will have transformed the environmental, socio-economic, political and technological conditions of the world in unpredictable ways.

The appropriate time frame for assessing the permanence of CO_2 storage is evidently longer than 10 years, but shorter than 100,000 years: the critical question is how much longer than 10 years and how much shorter than 100,000 years. The answer to this question depends in part upon how long ahead into the future present day society feels that it should shoulder responsibility for. For example, if we assume that we are only responsible for the next 100 years (the view of at least one expert in the field), then leakage beyond that time is not regarded as something that should influence our decision-making in the present. The rationale might be that in 100 years time, new low or zero-carbon technologies will have emerged that will have replaced fossil fuels, or else that societies in 100 years time from now will have devised other ways of managing or coping with global climate change. From this perspective things are just too uncertain to be able to say anything confidently about

100 years into the future. On the other hand, many other experts tend to look to a longer time period for establishing this cut-off point between taking responsibility and passing on that responsibility to the future. A typical time period quoted is 1000 years, implying a leakage rate of 0.1 per cent. This time period fits quite well into our understanding of the climate system, since we anticipate that global CO_2 emissions will peak well before the year 3000 under all scenarios, allowing the slow move towards equilibrium between the major carbon sinks.

Under the IPCC's highest CO_2 emissions scenario (SRES A1F), a peak in atmospheric CO_2 concentration occurs in about 2250 (Lenton & Cannell, 2002). In most other scenarios, the peak occurs earlier, e.g. between 2020 and 2100 (IPCC, 2001). This suggests that a risk averse strategy would avoid contributing more CO_2 emissions to the atmosphere until after 2250. With a leakage rate of 0.1 per cent, a quarter of the reservoir would have leaked out by 2250, which could be a significant additional contribution to emissions if the high emissions scenario were followed. There are inconsistencies in this simple analysis, however, because the high emissions scenario assumes that all the fossil fuel being used results in the release of CO_2 to the atmosphere, whereas this is clearly not the case if CCS is employed. Therefore, the atmospheric CO_2 concentration under the SRES A1F scenario would be higher than in the case of a scenario which includes reliance on fossil fuels but with CCS.

Nevertheless, given uncertainties in our understanding of the carbon cycle and of the climatic effects of CO_2 and its various positive and negative feedbacks, it would seem prudent to avoid a leakage rate that would add more than a small amount, say 1 or 2 per cent, of the stored CO_2 back into the atmosphere before the next several hundred years. This might imply a leakage rate which is an order magnitude smaller than 0.1 per cent, say between 0.01 per cent and 0.005 per cent (or all the CO_2 leaking out over 10,000 to 20,000 years time). It is interesting to note that a 0.01 per cent leakage rate has been used by a number of commentators in the field (IPCC, 2005). Environmental groups such as WWF have mentioned retention times of 100,000 years which would imply a leakage rate of 0.001 per cent. Some commentators have expressed the view that leakage rates should be reduced to about 0.001 per cent, not primarily for scientific reasons, but in order to provide a sufficiently large margin of error that the public and stakeholders are provided with reassurance.

Establishing an 'acceptable' leakage rate is far more complicated than implied by the simple analysis above. Indeed, it may be so complex and uncertain that a reasonably simple analysis is in practice preferable at the present time. For example, the level of acceptable leakage will to some extent depend upon the extent to which we rely globally upon CCS as a mitigation strategy. Let us suppose that there was just one CO_2 reservoir used in the entire world, with other zero- and low-carbon options having been deployed for the bulk of carbon mitigation. Leakage from this single reservoir would be much less important than if there were hundreds of reservoirs being used for CO_2 storage all around the world. Therefore the significance of leakage from CO_2 storage has to be investigated on a global scale, rather than from the perspective of an individual reservoir, and the acceptable leakage rate will depend upon the extent to which CCS is deployed globally. To investigate the problem thoroughly, it would be necessary to make some assumptions regarding how many CO_2 storage reservoirs are envisaged on a global scale, their storage capacities and the potential leakage rates from each reservoir. CO_2 leakage from a particular geological reservoir has then to be examined in terms of its contribution to atmospheric CO_2 concentration, which in turn has to be translated through a climate model into a given quotient of global climate change.

As noted in Chapter 1, a number of comprehensive energy scenarios have been created at the global scale which include some treatment of CCS, though they are not very detailed (IPCC, 2001). Some integrated modelling studies of CO_2 leakage have been conducted, although this work is very preliminary and unable to provide definite answers. Below we briefly review (and critique) the main modelling approaches taken to address the leakage problem and their key findings to date.

Herzog *et al.* (2003) use a concept of 'sequestration effectiveness' which they define as the net benefit from temporary storage compared to the net benefit of permanent storage. Their conceptual framework is borrowed from cost-benefit analysis, the costs in this case being the economic resources required for removing a unit of CO_2 (plus any monitoring costs for CO_2 sinks) over a given time period and the benefits being the avoided costs incurred by climate change impacts (frequently expressed as the value of a tonne of CO_2 abatement) over the same time period. As is usual in cost-benefit analysis, a discount rate is applied each year to the costs and benefits to reflect the extent to which money is preferred now rather than in the future. The effect of discounting is to reduce the monetary value of costs and benefits into the future. For example, if a typical commercial discount rate of 10 per cent is selected, then after ten years the value of $1 is $0.386, $0.149 after twenty years and $0.009 after fifty years. If a very low discount rate is applied, say 1 per cent, then the value of $1 after ten years is $0.905, $0.820 after twenty years and $0.608 after fifty years. Hence, in the longer-term the costs and benefits are largely discounted however low the discount rate is.

Herzog *et al.* (2003) argue that sequestration effectiveness depends upon the leakage rate, the cost of a unit of carbon emissions and the discount rate. It is generally assumed that the cost of a unit of carbon emissions will increase over time due to the acceleration of anthropogenic climate change over the next century. Hence, the benefits of sequestration will increase over time. This effect, however, is counter-acted by leakage of CO_2, which clearly reduces the quantity of CO_2 abated, and by the discount rate, which acts to reduce the costs and benefits associated with future CO_2 reduction. Imagine a situation where there are two different curves representing the increase in value of a tonne of carbon abatement – a low rate of increase and a high rate of increase. For a given discount rate, a high rate of leakage becomes more of a problem when there is a relatively low rate of increase in the value of carbon abatement. Provided that the value of carbon abatement increases more rapidly than the effects of discounting, CCS is still worth doing, provided that the leakage rate is not too high. However, if the value of carbon abatement remains static over time, the effect of the discount rate will fairly quickly reduce the rationale for CCS irrespective of the leakage rate.

Herzog *et al.* (2003) have also promoted the argument that CO_2 emissions which have been sequestered are no different from carbon locked in un-mined fossil fuels in the ground. The rationale here is that zero-carbon technologies such as renewables are not necessarily reducing CO_2 emissions permanently, but may simply be delaying the use of fossil fuels to some point in the future. They argue that society will eventually use up all the earth's accessible fossil fuels unless a new zero-carbon technology emerges which replaces fossil fuels. In making this argument Herzog *et al.* are straying well away from scientific analysis, into an assumption about future political and policy decisions which are highly contested. Clearly, there are many reasons to do with climate change, but also to do with human health and safety, potential future technological change and economic and political circumstances, which militate against full utilization of known fossil fuel reserves globally. If we assume, *contra* Herzog *et al.*, that renewables are in effect reducing CO_2 emissions on a permanent basis, then the occurrence of leakage reduces the economic efficiency of CCS compared to renewables (or other carbon abatement options such as energy demand reduction) over time (Shackley *et al.*, 2003).

Ha-Duong and Keith (2003) have also explored the trade-offs between leakage and discount rates, costs and the energy penalty. They argue that for an optimal mix of CO_2 abatement and CCS technologies, 'an (annual) leakage rate of 0.1 per cent is nearly the same as perfect storage while a leakage rate of 0.5 per cent renders storage unattractive' (Ha-Duong and Keith, 2003, p. 181). There are a number of criticisms of economic-based approaches such as those of Herzog *et al.* and Ha-Duong and Keith. These are discussed below:

1. Cost-benefit analysis assumes that the only concern of decision-making is economic efficiency. Many authors, including Baer (2003), have pointed out that concerns regarding ecological risk, financial risk and political risk need to be incorporated into a decision-making framework along side economic efficiency.

2. The ecological and geological risk relates to uncertainties associated with CCS, our knowledge and understanding of the global climate system and of the impacts of CO_2 leakage. A precautionary approach would imply selecting a very low leakage rate because we are unlikely to know enough with a high level of confidence prior to having to make decisions about CO_2 storage. Dooley and Wise (2003) explored CO_2 leakage on a short (100 year) simulation and found that relatively high rates of release (1 per cent per annum) from storage made it impossible to stabilize atmospheric CO_2 concentration at 450 ppmv. The leakage rate in their analysis had to be reduced to at least 0.1 per cent per annum in order to stabilize at 450 ppmv.

3. Financial risk refers (*inter alia*) to the possibility that the cost of a tonne of carbon emissions might increase much more rapidly than anticipated. In that case, there may be large liabilities arising from leakage which have not been properly accounted for at the project development stage.

4. Political risk refers to the uncertain behaviours and policies of governments and other regulatory institutions in the long-term with responsibility over CO_2 storage. For example, could the deliberate release of CO_2 from a nation's geological storage sites represent a future weapon of mass destruction?

5. The use of discount rates has been roundly criticized by some economists who have argued that they are not an appropriate conceptual framework for analyzing sustainability issues (Hanley and Owen, 2004). This is because climate change requires a very long-term time frame for analysis, stretching to hundreds of years as argued above. Discount rates are inappropriate for use over such long time periods.

6. The value of a tonne of carbon abatement into the future is not known, nor is there any reliable way of calculating what this value is (Pearce, 2005). Without knowledge of this variable, it is not possible to calculate 'optimal' leakage rates. All that can be done is scenario analysis, based upon different assumptions and covering a wide range of potential carbon abatement costs.

7. It is not possible to answer questions over acceptable leakage without recognizing that there are both political and moral dimensions to the issue. As argued above, the time scale over which we consider ourselves to be 'guardians' is an inherently moral and political issue. Some argue that we are responsible only for a short period of time into the future, whilst others think that our responsibility stretches much further into the future. These are not questions of science, but rather of political values.

8. In response to disagreement concerning the above points, it can be argued that in order to secure support from stakeholders and members of the public, it will be necessary to select a very low leakage rate to demonstrate that the implementation of CCS is being undertaken with the highest degree of care.

In its Special Report, the IPCC conclude on the issue of leakage that:

> within this purely economic framework, the few studies that have looked at this topic indicate that some CO_2 leakage can be accommodated while still making progress towards the goal of stabilizing atmospheric concentrations of CO_2. However, due to the uncertainties of the assumptions, the impact of different leakage rates and therefore the impact of different storage times are hard to quantify.   (IPCC, 2005, p. 359)

Elsewhere the report acknowledges that: 'Ultimately, political processes will decide the value of temporary storage and allocation of responsibility for stored carbon' (IPCC, 2005, p. 365). We believe that this is appropriate, given the role of moral and political values in considering the acceptability of leakage. However, the critical question is what such a political process should look like and how should it be designed?

9.5 Is CO_2 Storage in Geological Reservoirs Legal?

Legal impediments to storage are described in Chapter 4. Nearly all the UK's identified CO_2 storage potential is offshore. Amendment of the London Convention and OSPAR treaties may be necessary before CO_2 storage in saline aquifers can take place offshore and this could take years, especially as little is known about the impact of leaks of CO_2 into the marine environment. The provisions of the OSPAR Convention do seem somewhat strange and anomalous. For instance, amendments to the Convention may be required if CO_2 is sourced from onshore facilities and injected from a ship, an offshore installation or other structure in the maritime area (excluding pipelines) for the purposes of mitigating climate change. Yet, if the injection of CO_2 occurs from offshore installations produced on the same or other offshore installations (e.g. together with natural gas), it would not require modification of the Convention, even though the purpose would once again be climate change mitigation. (This is the conclusion of a report of June 2004 by the OSPAR Group of Jurists and Linguists.) Clearly, clarification and modification of the legal framework will be necessary before CCS can become an established carbon abatement technology. Companies are unlikely to initiate major investments in CCS technology where there are potential conflicts with the provisions of the OSPAR Convention.

9.6 Is CCS Technically and Economically Feasible?

There are no technical barriers to CO_2 capture and storage. All the technological steps throughout the entire CCS chain are proven and there are major drives to reduce the cost of CO_2 capture. Pilot power stations fitted for CO_2 capture, with outputs of tens of megawatts, are being constructed, and CO_2 storage is already taking place at an industrial scale at the Sleipner CO_2 storage site in the North Sea, just a few kilometres from the UK Continental Shelf (see Table 2.1). Moreover, steps have been taken to implement a full chain project in the UK, namely the Miller Field/Peterhead project led by BP and described in Chapter 8. The commercial attractions of EOR at today's oil prices make wider scale deployment in the North Sea oilfields a distinct possibility in the short term. Nonetheless there are several economic impediments to the wide scale deployment of CCS in the UK.

The wholesale electricity market in the UK is very competitive and no operator will be able to remain competitive when deploying CCS unless the correct fiscal incentives are in place. These incentives would have to include a price for the CO_2 saved through utilizing CCS, for example through extension of the current Renewables Obligation into a 'Zero Carbon Obligation' or 'Decarbonised Electricity Certificates' (potentially including CCS, nuclear and renewable energy), or perhaps through a separate 'CO_2 Storage Obligation' put upon electricity suppliers. There would need to be some kind of guarantee of long term stability in such a support mechanism that would allow investment in new power plant or retrofits to take place. Ultimately the market penetration of CCS will depend on the relative fuel prices for gas and coal and the way in which any fiscal incentives are distributed between nuclear, renewable and CCS power generation technologies. We discuss in more detail the issue of how to provide incentives for CCS given current UK energy policy in section 9.11.

A further economic issue is the impact of increasing fossil fuel energy consumption arising from CCS upon demand for coal and gas. In relative terms, the UK is a small market and therefore the salient issue is the impact on demand from the wider international implementation of CCS. In its analysis the IEA (2004) found that CCS would act to increase the use of coal, especially in regions with plentiful supplies such as North America, China and India, thereby increasing reliance on domestic energy sources. Depending upon market factors, ease of obtaining planning permission and technological innovation in new coal extraction technologies, implementation of CCS might stimulate demand for domestic supplies of coal in the UK.

9.7   Is CCS Acceptable to the Public?

The evidence in Chapter 5 suggests that CCS may be more acceptable to the public than some other low-carbon options, such as nuclear power, and higher energy bills (assuming that far-reaching CO_2 reduction targets have to be met). On the other hand, CCS is considerably less attractive to the public than renewable energy and energy efficiency. Two independent surveys and the Citizen Panel work in the UK context have confirmed these broad brush findings, though the surveys differ in the extent to which CCS was perceived positively. In the Tyndall survey of 2003 there was considerably more support for CCS than in the Curry *et al.* (2005) survey. The Tyndall survey also illustrated a stronger contrast between CCS (perceived positively) and nuclear (perceived negatively) than Curry *et al.* (in which survey nuclear was still perceived more negatively than CCS though there was little difference in the extent of positive perceptions between CCS and nuclear). The most striking difference between our survey and that of Curry *et al.* (2005) is the high percentage of respondents in the latter who replied 'don't know' (at 50 per cent) when asked about the desirability of using CCS compared to other low- or zero-carbon mitigation options compared to an equivalent question in the Tyndall survey, where only 10 per cent responded that they did not know (and a further 13 per cent were 'neutral', for which there was not an equivalent category in Curry *et al.* (2005)).

A likely explanation of this difference in the results from the two surveys is that the Tyndall survey provided more information about what CCS is and how it works than the Curry *et al.* survey. A limitation of both of the surveys is that it was not possible to present very much technical information about the various carbon mitigation options, e.g. regarding their environmental and other impacts and costs. With such limited survey work it is not possible to make any strong conclusions, especially regarding the comparison of CCS with other CO_2 mitigation options. As discussed in Chapter 8, public opinion in the UK regarding nuclear power is not static, and there is some evidence that it has been shifting over the past few years towards a more favourable perception. One survey of public perceptions of

CCS in the USA found that the respondents actually preferred nuclear power to coal with CCS (Palmgren *et al.*, 2004). Other survey work has illuminated that many respondents are unaware of the low-carbon status of nuclear power, or of the relatively higher costs of many renewables (Curry *et al.*, 2005a), and it is likely that this additional information would change the respondents' comparison of carbon mitigation options. The work in the Citizen Panels suggests that CCS, in order to win public support, has to be presented in the context of global climate change and that there needs to be some recognition of the scale of the problem and the urgency of reducing CO_2 emissions.

9.8 What Do Stakeholders Think About CCS and How it Compares with the Other Options for Sustainable Energy Futures?

Chapters 6 and 7 present detailed regional case-studies of the potential implementation of CCS in the electricity sector over the next fifty years. Both case-studies revealed a large disparity of opinion even amongst small samples of regional stakeholders who are knowledgeable about energy. Opinion was especially divided over the Fossilwise, Nuclear Renaissance and Renewable Generation scenarios in both regions. The three scenarios all scored well against some criteria and poorly on others. Nuclear Renaissance was generally more disliked than the other scenarios in both regions, though there were two nuclear power advocates. Opinion on the Spreading the Load (high) and (low) scenarios in the North West, and of the Spreading the Load and Capture as a Bridge scenarios in the East Midlands/Yorkshire & Humber regions was much more consensual across the stakeholder samples.

Nuclear does well with respect to reliability, security and infrastructure but poorly with respect to costs, lock-in, environmental impacts and public opposition. Renewables do well with respect to security, environmental impacts, resilience to disasters and lock-in but poorly with respect to costs, infrastructure, public opposition and reliability. Finally CCS does well with respect to costs, infrastructure, reliability, security, public opposition and resilience to disasters. It does less well with respect to environmental impacts and lock-in. One interesting finding is that some of the strengths of each of the three main generation types (nuclear, fossil CCS and renewables) are also the weaknesses of one of the other energy supply types. As the Multi-Criteria Assessment (MCA) illustrated, renewable energy generation is a key component of a low- or zero-carbon energy system, and has many significant advantages in terms of sustainability. Hence, a fairly good argument could be made for combining either or both of nuclear and fossil CCS with renewables, in designing a future sustainable energy system.

9.9 What are the Main Differences Between the Two Case-Study Regions?

There are some differences in the methods that were used in the two regional case-studies:

1. One of the five scenarios used in the North West (NW) region was changed when conducting the East Midlands / Yorkshire & Humber (EMYH) regional study.
2. One of the criterion used in the North West case study was dropped and replaced by a new criterion which seemed more appropriate.
3. The scenarios for the East Midlands/Yorkshire & Humber regional study were more detailed and more quantified than those for the North West.

These changes were made in the light of the experience of conducting one regional case-study and the lessons that were thereby learnt and which could be transferred. Whilst the above differences mean that a direct comparison is thereby rendered more difficult, the two case-studies are sufficiently similar in design that a comparison between the regions is broadly feasible with respect to criteria weightings and for the scoring of three scenarios – Fossilwise, Nuclear Renaissance and Renewable Generation.

With respect to weightings, the NW values tended to illustrate a slightly wider range for most criteria. For infrastructure, reliability and disaster, weightings were broadly similar in both regions, though with the NW respondents presenting a somewhat wider range of values. Cost-effectiveness, security and environmental impacts enjoyed a wider range of values for NW than for EMYH respondents. The greater variance in the weightings is unlikely to be explained by the fact that the NW respondents were a more diverse grouping of stakeholders than those in the EMYH region, since both groups were composed of broadly similar types of stakeholders. An alternative explanation is that there is greater consensus in EMYH region concerning the relative importance of criteria than in the NW region.

One possible reason for such consensus is that the EMYH respondents have been more actively engaged in thinking about energy issues than have those in the NW. Both the YH and EM regions certainly have a strong history of involvement and engagement in energy issues, having been for many years the centre of English coal mining and fossil fuel based electricity generation. There are reasonably well established energy fora in both regions at which stakeholders such as those we interviewed meet-up frequently. Such a tradition does not exist in the North West, and whilst there is a North West Energy Council it is a much more recent creation (established in 2003) and is less inclusive than those in EMYH. On the other hand, the YH and EM are two distinct governmental regions in terms of stakeholder networks and interactions so it is somewhat difficult to understand why a consensus should necessarily have emerged between them.

Looking at the Fossilwise scenario first, the EMYH respondents illustrated a somewhat greater preference for this scenario across most criteria than the NW respondents. The only criteria for which this was not the case was the environmental impacts of Fossilwise, which were rated more negatively in the EMYH scenario than in the NW scenario. Given the long history of coal mining and utilisation of coal for electricity generation in the EMYH region it is perhaps not surprising that there is a some what more favourable evaluation of Fossilwise compared to its evaluation in the NW region. It is also interesting to note that the environmental impacts of coal mining and of the utilisation of coal are perhaps more readily appreciated (and scored rather more negatively) in the EMYH region than in the NW (where coal mining has occurred in the past, but which mostly came to an end in the 1950s and 1960s).

The scoring of the Renewables Scenario is broadly similar in both regions. The main difference lies in the scoring of the cost-effectiveness criterion, for which the NW respondents tended to give a lower score than those in EMYH. There was a slightly more positive evaluation of the Renewables Scenario in EMYH than in the NW but, by and large, the evaluations were similar.

Turning to the Nuclear Renaissance Scenario, the scorings are once again broadly similar across the criteria. There is, however, a slightly more favourable evaluation of Nuclear Renaissance in the NW than in EMYH with respect to costs, infrastructure and security. There are also noticeable (positive) outliers with respect to costs for both regions. The NW is the centre of the nuclear power industry in the UK and home to some of its key installations. This may help to explain why there is a somewhat more positive evaluation of the Nuclear Renaissance scenario in the NW than in the EMYH region, which has no nuclear power plants.

9.10 Summary of Key Findings

Below we summarise the key findings reported in this book.

1. The extent to which CCS is implemented in the UK is likely to depend on a number of factors, including the degree to which CO_2 emissions reduction is regarded as a priority by government, stakeholders and, possibly, the public – should they choose to engage in the debate. This may in turn depend upon the perceived risks of climate change impacts, both in the UK and globally. The more that extreme events occur (Hurricane Katrina, European droughts of 2003, etc.), and the more that they are associated with anthropogenic climate change, the more attention and policy focus may be applied to the climate change issue and the concomitant need for CO_2 reduction. CCS is one of the few technologies which could deliver large CO_2 cuts in a short time-period (i.e. peak-shaving the future atmospheric concentration of CO_2) and this may be perceived as being a necessary response should sufficient disasters mobilize political concerns.

2. Also important are the perceived risks associated with different power generation technologies that can offer carbon reduction opportunities. The main zero-/low-carbon options in front of policy makers at present (not including demand side changes) are nuclear, a wide range of renewables and CCS. The following risks have to be evaluated in comparing the options: technological, economic and financial, environmental, socio-political, fuel security, reliability, flexibility and reversibility (lock-in). These are broadly the same as the criteria which have been used in our multi-criteria assessment (MCA) of energy scenarios in Chapters 6 and 7.

3. The costs of CCS in our techno-economic model for UK power stations in the East Midlands and Yorkshire to reservoirs in the Southern North Sea are between £25 and £60 per tonne of CO_2 captured from new build PF and IGCC coal power plant, transported and stored. This is between 2 and 4.5 times the current traded price of a tonne of CO_2 in the EU Emissions Trading Scheme. It is about 1/3 to 3 times the value to society (according to one estimated range by the UK Government) of avoiding an additional tonne of CO_2 from entering the atmosphere.

4. Hence, CCS is not yet a cost-effective way of reducing CO_2 emissions (compared to other options, some of which (such as renewables) are supported through government incentives) though it is potentially still a beneficial carbon mitigation option from a societal perspective. To become cost-effective, it will probably be necessary to provide some additional public subsidy or incentive similar to the Renewables Obligation, such as a Zero-Carbon Obligation or Carbon Storage Obligation, at least in the short to medium-term or until such as time as the market value of a tonne of CO_2 abatement is considerably higher.

5. In the medium-term (2010 to 2020) there will be an 'energy generation gap' in the UK, arising from the decommissioning of both existing nuclear and fossil fuel plant. Nuclear decommissioning results in an increase in CO_2 emissions from the electricity sector if nuclear is replaced by CCGT. It is generally accepted that new renewable generation cannot be introduced sufficiently quickly to fill the generation gap, and nor is it clear that current renewable technologies would be able to replace base and peak load from nuclear and/or fossil fuel generation in the longer-term (to 2050). Hence, a vital question for energy policy in the UK and other industrialized countries is how new nuclear and retrofitted and new fossil CCS compare as constituents of the generation mix and what is an appropriate balance.

6. New fossil CCS investment has some similarities with new nuclear plant investment. As with any capital-intensive option, there is a danger of becoming 'locked-in' to a CCS system. Renewables tend to avoid lock-in because of their smaller scale, shorter life-time and lower upfront capital costs but their distributed character tends to imply numerous environmental and aesthetic impacts in addition to raising issues for the design of power system infrastructure. Fossil CCS has fewer financial risks, and is more flexible, than nuclear if it is indeed the case that 10 new nuclear plants are required for a new nuclear build programme to be viable. Nevertheless, new fossil CCS is still much less flexible than CCGT plant, though the latter remains highly exposed to potentially volatile future gas prices.

7. The storage of CO_2 probably poses fewer health and safety and environmental risks than does storage of nuclear waste. Furthermore, the potential for catastrophe of a fossil CCS plant is much less than that arising from a nuclear

plant. On the other hand, there are more 'every day' hazards arising from coal mining than from the nuclear energy supply-chain, though the accident record for coal mining depends to some extent on where in the world coal is sourced.

8. Some preliminary evidence suggests that CCS may be a more acceptable decarbonisation option to the public than new nuclear fission. Public and stakeholder support for CCS will depend, however, on evidence that government and industry are also vigorously pursuing energy demand reduction, energy efficiency and renewables. It is the portfolio of options which is critical for public and stakeholder sanction, rather than any single option being privileged.

9. For many stakeholders one of the benefits of CCS compared to distributed generation is that it fits readily into the existing infrastructure of power plants and the electricity grid. It was also perceived that there would be fewer planning problems and less public opposition than for a large scale deployment of renewable energy technologies onshore.

10. The notion of CCS as a 'bridging' or 'stop-gap' technology (i.e. whilst we develop 'genuinely' sustainable renewable energy technologies) needs to be examined somewhat critically, especially given the scale of global coal reserves. If CCS plant is built, then it is likely that technological innovation will bring down the costs of CO_2 capture further, such that it could become increasingly attractive, though of course the costs of renewable energy are also likely to come down. It is not possible with currently available evidence to conclude that development of fossil fuel CCS plant in the UK would have any material impact on the availability of finance for either or both of new nuclear plant or renewables.

11. The UK has sufficient storage capacity to store CO_2 from the power sector (at current levels) for a least half a century, using both well understood, and therefore likely to be lower-risk, depleted hydrocarbon fields and less well characterized and therefore somewhat higher risk aquifers. It is very difficult to produce reliable estimates of the storage capacity of aquifers and no generic methodologies exist. A detailed case-by-case geological assessment of individual formations is required, and even then it is likely that only experience of pilot CO_2 storage reservoirs will provide greater certainty.

12. The greatest uncertainty with respect to CCS is whether the CO_2 will leak from the reservoirs and, if so, how quickly. Leakage does not negate the value of CCS as a carbon mitigation option provided that the leakage rate is very low. As with capacity estimation, the issue of leakage has to be evaluated on a site-by-site basis: no generic assumptions are feasible. Establishing an 'acceptable' leakage rate is highly complicated and depends on a large number

of scientific, economic, financial, social, political and even moral factors. The second most important uncertainty is the potential impacts of leakage from sub-sea reservoirs upon marine ecosystems.

9.11   Recommendations to Policy Makers

Our key recommendations to Policy Makers are provided below.

Legal Framework and the Post-Kyoto Phase

1. The legal framework requires clarification and appropriate modification rapidly. In particular, the OSPAR Convention needs to be modified in order that CO_2 storage for the purposes of climate change mitigation is clearly permitted under the terms of the Convention. The private sector will not undertake large scale investment in CCS until such time as a clear legal framework is in place.

2. The uncertainty regarding the post-Kyoto (> 2012) phase of the United Nations Framework Convention on Climate Change raises questions regarding the future of the EU Emissions Trading Scheme. European Union Governments need to agree on the long-term future of the EU ETS and on the emission allocation such that the power sector has some greater certainty in making its investment decisions. Tighter emission caps will be required in order to stimulate investment in low-carbon options such as CCS.

3. China, India and the USA all use large amounts of coal, and are likely to use increasing amounts in the future. From a climate change perspective, there is an urgent need to design new coal power plants in all major coal-using countries such that they can be modified to capture CO_2 in future (capture-ready).

Regulatory Framework

4. A regulatory framework for CCS will need to include risk assessment, legal context, accounting and monitoring, potential environmental and health and safety impacts. A new regulatory agency, the UK CO_2 Capture and Storage Agency, will be required to regulate and monitor the CCS industry. The monitoring and verification aspects are also necessary for inclusion of CCS as a CO_2 reduction option under the EU ETS.

5. In the long-term, it is likely that once a storage site has closed (following agreement between the operator and the regulator) liability for monitoring and managing the site will need to transfer from the operator to the state.

6. UK power generators should source their coal only from those countries with a proven health and safety track record and with a favourable record

of environmental protection. In this way, the wider impacts associated with fossil CCS in the UK through use of imported coal can be limited.

Public Engagement

7. There should be public involvement in the evaluation and on-going monitoring of CCS proposals and projects. This would help to address, and if appropriate, diffuse potential opposition, but also would allow developers and regulators to modify their plans on the basis of information on public perceptions. A consultative committee made up of stakeholders and local residents attached to each CCS project is one possibility (e.g. see suggestions in Shackley *et al.*, 2006).

Economic Incentives for CCS

8. Government needs to ensure that support for energy demand reduction, energy efficiency, low-carbon and renewable energy are not only maintained but accelerated. A 'level-playing' field has been promoted by the House of Commons Science and Technology Committee: 'A technology neutral incentive framework would better reflect the overall objective, which is to reduce CO_2 emissions. It would also be more efficient to let the market decide which technologies provide the best solution to meet this challenge' (HoC, 2006, p. 54). Such a level playing field is in principle provided by the EU Emissions Trading Scheme (EU ETS) but at present this does not apply to all sources of CO_2 emissions and to all mitigation options, has an uncertain future and is at too early a stage of operation to rely upon as a way of valuing the social benefits of carbon mitigation (e.g. the scheme currently lacks tight national caps on emissions). Therefore some other form of government support is required until such time as the EU ETS or an equivalent mechanism is fully operational.[1]

9. Decarbonised Electricity Certificates (DECs), an extension of the current Renewables Obligation Certificates, are one possibility and extend the range of options to include nuclear and fossil fuel with CCS in addition to renewables (HoC, 2006). The disadvantages of DECs include: a focus only upon electricity rather than decarbonised energy; the uncertainty created by fuel price volatility (meaning that a subsidy scheme based on generation would not guarantee revenue for the developer); disruption of investor confidence in the existing

---

1 The Sustainable Development Commission (2006) has argued against any additional incentives for the development of nuclear power. The implicit argument seems to be that the EU ETS is sufficient but in that case it is unclear as to why renewables should receive public subsidy. Most analysts agree that long term certainty in the value of carbon reduction will be required in order to stimulate the necessary innovation and that individual technologies should not be favoured or discriminated against.

Renewables Obligation market; and the danger of CCS derailing renewable technology development given substantive differences in scale and investment levels. For example, DECs may encourage investment in just a few large power plants rather than in a wider-range of renewable energy technologies. In terms of CO_2 abatement the Peterhead/Miller CCS scheme is estimated to be equivalent to 40 per cent of all the wind turbines in the UK which have been developed over the past 20 years. A few large-scale schemes face fewer transaction costs as incurred in project development, financing and planning, hence large companies may find it easier to meet CO_2 reduction targets through CCS rather than through renewables, especially if the technology is proven. This would not assist in the diversification of the supply-side which is desirable for a number of reasons in addition to CO_2 reduction, including fuel security, resilience, and avoidance of lock-in. A distinct support mechanism for large-scale zero- and low-carbon technologies is a potential option, perhaps allowing nuclear and fossil CCS to compete for investment. A capital grants scheme may be necessary to encourage the first CCS demonstration plants but does not provide a long-term incentive (HoC, 2006).

10. An alternative support mechanism is the idea of a 'carbon contract' whereby Government sets a target for CO_2 reduction from certain sectors or from the economy as a whole over a period of say 5 to 20 years and then invites companies to submit bids to deliver a certain quantity of CO_2 abatement in a given time period (Helm and Hepburn, 2005). A competitive bidding process could be adopted, or alternatively an auction could be held (*ibid.*). Government would then pay for the carbon mitigation component of the selected schemes, thus providing long-term security of income revenue. The benefits of carbon contracts are that they overcome the problem of the uncertainty of the future value of carbon abatement, thus allowing companies to have confidence in their future revenue streams and enabling them to seek finance to develop zero-/low-carbon technologies rapidly. On the downside, it is not clear where the Government itself would find the money to pay for such contracts from, especially given their long-term nature. Because of the uncertainty of the value of CO_2 in the EU ETS it is also possible that the Government might end up paying too much for carbon abatement over the life time of a project, though there are good reasons for believing that the value will increase over time, not decrease. Carbon contracts overcome the problem of market uncertainty by the Government effectively agreeing a price bilaterally or through an auctioning process, but this cannot overcome the real uncertainty attached to future abatement costs and benefits. There would also be political uncertainty attached to a long-term agreement, since a future government might renege on past commitments.

11. A more radical proposal than carbon contracts is that of Domestic Tradeable Quotas (DTQs), which extend CO_2 emissions trading to all households, and

with the expectation of a ratcheting-down of total emissions over the next 50 years (Starkey and Anderson, 2005). The advantages of DTQs over other policy instruments include facilitating liquidity in carbon reduction markets and maintaining flexibility in responding to different target levels. More specifically DTQs would: a) encourage a larger number of participants in the market place, thus encouraging more trading activity (i.e. liquidity); b) avoid making any distinction between CO_2 reduction through supply-side or demand-side change or through use of a particular technology (side-stepping supply-side capture or technology capture); and c) the ability to adjust the total quota available on a yearly basis to reflect progress (or lack of it) in international target setting (hence the UK would not have to commit itself to 60 per cent reduction if no other country intended to do likewise but emissions reduction could also be ratcheted-up if tougher targets were put in place (in this regard long-term carbon contracts are less flexible)). One problem with the DTQs approach is that it has a much higher transaction cost than carbon contracts or other traditional incentive schemes. Still, even the high end of the cost range does not appear to be unreasonable given the importance of climate change as an issue for public policy (Starkey and Anderson, 2005). It is also possible that there would be negative perceptions of a DTQs scheme amongst the public, though it could be argued that introduction of such a scheme would require Government to engage in a dialogue with the public as to why a DTQs scheme was necessary and desirable.

Learning-by-doing

12. Given the large uncertainties associated with CCS, a learning-by-doing approach is required. It is likely that many uncertainties regarding the behaviour of CO_2 in reservoirs and aquifers, its possible leakage out and the impacts if it should leak, will not be resolved with certainty until field trials are underway. This means that CO_2 storage trials need to be designed so that as much as possible can be learnt from them as soon as possible. If CCS becomes more of a routine practice it will still be essential to design 'learning systems' given the uncertainty which will persist regarding the long-term fate of CO_2 in geological reservoirs.

13. The case-studies provide us with some initial knowledge of how energy stakeholders trade-off different criteria with respect to the current array of energy technologies. However, our knowledge is limited by a small sample size and the fact that we did not interview national-level energy policy makers. It would be useful to repeat the MCA process with more stakeholders and with members of the public to assess the extent to which the polarisation amongst stakeholders is found more widely.

9.12   References

Anderson, K., Shackley, S., Mander, S. and Bows, A. (2005), *Decarbonising the UK: Energy for a Climate Conscious Future*, Tyndall Centre, Manchester.

Baer, P. (2003). 'An Issue of Scenarios: Carbon Sequestration as an Investment and the Distribution of Risk: An Editorial Comment', *Climate Change*, **59**, pp. 283–291.

Curry, T., Reiner, D., de Figueiredo, M. and Herzog, H. (2005), *A Survey of Public Attitudes towards Energy and Environment in Great Britain*, Unpublished MS, MIT, Cambridge, MA.

Curry, T., Reiner, D., Ansolabehere, S. and Herzog, H. (2005a), 'How Aware is the Public of Carbon Capture and Storage?', in E. Rubin, D. Keith and C. Gilboy (eds), *Proceedings of 7th International Conference on Greenhouse Gas Control Technologies, Volume 1: Peer-Reviewed Papers and Overviews*, Elsevier Science, Oxford, pp. 1001–1009.

Dooley, J.J. and Wise, M.A. (2003), 'Retention of CO_2 in Geologic Sequestration Formations: Desirable Levels, Economic Considerations, and the Implications for Sequestration R&D', in J. Gale and J. Kaya (eds), *Proceedings of the 6th International Conference on Greenhouse Gas Control Technologies*, Pergamon, Oxford, pp. 273–278.

DTI (2003), *Review of the Feasibility of Carbon Dioxide Capture and Storage in the UK*, DTI Report URN 03/1261, Department of Trade and Industry, London.

DTI (2005), *Carbon Abatement Technologies*, Department of Trade and Industry, London.

Ha-Duong, M. and Keith, D.W. (2003), 'Carbon Storage: The Economic Efficiency of Storing CO_2 in Leaky Reservoirs', *Clean Technologies and Environmental Policy*, **5**, pp.181–189.

Hanley, N. and Owen, A.D. (eds) (2004), *The Economics of Climate Change*, Routledge, London.

Helm, D. and Hepburn, C. (2005), 'Carbon Contracts and Energy Policy: An Outline Proposal', Unpublished MS, University of Oxford.

Herzog, H.J., Caldeira, K. and Reilly, J. (2003), 'An Issue of Permanence: Assessing the Effectiveness of Temporary Carbon Storage', *Climatic Change*, **59**, pp. 293–310.

HoC (2006), *Meeting UK Energy and Climate Needs: The Role of Carbon Capture and Storage*, First report of session 2005–06, Vol. 1, House of Commons Science and Technology Committee, HC 578-1.

Holloway, S. and Baily, H.E. (1996), 'The CO_2 Storage Capacity of the United Kingdom', in S. Holloway (ed.), *The Underground Disposal of Carbon Dioxide, Final Report of Joule 2 Project*, No. CT92-0031, British Geological Survey, Nottingham, pp. 92–105.

Holloway, S., Vincent, C.J., Bentham, M.S. and Kirk, K.L. (in press). 'Top-down and Bottom-Up Estimates of CO_2 Storage Capacity in the UK Sector of the Southern North Sea Basin', in press, *Environmental Geoscience*.

IEA (2004), *Prospects for CO$_2$ Capture and Storage*, International Energy Agency, Paris.

IPCC (2001), *Climate Change 2001: Synthesis Report*, Cambridge University Press, Cambridge.

IPCC (2005), *IPCC Special Report on Carbon Dioxide Capture and Storage*, prepared by Working Group III of the Intergovernmental Panel on Climate Change, B. Metz, O. Davidson, H.C. de Coninck, M. Loos and L.A. Meyer (eds), Cambridge University Press, Cambridge.

Lenton, T.M. and Cannell, M.G.R. (2002), 'Mitigating the Rate and Extent of Global Warming', *Climatic Change*, **52**(3), pp. 255–262.

Palmgren, C., Granger Morgan, M., Bruine de Bruin, W. and Keith, D. (2004), 'Initial Public Perceptions of Deep Geological and Oceanic Disposal of CO$_2$', *Environmental Science and Technology*, **38**(24), pp. 6441–50.

Pearce, D. (2005), 'The Social Cost of Carbon', in D. Helm (ed.), *Climate-change Policy*, Oxford University Press, Oxford, pp. 99–133.

Shackley, S., Cockerill, T. and Holloway, S. (2003), 'Carbon Capture and Storage: Panacea or Long-Term Problem?', *Climate Change Management*, September Issue, p. 11.

Shackley, S., Mander, S. and Reiche, A. (2006), 'Public Perceptions of Underground Coal Gasification in the United Kingdom', *Energy Policy*, in press.

Starkey, R. and Anderson, K. (2005), *Domestic Tradeable Quotas: A Policy Instrument for Reducing Greenhouse Gas Emissions from Energy Use*, Tyndall Technical Report 39, Tyndall Centre, Manchester.

The Sustainable Development Commission (2006), *The Role of Nuclear Power in a Low Carbon Economy*, SDC Position paper, March 2006.

Vendrig, M., Spouge, J., Bird, A., Daycock, J. and Johnsen, O. (2003), *Risk Analysis of the Geological Sequestration of Carbon Dioxide*, Report No. R246, DTI/Pub URN 03/1320, Department of Trade and Industry, London.

Index